T5-BAP-075

POPULATION REGULATION AND DYNAMICS

POPULATION REGULATION AND DYNAMICS

PROCEEDINGS OF
A ROYAL SOCIETY DISCUSSION MEETING
HELD ON 23 AND 24 MAY 1990

ORGANIZED AND EDITED BY
M. P. HASSELL AND R. M. MAY

LONDON
THE ROYAL SOCIETY
1990

Printed in Great Britain for the Royal Society
by the
University Press, Cambridge

First published in *Philosophical Transactions of the Royal Society of London*,
series B, volume 330 (no. 1257), pages 121–304

(∞) The text paper used in this publication meets the minimum requirements of American National Standard for Information Sciences Permanence of Paper for Printed Library Materials, ANSI Z39.48-1984.

British Library Cataloguing in Publication Data

Population regulation and dynamics: a discussion organized and
edited by M. P. Hassell and R. M. May
1. Organisms. Population Dynamics
Hassell, M. P. II. May, R. M. III. Royal Society
574.5248

ISBN 0-85403-424-2

Published by the Royal Society
6 Carlton House Terrace, London SW1Y 5AG

PREFACE

It is over a century since Darwin provided the essentials of the explanation for how species originate. One cornerstone for his argument was the observation that all natural populations have an inherent capacity to increase from generation to generation in the absence of checks and balances. A number of the fundamental questions of ecology follow on from this simple observation. What are the regulatory factors that, over the long run, prevent a population from realizing its potential for unbounded increase? How do these regulatory factors combine to produce observed patterns in the relative abundance of species? How do these dynamical factors influence the structure of plant and animal communities? And, ultimately, how does all this add up to determine the number of species, either locally or globally? These questions remain largely unanswered. Some patterns of relative abundance have been documented and some remarkable regularities noted, but these regularities remain largely unexplained, and much controversy surrounds efforts to determine the (often markedly nonlinear) mechanisms that regulate the abundance of particular populations. We have even less information on the dynamics of multispecies interactions, and only know the total number of species on earth to within a factor of 10 or so. The questions above, moreover, have important practical implications. A range of current problems, from the ecological effects of global climate change to protocols for the release of genetically engineered microorganisms, require a much better understanding of the factors regulating population abundance and determining community structure.

The past decade has, however, seen important advances, particularly in providing an agenda for research that brings closer the integration of ecological theory and empirical studies on natural systems. The papers in this volume show this by focusing on the factors affecting the dynamics of plant and animal populations, and examining patterns of species abundance on a much grander scale.

We thank all the authors for their contribution to the meeting and to this volume. We are also very grateful to the Royal Society in general, and to Christine Johnson (who organized the meeting) and Miss Funda Suleyman (who saw the papers into print) in particular, for making it all possible.

November 1990

M. P. Hassell
R. M. May

CONTENTS

Introductory remarks

T. R. E. SOUTHWOOD

Department of Zoology, University of Oxford, Oxford OX1 3PS, U.K.

These proceedings, like the meeting from which they sprang, bring together two topics of long-standing interest to ecologists: the regulation of individual species and the relative abundance of different species in a community. Indeed, the explicit quest for understanding population regulation can be traced to Gilbert White who, in 1778, queried why year after year there were always eight pairs of swifts in Selbourne. In view of the amount of study which has already been expended on these topics, one may ask 'what is new?' As is clear from this symposium, we are now more in a position to achieve a synthesis of these two topics. Another difference from the situation of some 40 years ago is the fact that plants are now being considered alongside animals and we are gaining new insights from their study.

Coming to more recent history, from about 1920 until 1970 the emphasis was on the dynamics of single- and two-species population interactions. Several models were produced, of which the Lotka–Volterra and Nicholson–Bailey models have been the most enduring. Quite properly, ecologists searched for meaningful field tests of the theories which underlay these models. Such tests involved the measurement of populations in the field, to constructing precise life tables. This work was most straightforward in certain insects, where discrete generations imposed by the seasonal cycle enabled the complexities of overlapping generations and the problems of the determination of the precise age of an individual to be circumvented. Some of the leading work of that time was undertaken in Australia by Davidson & Andrewartha, in Canada by R. F. Morris and his colleagues, and in the United Kingdom by O. W. Richards, N. Waloff at Silwood Park and by G. C. Varley in Oxford. Many techniques were developed for enumerating populations and for analysing them, among the best known being key factor analysis. This technique was devised to detect the stage at which variations in survival contributed most to generation-to-generation fluctuations in population size. As modified by Varley and Gradwell, the technique could also be used in the search for regulating factors, but there were many complications. Some of these factors were statistical arising from a lack of independence of data, but others were because of the sheer mechanical labour of analysing complicated sets of data when all one had at one's disposal were hand-turned calculators, a situation that it is hard to envisage today.

As this work progressed, two further insights added to the complications, though at the same time refocusing our questions in a more constructive way. The first concerns a recognition that we are seldom dealing with individual, isolated populations. Varley always stressed that he chose the winter moth for his investigations because the flightless females ensured that he was handling an isolated population. At the same time as he was undertaking this work, I was involved in studies on the fritfly (*Oscinella frit*). This is a pest species and its study was a necessary condition of my employment. I quickly found that the population in any area was by no means isolated: not only did it change its location every generation (and there were generally three generations in a year) but frequently the adults would move over many miles every few days. Virtually all I was able to say about the metapopulation was that the total mass over England and Wales on a warm day in late summer would be well over 200 t! This underlines some of the basic ideas that had led Andrewartha and Birch to entitle their iconoclastic work '*The distribution and abundance of animals*'. As this book shows, the patchiness of nature and the dispersal and interplay of local populations within the mosaic of the environment are today recognized as central topics in ecology.

Underlying much of the early work on population change in the field was the belief that this change could be simply analysed to show whether or not there were regulatory mechanisms at work. However, theoretical advances in the 1970s (in which R. M. May played an important part) have shown that even simple regulatory mechanisms can, under certain conditions, give rise to apparently random fluctuations which in fact are deterministic 'chaos'. This is particularly true when there are time delays and strong nonlinearities, as probably often occur in populations. This theoretical understanding shows that much of the debate at the famous Cold Spring Harbor Conference in 1957 was based on a false premise. The superficial behaviour of a population over time be it steady, cyclic or fluctuating, does not, of itself, tell us a lot about the fundamental dynamical mechanisms: erratic fluctuations may be caused by unpredictable and density-independent environmental events, or they may equally well be caused by strong density-dependence.

The structure of this volume follows in many ways the development of the subject. This is entirely logical; it is important to build our present view of ecology on the fundamental understanding of simple population interactions. On this edifice, we can then turn to the wider view of communities, including genetic variation, as well as the variation in time as reflected through the fossil record.

With environmental problems dominating the news,

it is particularly important for ecologists to ask 'where are we now?' and 'what is needed next?' Looking back to my own career in 1950, the justification for ecological research came largely under two headings. First, there was what was then known as 'economic zoology', namely pest control or the harvesting of economically important animals (fish and fur-bearers). Such work was widespread in Australia, Canada and the United States. Secondly, and much less extensive, there was pure research, curiosity about our natural world. Today the situation is entirely different. Human impact on the environment is widely recognized and we stand on the threshold, 'if not somewhat beyond the threshold', of a global environmental change. Many resources are being made available for measurements of the physical environment, but I would suggest that it may well be easier, indeed more appropriate, to detect the integrated impact of such environmental changes through biological indicators. For this we need long-term and comprehensive studies: base line studies. We need to monitor our populations, to observe how they change. It is indeed ironic that just as we are embarking upon this great, unintended experiment with nature of human-induced change in the global climate, long-term studies in Britain which will provide basic information, such as the Rothamsted Insect Survey and fundamental taxonomy, are under tighter constraints than at any time during the past few decades, while abundant resources are available for study of distant worlds (i.e. astronomy) and for the physical conditions of our own world. Those funding biological work need to consider these parallels and to recognize both the need (that is, the understanding and recognition of environmental change) and the opportunity presented by this unique event, at a time when novel and 'high-tech' instrumentation is available to help with the required data. Against this background, the exciting developments in ecology that are reported in this symposium are especially important in helping us to gain fuller and much needed understanding of the world in which we live!

The population dynamics of plants

MICHAEL J. CRAWLEY

Centre for Population Biology, Department of Biology, Imperial College, Silwood Park, Ascot, Berkshire SL5 7PY, U.K.

SUMMARY

Long-term studies of plant populations are reviewed, and their dynamics summarized in three categories. Many short-lived plants have ephemeral, pulsed dynamics lasting only a single generation, with recruitment determined almost entirely by germination biology and by the frequency and intensity of disturbance. Such populations are not amenable to traditional population models. At the other extreme, some long-lived plants have such protracted tenancy of their microsites that it is impossible to establish what pattern of dynamics (if any) their populations exhibit. A relatively small number of species show what we would traditionally regard as population dynamics at a given point in space (i.e. reasonably predictable trajectories that can be modelled by $N_{t+1} = f(N_t)$). A major difficulty in generalizing about plant dynamics is that the majority of species are successional; their recruitment depends upon the death, through senescence or disturbance, of the dominant plants. Where we do have data spanning several generations, it is clear that: (i) the populations are regulated by density dependent processes; (ii) in contrast to some animal populations, numbers appear to vary less from year to year in places where mean density is higher, and less from place to place in years when mean density is high than when density is low; (iii) few, if any, plant populations show persistent cyclic or chaotic dynamics, but (iv) there are several robust generalizations that stem from the immobility and phenotypic plasticity of plants (the law of constant yield; self-thinning rules, etc.). These generalizations are analysed in the context of simple theoretical models of plant dynamics, and the patterns observed in long-term studies are compared with similar data from animal populations. Two important shortcomings of traditional plant demography are emphasized; (i) the dearth of simple manipulative experiments on such issues as seed limitation, and (ii) the tendency to locate study plots around existing mature individuals (the omission of 'empty quadrats' may introduce serious bias into the estimation of plant recruitment rates).

1. INTRODUCTION

The idea of counting plants has never had widespread appeal. There are two main reasons for this. First, the longevity of plants is such that one individual may survive for many human lifetimes. Second, the phenotypic plasticity of plants is so great that the fecundity of the same individual genotype may vary by four orders of magnitude, depending upon the circumstances of its cultivation. There is a further general problem about what precisely to count. In animal ecology is taken as axiomatic that counting refers to individual, free-living genets (Harper 1977). Plants, however, are modular in construction and for most practical purposes it may make more sense to estimate the plant biomass rather than to count the number of individual genotypes. Again, it is unclear whether a count of the numbers of shoots per genet, or of the number of modules per shoot is the most appropriate currency for assessing plant abundance. The traditional distinction in population biology between plants and animals is not really taxonomic, but based on a comparison of mobile unitary organisms with sessile, clonal organisms. The population dynamics of sessile animals (like corals or sponges) and of mobile plants (like diatoms and non-rooted aquatics) have received very little attention.

(*a*) *History of plant dynamics*

The first published counts of plants refer to rarities, and to the documentation of the demise of much-loved species (e.g. successive presidents of the Tyneside Natural History Society, Carr (1848) and Burdon (1856), described the decline of the lady's slipper orchid in Castle Eden Dene, County Durham, to its local extinction in about 1850).

The idea of counting common plants is very much a 20th century pursuit. Work in the arid S.W. of the U.S.A. during the first decade of this century enumerated the decline of conspicuous plants like the saguaro cactus (*Carnegiea gigantea*) and the increase of invasive woody plants like mesquite (*Prosopis glandulosa*) with the advance of cattle ranching. The first permanent quadrats in which plants were routinely counted were set up by V. M. Spalding in 1906 at the Carnegie Institute's desert laboratory at Tumamoc Hill in Arizona (Hastings & Turner 1965). Results from these quadrats documented the gradual increase of grazing sensitive and the decrease of uncompetitive plants following fencing to exclude the cattle (Goldberg & Turner 1986). The discrete morphology and wide spacing of these desert plants may have been conducive to counting. None of the species exhibited any dramatic ups or downs, and none showed any signs of cycling.

The famous Park Grass and Broadbalk experiments at Rothamsted, England, were set up much earlier than this (in 1840s and 1850s), but they were concerned with changes in plant biomass rather than with the determination of plant population density (Williams 1978).

(*b*) *Plant population biology*

It is worth cataloguing some of the important attributes of plant morphology because their impact on population dynamics is so profound. (i) Plant size is extremely plastic, as we have already seen. (ii) The sedentary, rooted habit means that neighbour relations are vital. Plant performance can often be accurately predicted simply from knowledge of the number of neighbouring plants that are bigger than the plant in question (Hutchings 1986). (iii) Successional change in plant communities may be the rule (rather than the exception as is implicitly assumed in many animal population models). Thus the probability of recruitment may change systematically through time, independent of the reproductive performance of the plant species in question. (iv) Recruitment may be infrequent and unpredictable. Some plant species, like seed-bank annuals, may have population dynamics that are determined more by their germination biology (the duration of dormancy and the stimuli necessary to break dormancy) than by the number of seeds produced in the previous generation. (v) Competition is asymmetric. Larger plants influence the growth and reproduction of smaller ones, but small plants rarely have measurable impact on the performance of larger ones. (vi) Mortality is size dependent. The plants that die as a result of plant competition come exclusively from the smallest size classes. (vii) The death of large plants from old age, accident, or herbivore attack, may open up conditions for recruitment (Grubb 1988).

2. PATTERNS OF DYNAMICS

There is no typical pattern of plant dynamics. Instead, we observe a continuum from virtual stasis (e.g. very long-lived desert shrubs that have shown no recruitment since records began) to violent year-to-year fluctuations in abundance (e.g. desert annuals whose recruitment is controlled by the intensity of rainfall). Within this continuum, there are habitat differences (associated with the frequency and intensity of disturbance) and differences associated with plant life history (principally longevity and the degree of iteroparity). The degree to which plant dynamics are affected by internal, over-compensating density dependent processes of the kind that might lead to cycles or to chaotic dynamics is unclear, and examples are few.

The role of herbivores in affecting plant abundance and the stability of plant populations is potentially great, and appears to be related to the body size, mobility and degree of monophagy of the herbivores involved. For example, large, mobile, polyphagous vertebrate herbivores tend to have more impact on plant dynamics than small, sedentary, monophagous insect herbivores (Crawley 1989).

(*a*) *Episodic recruitment*

Many plant populations consist of only one or two age classes, and any given age class may cover areas of many square kilometres. The forces responsible for creating these large, single-aged patches include fire (Noble 1989), tropical storms (Whitmore 1989), volcanic eruption (Mueller-Dombois 1981), drought (Hubbell & Foster 1990), unusually heavy rains (Epling *et al.* 1960), and epidemic disease among keystone herbivores (Crawley 1983).

Many annual plants of arable land will not produce any seedlings at all unless there is cultivation or other soil disturbance sufficiently severe to open up large gaps in the perennial plant cover. These annuals typically possess long-lived seed banks, and some of them have very precise germination requirements, such that the weed flora that develops in the first year after cultivation depends critically upon the timing of soil disturbance.

The agricultural practice of fallowing is intended to cause mass germination of dormant seeds, then to kill the plants by cultivating again before any of them have set seed. Fallowing has varying degrees of success with different species of annual weeds. Some like the poppy *Papaver rhoeas* have such large seed banks, that one year's germination causes no significant decline in seed bank numbers. Others like the grass *Alopecurus myosuroides*, have their numbers substantially depleted by fallowing, but are so fecund in the low competition environment that occurs after a successful fallow, that the few surviving plants grow so large that the seed bank is replenished in a single year (Brenchley & Warrington 1930).

(*b*) *Ephemeral populations*

John Harper has remarked that it is the fate of most populations studied by plant ecologists to go extinct during the course of the study. This is graphically shown on the cover of the thesis by de Jong & Klinkhamer (figure 1), and documented by the local demise of *Vulpia membranacea* (Watkinson 1990) shown in figure 2. In such cases, the conditions favouring plant recruitment are not met within the quadrats where the initial cohort of adult plants was selected. The only solution to this problem is to ensure that 'empty quadrats' are included from the outset of the project (i.e. potentially habitable sites from which the adult plants are currently missing). Ideally, the quadrats should be placed at random within suitable habitats, but this is rarely done in practice.

The notion of potentially habitable sites begs several questions. A sample of unoccupied sites is almost certain to contain an unknown proportion of potentially occupiable but currently unoccupied sites, plus a proportion of genuinely uninhabitable sites. If, for simplicity, we were to assume that plants were distributed over the suitable sites in a Poisson manner, then the whole set of quadrats would represent a

Figure 1. The cover from the doctoral thesis of Tom De Jong and Peter Klinkhamer (1986), showing the demise of plant populations singled out for detailed study.

'Poisson plus added zeros' and would be bound to exhibit an artificially aggregated pattern (Pielou 1977). Again, a quadrat that is empty of adult plants may contain countless dormant seeds within the seed bank beneath it, so 'local extinction' in these cases should strictly be read as 'current absence of vegetative plants'.

(*c*) *Self-replacing, equilibria*

Seed-limited dynamics have been studied by a number of theorists (Watkinson 1986; Pacala & Silander 1985) and described from a few field systems (Symonides *et al.* 1986) and experimental trials (Pacala & Silander 1990). One strong generalization appears to be that populations can get to their carrying capacity extremely quickly, and often in as little as one or two generations (e.g. the annual grass *Poa annua*; Law 1981). Slower increases may be indicative of microsite limitation rather than seed limitation (Rabinowitz *et al.* 1989). In many cases, the carrying capacity itself will fluctuate from generation to generation, so that the resulting dynamics may reflect more about shifts in recruitment and microsite availability than any internally generated, cyclic tendencies.

(*d*) *Model populations*

The model of plant population dynamics that forms the central organizing theme of this paper was described by Crawley (1986). We depict the rate of change in plant numbers, dN/dt, as a function of six controlling processes, and we require that the rate of change is positive, at least when the plant is rare:

dN/dt = intrinsic rate of increase of the plant
− resource-limitation effects
− interference effects
− herbivore and pathogen effects
− mutualist-limitation effects
+ refuge effects

The intrinsic rate of increase is known for rather few plant species. The potential fecundity of plants is extremely high, of course, but the minimal level of mortality that is to be expected under the best possible conditions is unknown (see Crawley 1989). The intrinsic rate of increase of crop plants has been known since Biblical times to be between 30-fold and 100-fold per generation (St Matthew 13). There is no information at all on the intrinsic rate of increase of longer lived plants like trees, although some simple approximations may be possible (Crawley 1983, p. 28).

The effects of competition for light, water and nutrients have been studied intensively, and competition-based models of dynamics are quite numerous (Watkinson 1986; Pacala & Silander 1985; Crawley & May 1987). It has long been implicit among plant ecologists that asymmetric interspecific competition for light is the driving force in the dynamics of most plant communities (and competition for water in desert environments).

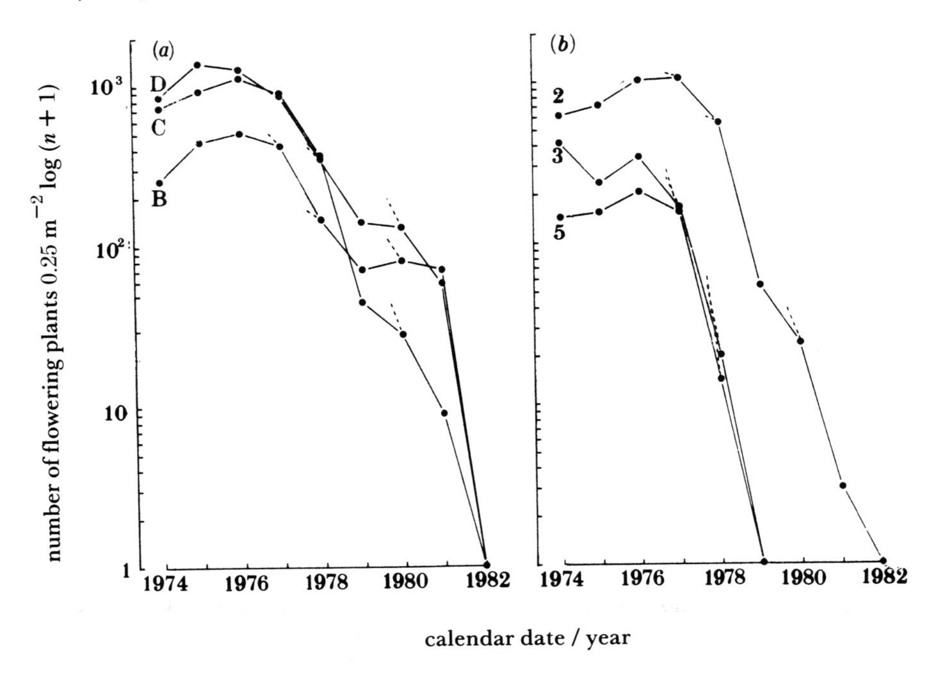

Figure 2. The demise of local populations of the annual grass *Vulpia fasciculata* on sand dunes in North Wales. Conditions for recruitment deteriorated over the course of the study period as a result of changes in the pattern of movement of surface sand (from Watkinson 1990). The number of flowering plants of *Vulpia fasciculata* on (*a*) plots *B*, *C* and *D* at Aberffraw and (*b*) plots 2, 4 and 5 at Newborough Warren between June 1974 and June 1982. The number of individuals dying at the stage of inflorescence development is shown by the dashed lines.

Behaviourial interactions between plants in the form of chemical warfare (allelopathy) have been as comprehensively ignored by modellers as they have been enthusiastically documented by field workers (Rice 1984). There has been a view (Harper 1977) that it is so difficult to show the importance of allelopathy unequivocally under field conditions, that it can safely be ignored. While it is true that allelopathy experiments need be carefully designed to eliminate other possible causes of differences in plant performance (herbivory, resource competition), the role of allelopathy in plant dynamics, and in successional dynamics in particular, deserves a reassessment.

The effects of herbivory on plant dynamics are reviewed elsewhere (Crawley 1983, 1989). The important point is that relatively small changes in competitive ability may be sufficient to tip the balance in favour of one plant species over another, and that selective herbivory (taking 5–15 % of leaf tissue) may be just the kind of mechanism to bring about such changes (Crawley & Pacala 1990). The relative importance of vertebrate and invertebrate herbivory is readily seen from a comparison of fencing exclosures against vertebrates and insecticide spraying against invertebrates. Crawley (1989) found that the identity of the dominant plant was almost always changed after the exclusion of vertebrate herbivores, and that measurable change in plant relative abundance occurred in more than 90 % of studies. Of the 30 or so studies of insect exclusion in natural vegetation, only half produced measurable impact and the effects on plant abundance were often subtle. Thus, while fence-lines excluding vertebrate of often conspicuous landscape features, the edges of insecticide-sprayed plots are seldom so clear-cut. Plant pathogens can inflict highly selective mortality on seedlings, and may have a substantial impact in determining the identity of recruits, especially in tropical plant populations (Burdon *et al.* 1981: Augspurger & Kelly 1984). Long-term data on pathogen-exclusion trials in natural vegetation are not yet available.

Mutualists are often omitted from models of plant dynamics. This is despite the fact that a number of convincing data sets show that recruitment may be limited by a shortage of pollinators (see below). The absence of specific mycorrhizal fungi may be the reason that orchids have such low rates of establishment in many habitats. In general, it is preferable to include mutualism in models as a negative rather than a positive factor, since the benefits of mutualists saturate very quickly. This avoids, in Bob May's graphic phrase, the 'orgy of mutual benefaction', that would result from including mutualists and hosts as being reciprocally beneficial in a Lotka–Volterra-style model of mutualism.

Refuge effects include a variety of stabilizing influences that provide a positive input of propagules, even when the local above ground plant population has gone extinct. There may be subterranean reserves of ungrazable meristems (Noy Meir 1975), long-lived seed banks (Roberts 1981) or immigration of wind borne or animal carried seeds from elsewhere (Rabinowitz *et al.* 1989). In some cases, recruitment from these refugia may be greater than from current reproduction, so that traditional models of the form $N_{t+1} = \lambda N_t f(N_t)$ become wholly inappropriate.

(*e*) *Dynamics*

For the species and the environment in question we have defined that $dN/dt > 0$ when rare, so that exponential increase in plant abundance is to be expected. The question of interest, therefore, is what determines the average level of abundance, and what determines the pattern of fluctuation about that average? What is the role of endogenous, time-lagged density dependent processes within the plant population (e.g. the tendency to cyclic or chaotic dynamics), and what is the relative importance of exogenous variation in producing catastrophic plant mortality or causing random fluctuations in plant recruitment (environmental stochasticity)?

Two aspects of plant population biology have a profound bearing on these questions. First, the size-specific fecundity of plants is such that in most reported cases, seed production is directly proportional to shoot mass, and there is no evidence for a threshold plant size below which reproduction is impossible (Rees & Crawley 1989) (figure 3*a*). This means that substantial changes in the size structure of the plant population (e.g. as a result of intraspecific competition) have no measurable effect on total seed production (see below for exceptions to this). Secondly, intraspecific plant competition tends to take the form of a contest rather than a scramble for resources (figure 3*b*). Thus the number of plants surviving intense self-thinning is more or less independent of the precise form of the size distribution (J. Weiner & S. Pacala, unpublished results).

Taken together, these two processes mean that the number of recruits is independent of the size of the parent plant stock over a wide range of plant densities. This effect is embodied in one of plant ecology's most robust empirical generalizations; the Law of Constant Final Yield (Harper 1977).

3. RECRUITMENT

The biology of recruitment has been understood since biblical times (see The Parable of the Sower; St Matthew 13: 4–8). The 'fowls of the air' (herbivory), 'stony places' (unsuitable microsites) and the 'thorns which sprang up and choked them' (interspecific plant competition) accounted for most of the seeds, most of the time. However, the question as to whether or not plant recruitment is seed-limited remains controversial.

(*a*) *Seed-limited recruitment*

It is important to establish whether or not recruitment in a given plant population is seed-limited or not. If it is not seed limited, then herbivores that cause only moderate reductions in plant fecundity may have no measurable impact on plant abundance or on population stability. It is curious that the simple experiment of sowing extra seeds and recording the numbers of recruits has been carried out so seldom. If a population is not seed-limited, then recruitment may be limited by microsite availability, by competition with mature plants of the same or of other species, or by seed and seedling predators. In very few cases have the necessary factorial experiments been carried out to discover which of these factors limits recruitment in a given circumstance. Presumably, the result will depend both upon the kind of habitat (e.g. seed-limited recruitment will be more likely in plant communities that have high equilibrium proportions of bare ground, and competition-limited recruitment will be more likely in forest than in grassland communities), and upon the growth form of the plant species (e.g. its longevity, its fruiting pattern and the degree of iteroparity exhibited).

M. Johnston (unpublished results) sowed 1000 seeds per square metre of 20 plant species growing in acid

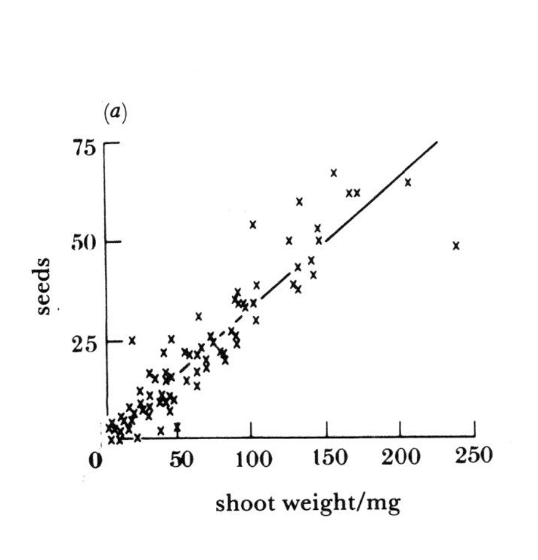

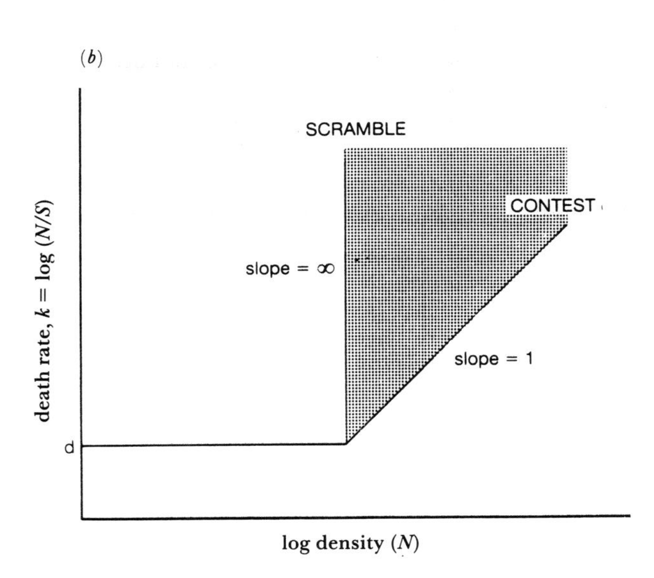

Figure 3. (*a*) The relation between shoot weight and fecundity in the annual crucifer *Thlaspi arvense*, showing direct proportionality and no threshold size prior to reproduction. (*b*) The relation between mortality (k) and log population density for plants that compete by scramble or contest; most plants have slopes close to 1, suggestive of contest competition.

grasslands in Nash's Field, Silwood Park. Only two species produced any seedlings at all; the large-seeded legume *Lotus corniculatus* and the sheep's sorrel, *Rumex acetosella*, both appeared to be seed limited, but the remainder were not. He has yet to establish the factors limiting the recruitment of the other 18 species.

(*b*) *Experimental protocol*

In order to establish the factors responsible for the limitation of recruitment requires a very simple experimental protocol. The 4 potential limiting factors (seed, competition, microsites and herbivores) are varied in factorial combinations and the resulting plant recruitment is monitored. In the simplest case, each factor just has two levels. Extra seed is sown or not, the ground is cultivated or not, competing vegetation is clipped or not, and herbivores are excluded or not.

The role of seed recruitment in perennial herbaceous vegetation is controversial. Some authors argue that seed recruitment is rare in general, but this view is disputed by Erkison (1989). He found that in a review of more than 60 studies, seedlings were present in 40% of cases. Of course, the fact that seedlings are found within populations of mature plants does not mean that recruitment is necessarily seed limited.

Within any one year, recruitment may be seed limited in some microhabitats but not in others (e.g. Keddy 1981). This argues for spatially replicated, long-term studies of the kind so rarely carried out (e.g. they are logistically impossible for Ph.D. students working on their own).

(*c*) *Data on recruitment limitation*

We have done the recruitment-limitation experiment on four species so far: the crucifer *Cardamine pratensis* (Duggan 1986), the umbellifer *Heracleum sphondylium* (Sheppard 1987), the composite *Senecio jacobaea* (Crawley & Nachapong 1985) and the oak tree *Quercus robur* (Forrester 1990). All the studies were done in acid grasslands in Silwood Park, Berkshire.

Cardamine pratensis growing in wet grassland was found to be competition-limited, with slugs and rabbits acting as important factors in reducing recruitment following cultivation. Recruitment was increased by seed addition only on the cultivated plots protected from herbivory (Duggan 1986).

Ragwort *Senecio jacobaea* showed no recruitment in grassland when extra seeds were sown, but recruitment was seed-limited following cultivation. Ragwort produces smaller seeds on regrowth shoots (following defoliation and destruction of the primary flower heads by cinnabar moth). Large seeds from primary capitula were capable of recruitment when sown into grassland cut to 2 cm in height, whereas the smaller seeds from regrowth shoots produced no recruits in cut grassland. Small and large seeds were equally likely to produce recruits in cultivated ground (Crawley & Nachapong 1985).

When the acorns of oak *Quercus robur* were placed on the surface of grassland, they were all consumed within a matter of days by mice, rabbits and larger seed-feeding birds (jays and wood pigeons). If piles of 1000 acorns were put out, only the dozen or so weevil-containing acorns were left after 48 h (M. J. Crawley, unpublished results). If the acorns were buried beneath the grassland, however, their survival rate was very high, and the seedlings were competitive with intact grassland canopy, whether or not they were exposed to rabbit grazing. The large acorn gives them both substantial competitive ability and substantial powers of regrowth (Forrester 1990). Recruitment of the oak, therefore, is seed-burial limited. Recruitment is herbivore-limited if the acorns are placed on the ground surface, and only seed-limited if the seeds are buried. Levels of oak recruitment from acorn caches buried by jays and squirrels are sufficiently high for the plant to be considered a flower-bed weed in gardens frequented by domestic cats. Apparently, the cats reduce the cache retrieval efficiency of the jays and squirrels, so that tree recruitment in this case is predator-mediated. This parallels the kind of three-trophic-level effect that led to the regeneration of large areas of oak woodland following the extermination of the rabbit population after the introduction of the myxoma virus (above).

The umbellifer *Heracleum sphondylium* is the only species so far that has proved to be seed-limited (and then in only one site out of two). Adding hogweed seeds to moist grassland in 1981 initiated a population which has persisted and is still expanding in 1990. Adding seed in a drier part of the same grassland led to no recruitment even after cultivation, and it appears that in this second site recruitment was limited by a shortage of water.

Fowler (1986) was able to increase recruitment in the grass *Bouteloua rigidiseta* by a factor of 3.4 when she sowed seeds at a rate of 4500 m^{-2} in arid grasslands in the southern United States.

4. EXAMPLES

The search for general principles of plant population dynamics is unlikely to produce robust generalizations unless we distinguish the habitat and the life forms involved. Hairston (1989) has reviewed the limitation of plant populations in a range of habitats (forests, successional communities, deserts, freshwater and marine environments), and I shall concentrate here on differences between plant growth forms. These, of course, are often closely correlated with habitat differences (trees in forests, shrubs in deserts, annuals in early successional communities, herbaceous perennials in grasslands, and so on).

(*a*) *Trees*

As a consequence of their great longevity, we have little direct evidence on tree population dynamics. Most studies have relied on indirect data, of the kind that are obtained from age structure studies (based on ring-counts from increment borers; Ogden 1985), or by observation of the relative abundance of saplings of different species beneath various canopy species (Horn 1975). The assumptions of constant recruitment and

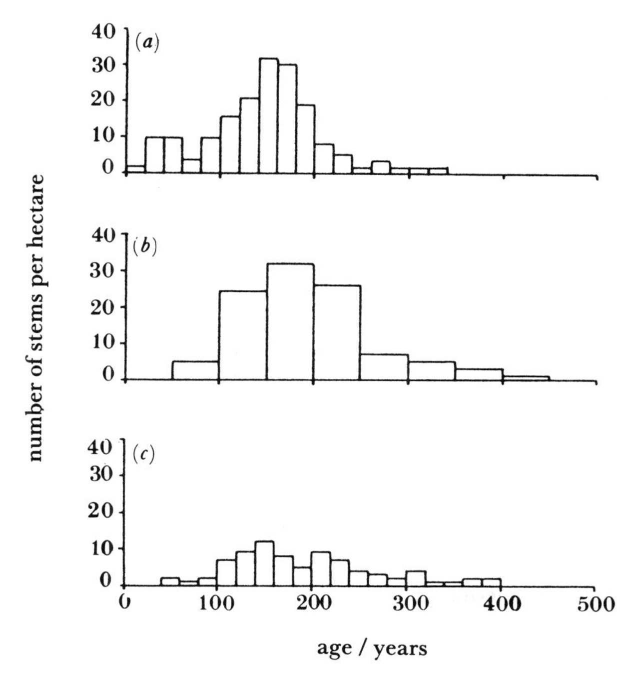

Figure 4. The age structure of three populations (*a–c*) of Japanese beech, *Fagus crenata* (from Nakashizuka 1987), showing a complete absence of recruitment in recent years.

time-invariant survivorship that underlie many of these reconstructions of forest dynamics are unlikely to be realistic in the majority of cases.

For example, the Japanese beech *Fagus crenata* exhibits large patches of uniform age, resulting from disturbance from typhoons, and dominance of large areas of the forest floor by dwarf bamboos (*Sasa* spp.), with the result that most age structure diagrams lack any recruiting juvenile age classes (figure 4). The absence of recruits in this case is due to asymmetric intraspecific competition between seedlings and adult trees and to intense interspecific competition with the perennial ground cover bamboos (Nakashizuka 1987).

Wildfire and wind are the major disturbances in the coniferous forests of the N.W. United States. There have been 16 major fires in Mount Ranier National Park since the year 1260 (Hemstron & Franklin 1982) with an average fire rotation of 434 years. The largest fire destroyed over 150 square miles, almost half the area of the Park.

Hubbell & Foster (1990) re-surveyed their permanent plot on Barro Colorado island in 1985, 5 years after the initial survey. They discovered greater than 10% change in the absolute abundance of more than 40% of the plant species. Catastrophic mortality of moisture loving species occurred during the drought year of El Nino. Rare species suffered greater declines, and common species more increases, than would be expected by chance alone. Hubbell & Foster conclude that the present forest is not an equilibrium assemblage, but that drought caused differentially high mortality among the smallest, shallowest rooted under-story plants, and the most intensely irradiated, emergent trees.

In a 21-year study of the Kolombangara rain forest in the Solomon Islands (Whitmore 1990) with 22 plots each of 0.63 ha† all trees greater than 30 cm girth were measured and all individuals counted for the 12 dominant species. This study stressed the importance of cataclysmic destruction in tropical forest dynamics. It catalogued the differences on permanent plots through time and from place to place, within and between floristic types of forest, and related these to canopy disturbance. Successional elimination of light-demanding tree species occurs under the present canopy disturbance régime; the more shade tolerant species increase in numbers until a sufficiently extensive canopy disturbance occurs to allow the establishment of more light-demanding species.

Coining the phrase 'shifting mosaic steady state' (Bormann & Lichens 1979) described the pattern of recruitment dynamics for deciduous forest in the N.E. United States. The average canopy turnover of 1–2% per year obscured big spatial variations of up to 10% in a single year.

While episodic disturbance appears to regulate recruitment in the majority of forest types, there are some relatively undisturbed forests that appear to have equilibrium species assemblages, in which the relative abundance of mature trees can be predicted on the basis of tree replacement probabilities. These are judged simply by counting the numbers of saplings of different species beneath the canopy of each kind of mature tree. The probability of replacement of canopy dominant of species A by a sapling of species B is taken to be the proportion of all saplings belonging to species B (of course, the equilibrium dominant species will have a high probability of self-replacement). Horn (1975) has documented the existence of such forests in the N.E. United States. There is no circularity in these predictions, since data gathered on replacement probabilities in one site can be used to predict the equilibrium tree assemblage in another site.

On balance, it appears that tree dynamics within most forests are governed by disturbance-limited recruitment, rather than by equilibrium successional dynamics. The impression gained from long-term study plots in tropical forests like Whitmore's (1989) long term study in the Solomon Islands and Hubbell & Foster's more recent work on Barro Colorado Island, is of much higher levels of dynamism and turnover than was previously believed. The fact that some of the individual trees are long-lived, does not mean that turnover within forests is slow.

While it is often assumed that the keystone herbivores in forest communities are the invertebrates feeding in the canopy, it is clear from recent work that seed and seeding-feeding mammals may be influential in determining the identity of the species that survive to the sapling stage, and eventually form the canopy dominants (Forrester 1990; C. D. Pigott, personal communication).

Stable age structures will not be found in the majority of tree populations because: (i) recruitment is too episodic; (ii) plant size has more impact on demography than plant age (sapling banks with negligible growth rates); (iii) predominance of strong

† 1 hectare = 10^4 m^2.

age classes in many populations; (iv) we know virtually nothing about the history of recruitment, but what little we do know suggests that recruitment is most unlikely to be constant from year to year.

(*b*) *Shrubs*

The dynamics of shrub populations have received rather little attention, except in desert regions where long-term studies have been carried out since the early years of this century (Goldberg & Turner 1986). A variety of recruitment limitation mechanisms has been reported, including insect herbivory (Parker & Root 1981), rainfall (Prentice 1986), limited seed dispersal (Brown & Archer 1897) and direct seed limitation (Louda 1982), but too few studies are available for any broad patterns to emerge.

Vegetation dynamics in some heathland shrubs are cyclical, driven by changes in the age structure of relatively large patches of the dominant plant (e.g. the so-called stand cycle of *Calluna vulgaris*; Watt 1964). Even with species that may exhibit local internal cycles, however, the more typical pattern of recruitment may be dominated by external, relatively large scale disturbance. Regeneration of heather *Calluna vulgaris*, for example, can occur from seed or by layering, but the relatively large scale patterns of recruitment tend to be dominated by fire (Hobbs & Gimmingham 1987).

(*c*) *Herbaceous perennials*

Herbaceous perennials may be evergreen or deciduous, mobile or fixed, tussock-forming or straggling, bulbous or rhizomatous, geophyte or hemicryptophyte, and all these different features of their growth form are likely to influence their population dynamics. They will affect their characteristic patterns of survivorship (depending upon the longevity of their perennating tissues and the ability to produce vegetative offshoots under a range of competitive conditions) and fecundity (whether they are monocarpic or polycarpic, alternate bearing or masting, etc).

We have detailed demographic information on only a small sample of herbaceous perennials. Several authors have published long-term studies on the dynamics of terrestrial orchids (Inghe & Tamm 1985; Wells 1981; Hutchings 1987). These studies show: (i) periodic recruitment; (ii) unequal age structures with strong age classes; (iii) irregular flowering. Orchids also exhibit the irritating tendency to come back from the dead. Plants that have disappeared for one or more years may reappear above ground, having spent the intervening period underground either dormant, or living a parasitic existence, obtaining sustenance from mycorrhizal fungi. This means that losses can not be equated with death until many years have elapsed without evident vegetative growth.

An interesting twist to orchid dynamics comes from Gill's (1989) 10-year study of the pink lady's slipper *Cypripedium acaule* in Virginia. Despite their extremely ornate flowers, these flowers appear to be virtually useless at attracting insect pollinators. The flowers 'cheat' the insects by producing neither nectar nor fragrance, and the insects appear to have learnt to avoid them. Cross pollinations by hand produce perfectly satisfactory seed set, and Gill's study provides one of the most convincing examples of pollinator-limitation of plant population density. More plants are produced if more seed is set, and more seed is set if more pollinia are transported to receptive stigmas. In only 1 year out of 10 did more than 5% of the plants produce seed and in 4 years out of 10 there was complete reproductive failure.

The question of how often long-lived herbaceous plants recruit from seed has been the subject of some debate. Everyone agrees that recruitment from seed is uncommon in these plants, but the impression has been given by some authors that seedlings are never found at all. Eriksson (1989) reviewed studies of 68 species and found that seedling recruitment in established populations was actually observed in 40% of cases. Seedling recruitment was more likely in grassland than in woodland species, in above-ground rather than below-ground clonal species, and in species lacking long-distance seed dispersal mechanisms. Seedling recruitment is obviously more likely in communities that have high average proportions of bare ground than in communities with high average cover values (Crawley 1986). Even when seedling recruitment does occur in these clonal communities, the death rate of seedlings is typically much higher than the death rate of daughter ramets. Even though the sample sizes are small, it appears that the possession of seed banks in herbaceous perennials is correlated with the turnover rate of ramets; plants with high turnover (like *Ranunculus repens*) tend to rely relatively more on seedling establishment and to possess seed banks, whereas the slow turnover species (like *Potentilla anserina*) have no seed bank and show little seedling recruitment

Table 1. *Depletion data for long-term studies on herbaceous perennials, showing the estimated half-life. Note that for some of the forest herbs, the plants may live through several generations of the dominant trees. Many of the orchids disappear temporarily then reappear in later years, making estimation of their half-lives more difficult. Data from Inghe & Tamm (1985) and references therein*

species	half-life/years
Hepatica nobilis	360
	59
	32
Sanicula europaea	221
	74
Primula veris	50
	6
	3
Dactylorhiza sambucina	20
	10
Dactylorhiza incarnata	
original cohort	5
1966–67 cohort	3
Orchis mascula	5
Listera ovata	56

(Eriksson 1986). In Tamm's long term study plots (above), there was no relation between the numbers of seedlings recruiting in any year, and the magnitude of seed production in the previous season (Inghe & Tamm 1985), suggesting that although seedlings were found in most years, seedling recruitment was not seed limited.

Clearly, these kinds of studies need to be done over protracted periods, especially in circumstances where microsite availability is likely to be episodic (e.g. in arid systems). It is entirely plausible, for example, that recruitment could be seed limited in that rare year in which a combination of adequate rainfall and sufficient disturbance to create plentiful microsites means that recruitment is possible.

Most of the information on clonal plants comes in the form of depletion curves, in which the mortality of permanently marked clones is monitored (table 1). It is striking that the half-lives of several woodland perennials are long enough that one genet of *Hepatica nobilis* or *Sanicula europaea* might outlive two or more generations of the dominant canopy trees (Inghe & Tamm 1985).

(*d*) *Annuals in dunes and deserts*

More is known about the population dynamics of annuals than about any other group of plants. Several data sets exist in which the dynamics have been followed for more than five years and a few data sets of 10 or more years have been collected. We need to distinguish the dynamics of desert and sand dune plants in which conditions for recruitment are available more or less constantly, providing there is sufficient rain to allow germination, with early successional communities, in which only one or two years following disturbance are suitable for recruitment. In both cases, we need to consider the relative importance of long-term seed dormancy (the size of the 'seed bank'; Leck *et al.* 1989) and of long-distance seed dispersal.

Given that they have the longest runs of data, they also have the widest variety of dynamic behaviours. Perhaps the commonest pattern is that exponential increase is so rapid that all microsites are occupied after only one or two generations (e.g. *Poa annua* on waste ground; Law 1981). Alternatively, long-term successional changes in the distribution and abundance of perennial plants may so change the local microsite conditions that recruitment becomes impossible, and the population declines to local extinction (e.g. *Vulpia* in sand dunes, figure 2; Watkinson 1990).

In a few cases, cyclic dynamics have been reported. In one case, the skeletons of dead *Salicornia* were thought to occupy microsites in sheltered places for a year after their death, taking up space that would be occupied by recruits in more exposed places, where the skeletons would be washed away. In the second year, the microsites became available, and a two-year cycle in population density was initiated (Wilkon-Michalska 1976, in Symonides (1984)). In a second case, Symonides (1984) described how the tiny annual crucifer *Erophila verna* underwent a two-point cycle in inland dunes in Poland. Fecundity was over-compensating, so that a single plant produced more seed per unit area than did a large number of small plants. At very high densities, many of the rosettes failed to flower, or flowered very late, so that their leaves had not withered at the time of fruit maturation. This meant that much of the seed was intercepted by the upper surface of leaves and desiccated or was eaten by seed predators before it could germinate. Thus extremely low recruitment followed extremely high adult-plant densities, but the large plants that grew at these low densities produced sufficient seed to ensure exceptionally high densities in the following year. She suggests that this is the origin of the observed two-year cycle. A climatic cause of the cycles is ruled out by the fact that different patches within the same dune system are one step out of phase with each other, so that high and low density patches can be found in any given year (Symonides 1984). Only some of the study plots exhibited these cycles, and there is no suggestion that cyclic behaviour is characteristic of *Erophila verna* populations in general.

Both these cases are unusual in that the physical presence of dead or dying parts of mature plants in high density populations ensured that only a small number of plants become established in the following year. It is more common that annual fluctuations in plant abundance are caused by variations in the weather conditions affecting seedling recruitment, and density-dependent variations in fecundity are rarely sufficiently intense as to cause seed-limitation in the years following high density.

(*e*) *Annuals in early successional communities*

The majority of early successional annuals possess long-lived seed banks (Roberts 1981; see also D'Angela *et al.* (1988) for exceptions). Those early successional ruderals that lack dormancy rely upon high fecundity and wide seed dispersal for their persistence (Egler 1954). The two types of plant have quite different dynamics.

In the case of seed bank annuals, there is virtually no relation between population density and seed production in the previous year. This means that classical population dynamics models are virtually useless. The factors determining the relative abundance of these plants in any given year depend almost entirely upon the extent and the precise timing of soil disturbance. We have found that the dominant weed in Pound Hill Field in Silwood Park can be altered from *Myosotis discolor* to *Solanum nigrum* solely by altering the timing of cultivation (M. J. Crawley, unpublished results). In the absence of disturbance these plants will not recruit at all.

Many species will not germinate beneath a canopy of green leaves, while others require several gap-sensing stimuli (e.g. white light and fluctuating temperatures) to initiate germination (Leck *et al.* 1989). In diverse communities of ruderals, a knowledge of the germination biology of the component species is vital to an understanding of their coexistence. For example, despite over 120 repeated annual cultivations, the species richness of the weed flora of the

Broadbalk experiment has not declined, and there is no evidence of competitive exclusion (Crawley 1986).

We know much less about the population dynamics of 'dispersal ruderals', mainly because their occurrence in permanent quadrats is sporadic. There is a well developed theory for the persistence of this kind of annual (Crawley & May 1987; Kadmon & Shmida 1990). Briefly, their fecundity must exceed the reciprocal of the equilibrium proportion of gaps in the canopy (the proportion of the total area over which seed is distributed that constitutes a suitable microsite). The smaller the number of gaps, the higher the fecundity must be to ensure persistence. It is axiomatic that dispersal must be sufficiently efficient to 'find' the available gaps with Poisson (or better) probability.

In a nine-year study of common and sparse prairie grasses, Rabinowitz *et al.* (1989) found that the uncommon species had lower variation in reproductive output and were less likely to show reproductive failure than the common species. The correlation between rainfall and reproductive output was good for the common grasses, but poor for the sparse species, suggesting that the pattern of environmental fluctuations buffers the reproductive output of the sparse species.

5. SPATIAL DYNAMICS

In concluding his paper on the population dynamics of invading *Poa annua*, Law (1981) wrote: 'It may prove to be more enlightening to take spatial pattern as an essential component of plant demography, rather than as an unfortunate barrier to reliable population estimates'. Rather few data sets are sufficiently extensive that they allow simultaneous estimation of spatial and temporal variance in population density in such a way that the relative magnitudes of spatial and temporal variance can be compared.

There are some fundamental problems in obtaining precise, unbiased estimates of spatial variance. Most plant ecologists place their long-term study quadrats around existing aggregates of plants (see above). This means that plant recruitment into previously unoccupied quadrats is rarely estimated in an unbiased way. It has the other undesirable property that most studies are of declining populations and tend, therefore, to give negative estimates of the intrinsic rate of increase. Studies in which the quadrats are placed within suitable habitat at random, and in which the 'empty' quadrats are given as detailed attention as the occupied quadrats, are very rare. I present two examples: one of the annual *Salicornia brachystachys* from Dutch salt marsh and the mobile perennial *Glaux maritima* from meadows on the shores of the Swedish Baltic.

There are several technical problems in any comparison of spatial and temporal variances. While it is relatively clear that one year is a sensible time unit for calculating the variance in population size of an annual plant, it is by no means clear what quadrat size should be used in calculating spatial variance. Plant ecologists have long known that the variance obtained depends upon the size of the quadrat employed, and that the same plants can exhibit spatial patterns on several scales simultaneously (Kershaw 1964).

Again, Hanski (1987) has pointed out that the slope of Taylor's Power Law for plots of spatial variance against spatial mean density is influenced by the degree to which the dynamics of local populations are correlated over large spatial scales. Do all local populations experience the same 'good years' or is there relatively low cross-correlation between local dynamics?

Note, however, that temporal variance is bound to be calculated for quadrats of a given size, and that changing quadrat size would alter the estimate of temporal variance. Quadrat size, therefore, is not just an issue for considerations of spatial variance, but is central in any discussion of plant population dynamics.

(*a*) *Salicornia dynamics*

These data relate to a study from 1977 to 1989 (with gaps in 1981 and 1988) along a transect of 42 quadrats, each measuring 50 × 50 cm (B. Koutstaal & A. Huiskes, unpublished data). The time series shows a fluctuating pattern of abundance that might best be described as a damped decline. There is evidence for density dependent regulation, and a plot of the change in log population size against population density has a significant negative slope (figure 5). Population variability is measured by the standard deviation of log population size. This population fluctuated 12-fold over the 13-year period, and gave a standard deviation (s.d.) in log density of 0.717.

Averaged over 13 years it is quite clear that there are good sites and bad sites for *Salicornia* along the transect (figure 5). Taking the total over all years, we can calculate the s.d. in log population size from quadrat to quadrat along the transect (s.d. = 1.481). It is clear that the population is more variable in space than in time.

Another way of presenting data on population density variation is in the form of plots of log variance against log mean density (Taylor's Power Law). *Salicornia* has a temporal power law slope of 1.24 and a spatial slope of 1.24. The intercept is also significantly higher in the spatial plots.

(*b*) *Glaux dynamics*

A 7-year study by Jerling (1988) involved the repeated census of 13 permanent quadrats, each measuring 2 × 2 m. The time series shows wide fluctuation, but no obvious upward or downward trend (figure 6). There is evidence of density dependent regulation, and although the scatter of points is greater than in the case of *Salicornia* there is a significant negative relationship between change in log population size and population density.

The spatial variation is shown in figure 6, and as with *Salicornia*, it is greater than the temporal variation. The power law plots have slopes of 2.09 for the temporal and 4.74 for the spatial variance.

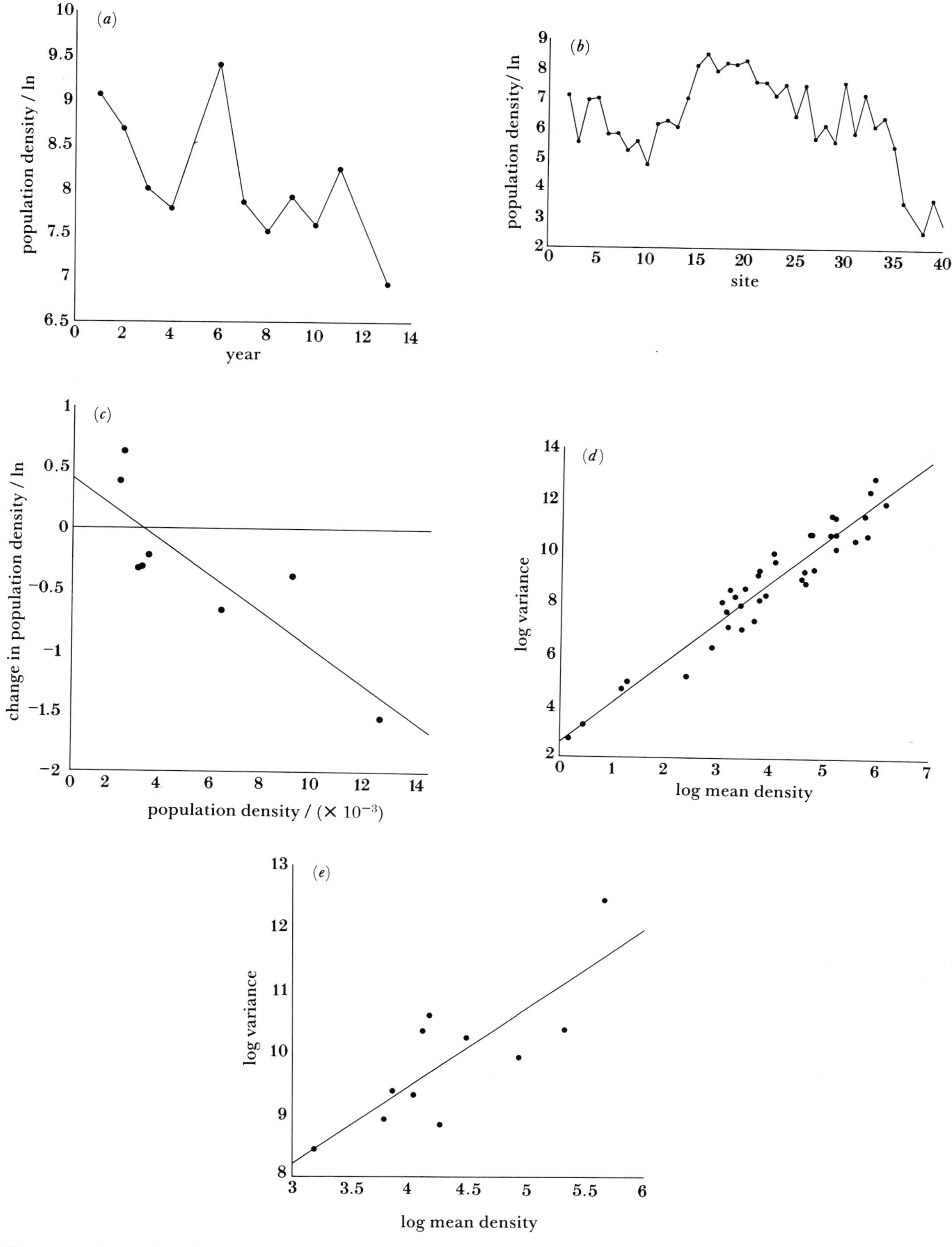

Figure 5. Population dynamics of *Salicornia brachystachys* in Holland. (*a*) Time series. (*b*) Spatial distribution. (*c*) Population regulation. (*d*) Temporal power law. (*e*) Spatial power law. For details see text.

(*c*) *Overview*

Drawing together the data for *Salicornia* and *Glaux* with the classic long-term studies of Epling *et al.* (1960) on desert annual *Limnanthes parryae* and Symonides (1984) on the dune annual *Erophila verna* with our own studies on ragwort *Senecio jacobaea* (Crawley & Gillman 1990) we can begin to compare the relative magnitude of temporal and spatial variation (table 2). In all cases but one, spatial variation exceeded temporal variation by a factor of between 1.7 and 4.4.

Even in the case of *Limnanthes*, however, the spatial

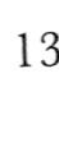

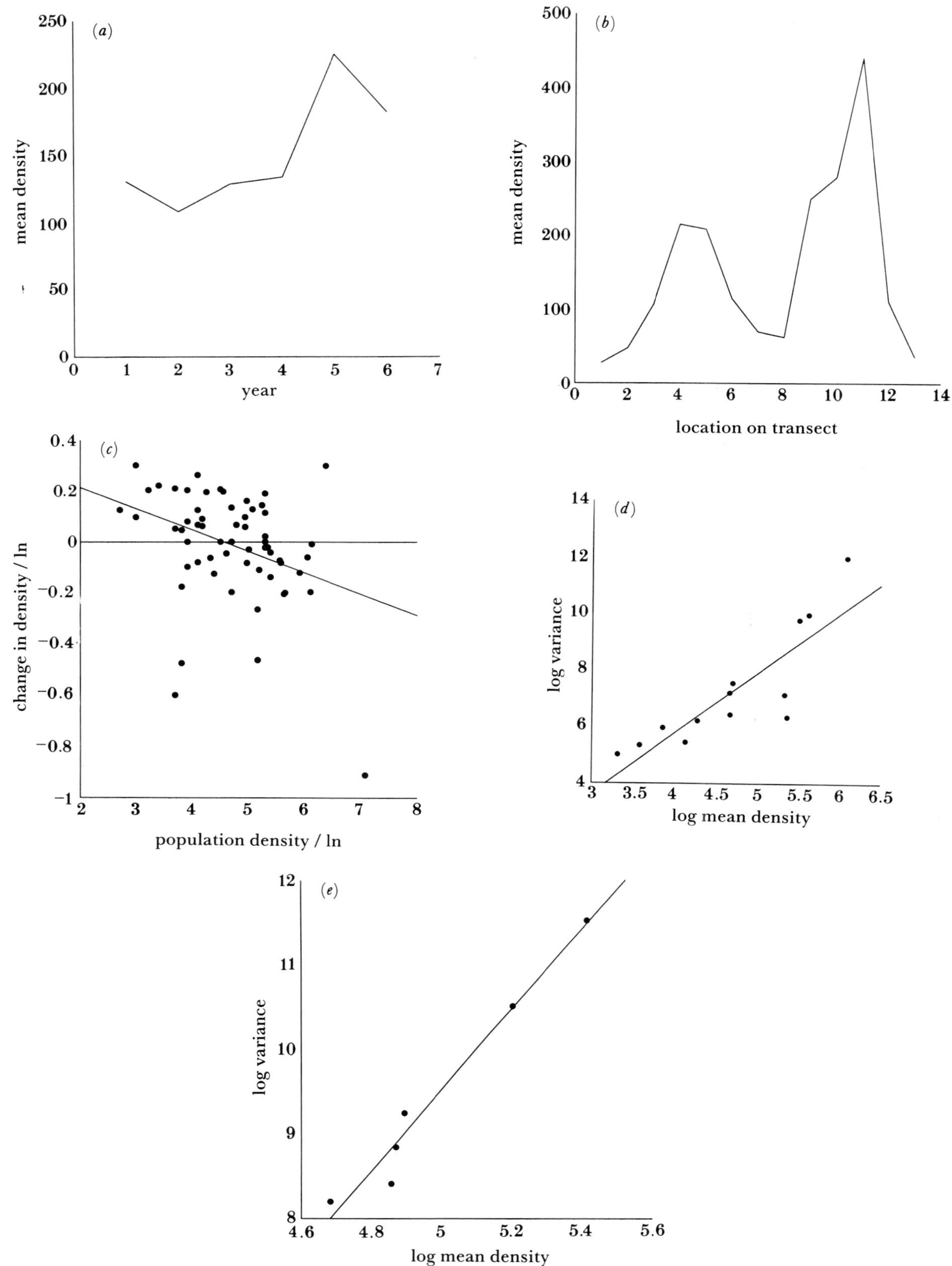

Figure 6. Population dynamics of *Glaux maritima* in Sweden. (*a*) Time series. (*b*) Spatial distribution. (*c*) Population regulation. (*d*) Temporal power law. (*e*) Spatial power law. For details see text.

dynamics were pronounced and consistent. Epling *et al.* (1960) write 'Within any year their dispersion is not uniform but follows a very local pattern of greater or less concentration. This pattern has persisted over the period of study'.

6. DISCUSSION

Our understanding of general patterns in plant population dynamics has been hampered by shortcomings in all the important scientific aspects of plant

Table 2. *The relative magnitude of spatial and temporal variation in population density for five plant species, measured as the standard deviation in log population size. In four cases the spatial variation is substantially greater than the variation from year to year. The exception (Limnanthes) is a widespread desert plant where recruitment is limited by widespread, unpredictable rain storms. For sources, see text*

species	time	space	multiple
Salicornia	0.717	1.48	2.07
Erophila	0.157	0.688	4.39
Limnanthes	0.754	0.313	0.41
Glaux	0.270	0.853	3.16
Senecio	0.793	1.340	1.69

ecology: in its theoretical development, in an unwillingness to carry out simple experiments, and in the short-sighted protocols of observational studies.

Plant populations dynamics theory has relied too heavily on models borrowed from animal ecology. Thus many of the essential attributes of plants have been played down (their longevity, tremendous size plasticity, the importance of neighbours, the constancy of yield per unit area), and the models have adopted the animal ecologists' relish for assumptions of equilibrium, predictable recruitment, and constant rather than successional environments.

The reluctance to carry out simple manipulative field experiments on recruitment has meant that the relative importance of seed, microsites, competition and herbivory remains unknown even in systems that have been studied over many years. The practice of sowing extra seeds and following their fate and the fate of any seedlings they may produce should be a routine element of any field study in plant dynamics. Seeds should be sown into a range of microhabitats, apparently unsuitable as well as apparently suitable, so that we increase our understanding of why plants do not occur in certain places.

Observational studies of plant populations are typically set up around existing mature plants and empty quadrats are ignored. This has two undesirable consequences: (i) most observed populations drift inexorably to extinction, and (ii) because recruitment may occur in empty rather than occupied quadrats, such studies lead to serious underestimates of the population's actual rate of increase. Observational studies must include empty quadrats, and quadrats should be sited at random within the chosen habitat, unless persuasive arguments to the contrary can be presented. Bias in siting the quadrats is one of the most pernicious problems in field ecology.

Incorporating spatial dynamics into the mainstream of plant population dynamics studies is important because spatial variation is as great or greater than temporal variation in most studies for which suitable data are available. Spatial variation in density should not be regarded as a nuisance, but rather as something that is interesting and important to study in its own right. The 'empty' quadrats are important: are they habitable or not; are they really empty, or do they contain dormant seeds; are they empty because they have not received propagules or because they have undergone successional change rendering them uninhabitable until the next major disturbance?

Spatial heterogeneity also exists in the kind of dynamics exhibited by a given species. For example, there may be within-species differences in life histories that have both genetic and environmental causes. Thus *Plantago lanceolata* sets more seed on dry soils than on wet; there is variation in the production of side shoots, the size of seed bank; the mass of individual seeds and seedling competitiveness; mature plant size and longevity also vary from site to site (van Groenendael & Slim 1988). Adding the complexity associated with overlaying spatial heterogeneity in genotypic composition on spatial heterogeneity in environmental conditions represents a major challenge to theoreticians and experimentalists alike.

Plant population dynamics are relatively tame compared to animal dynamics, and we have few convincingly documented examples of cyclical plant populations and none of chaotic dynamics among plants (exceptions include certain annuals (Symonides 1984), vines (Mueller-Dombois 1981), some shrubs (Watt 1964) and trees exhibiting 'wave dynamics' (Sprugel 1976), but in most cases these are age-structure effects, rather than generation-to-generation delayed density dependence effects). There are at least three reasons for this. The first is morphological. Because of their iterated, modular construction, plants are immensely plastic in their growth form. There is typically a linear relation between size and fecundity and there is rarely any threshold size below which reproduction is impossible (Rees & Crawley 1989). This means that the tendency for over-compensating density dependence in fecundity is reduced (but see the special case of *Erophila verna*, above). The second reason is that the nature of plant competition is essentially asymmetric. Plants contest rather than scramble for light, so that there are winners and losers. The relatively large plants prosper, no matter what the details of the size distributions among the smaller plants. The smallest plants are almost always the most likely to die. This combination of a linear size-fecundity schedule followed by contest competition among the recruits is sufficient to guarantee extremely stable dynamics in a constant environment, and rapid recovery from catastrophe in a fluctuating environment. Thirdly, recruitment is often not seed limited. Thus population fluctuations tend to be due mainly to exogenous controls on recruitment, rather than to the kind of over-compensating density dependence that can cause cycles or chaotic dynamics in model systems.

It is possible to develop mathematical models of plant dynamics that exhibit 'interesting dynamics' like cycles or chaos (Pacala & Silander 1985), but the consensus is that these models are more animal-like than plant-like in their behaviour (Watkinson 1985). More realistic plant models, incorporating linear size–fecundity relation, constant final yield and contest competition, tend to exhibit much more stable dynamics (Crawley & May 1987).

The role of herbivores in plant dynamics has been the subject of controversy (Crawley 1983; Hairston

1989). As so often, the protagonists are arguing about different sides of the same coin. Herbivores as a group appear to have relatively little impact on system level properties of the plant community like total biomass (the world is green despite herbivory). This does not mean, however, that herbivores have no impact on the abundance of individual plant species. Small changes in herbivory can change the identity of the dominant plant species (Crawley & Pacala 1991). The importance of herbivory is implicit in the recognition of certain plant species as increasers or decreasers under grazing. It is important to bear in mind, however, that the classification of a given plant is seldom unequivocal, and there is considerable inconsistency in the behaviour of increasers and decreasers. For example, a plant that might increase under moderate grazing intensities, might decrease rapidly under more intense grazing pressure (Crawley 1983).

There are a number of important plants traits that make plant dynamics (or more strictly the dynamics of sessile modular organisms) fundamentally different from those of animals (and other mobile, unitary organisms). (i) They exhibit extraordinary phenotypic plasticity in size, so that the fecundity of an individual genotype can vary by four orders of magnitude depending upon the conditions it experiences during development. (ii) Plants are sedentary, and so neighbour relations are extremely important (performance is well predicted by the number of neighbours larger than a given individual; smaller neighbours are typically of little consequence). (iii) The importance of competition in animal population dynamics is a matter of debate; in plant population dynamics it is quite clearly of over-riding significance in the majority of cases. (iv) Stock recruitment curves are often useless in understanding plant population dynamics, because of seed banks, immigration, repeated reproduction and the existence of strong age classes (in this, plants are like fish). (v) Plant populations have more stabilizing refugia than most animal populations, in the form of long-lived seed banks, very long life, ungrazeable perennating organs, repeated reproduction and wide seed dispersal.

The directions for future research in plant populations dynamics are reasonably clear. (i) We need to carry out simple experiments on seed, microsite and herbivore limitations on recruitment. These experiments should be carried out on a variety of plant growth forms in a range of different habitats. (ii) Observational studies should place their quadrats at random within suitable habitat, instead of around selected mature individuals. If we want to understand recruitment, we need to observe empty quadrats. (iii) We should develop theoretical models specifically for plants, incorporating the key aspects of plant biology outlined above, and attempting to make these models spatially explicit.

7. SUMMARY

(i) Plant populations are regulated by density dependent processes. This may involve competition between seeds for access to limited microsites, or strongly asymmetrical competition between seedlings and established adult plants. Herbivory can regulate plant density below the level that would be set by plant competition (but note that when it does so, unpalatable plants often increase in density to take up the available space).

(ii) In terms of their overall relative importance in plant population dynamics, this review suggests that the various processes be ranked as follows: interspecific competition > herbivory > intraspecific competition for microsites > seed limitation.

(iii) Plant competition is fundamentally asymmetric. Plants contest rather than scramble for resources and this has a major stabilizing effect on plant population dynamics.

(iv) Plant population dynamics are relatively tame. We have very little evidence for cyclic dynamics caused by over-compensating density dependence, and no evidence for chaotic dynamics.

(v) Spatial dynamics of plants are very important, and plants conform to Taylor's Power Law in their patterns of both spatial and temporal variation. In four studies out of five where it was possible to contrast the relative variation of temporal and spatial dynamics, the spatial component of variation was greater by a factor of between 1.7 and 4.4.

(vi) Protocols for theoretical, experimental and observational studies of plant population dynamics are proposed.

8. REFERENCES

Augspurger, C. K. & Kelly, C. K. 1984 Pathogen mortality of tropical tree seedlings: experimental studies of the effects of dispersal distance, seedling density and light conditions. *Oecologia* **61**, 211–217.

Bormann, F. H. & Lichens, G. E. 1979 *Patterns and process in a forested ecosystem.* New York: Springer–Verlag.

Brenchley, W. E. & Warrington, K. 1930 The weed seed population of arable soil. I. Numerical estimation of viable seeds and observations on their natural dormancy. *J. Ecol.* **18**, 235–272.

Brown, J. R. & Archer, S. 1987 Woody plant seed dispersal and gap formation in a North American subtropical savanna woodland: the role of domestic herbivores. *Vegetatio* **73**, 73–80.

Burdon, J. J., Groves, R. H. & Cullen, J. M. 1981 The impact of biological control on the distribution and abundance of *Chondrilla juncea* in south-eastern Australia. *J. appl. Ecol.* **18**, 957–966.

Callaghan, T. V. & Emanuelson, U. 1985 Population structure and processes of tundra plants and vegetation. In *The population structure of vegetation* (ed. J. White). Dordrecht: Junk.

Crawley, M. J. 1983 *Herbivory. The dynamics of animal-plant interactions.* Oxford: Blackwell Scientific Publications.

Crawley, M. J. 1986 The population biology of invaders. *Phil. Trans. R. Soc. Lond.* B**314**, 711–731.

Crawley, M. J. 1989 Insects and plant population dynamics. *A. Rev. Entomol.* **34**, 531–564.

Crawley, M. J. & Gillman, M. 1990 The population dynamics of ragwort and cinnabar moth in acid grasslands. *J. Anim. Ecol.* **58**, 1035–1050.

Crawley, M. J. & May, R. M. 1987 Population dynamics and plant community structure: competition between annuals and perennials. *J. theor. Biol.* **125**, 475–489.

Crawley, M. J. & Nachapong, M. 1985 The establishment of seedlings from primary and regrowth seeds of ragwort (*Senecio jacobaea*). *J. Ecol.* **73**, 255–261.

Crawley, M. J. & Pacala, S. 1991 Herbivores, plant parasites and plant diversity. In *Parasitism: conflict or coexistence* (ed. C. Toft), pp. 157–174. Oxford University Press.

D'Angela, E., Facelli, J. M. & Jacobo, E. 1988 The role of permanent soil seed bank in early stages of a post-agricultural succession in the Inland Pampa, Argentina. *Vegetatio* **74**, 39–45.

de Jong, T. & Kinkhamer, P. 1986 Population ecology of the biennials *Cirsium vulgare* and *Cynoglossum officinale*. Ph.D. thesis, University of Leiden, Holland.

Duggan, A. E. 1986 Ph.D. thesis, University of London.

Egler, F. E. 1954 Vegetation science concepts: I. Initial floristic composition—a factor in old field vegetation development. *Vegetatio* **4**, 412–417.

Epling, C., Lewis, H. & Ball, F. M. 1960 The breeding group and seed storage: a study in population dynamics. *Evolution* **14**, 238–255.

Eriksson, O. 1986 Survivorship, reproduction and dynamics of ramets of *Potentilla anserina* on a Baltic seashore meadow. *Vegetatio* **67**, 17–25.

Eriksson, O. 1989 Seedling dynamics and life histories in clonal plants. *Oikos* **55**, 231–238.

Forrester, G. 1990 The population ecology of acorn weevils and their influence on natural regeneration of oak. Ph.D. thesis, University of London.

Fowler, N. L. 1986 Density-dependent population regulation in a Texas grassland. *Ecology* **67**, 545–554.

Gill, D. E. 1989 Fruiting failure, pollinator inefficiency, and speciation in orchids. In *Speciation and its consequences* (ed. D. Otte & J. A. Endler), pp. 458–481. Sunderland, Massachusetts: Sinauer.

Goldberg, D. & Turner, R. M. 1986 Vegetation change and plant demography in permanent plots in the Sonoran desert. *Ecology* **67**, 695–712.

Groenendael, J. M. van & Slim, P. 1988 The contrasting dynamics of two populations of *Plantago lanceolata* classified by age and size. *J. Ecol.* **76**, 585–599.

Grubb, P. J. 1988 The uncoupling of disturbance and recruitment, two kinds of seed bank, and persistence of plant populations at the regional and local scales. *Ann. Zool. Fennici* **25**, 23–36.

Hairston, N. G. 1989 *Ecological experiments: purpose, design and execution.* Cambridge University Press.

Hanski, I. 1987 Cross-correlation in population dynamics and the slope of spatial variance-mean regression. *Oikos* **50**, 148–151.

Harper, J. L. 1977 *Population biology of plants.* London: Academic Press.

Harper, J. L. 1975 Review of Allelopathy by E. L. Rice. *Q. Rev. Biol.* **50**, 493–495.

Hastings, J. R. & Turner, R. M. 1965 *The changing mile: an ecological study of vegetation change with time in the lower mile of an arid and semiarid region.* Tucson: University of Arizona Press.

Hemstron, M. A. & Franklin, J. F. 1982 Fire and other disturbances of the forests in Mt Ranier National Park. *Quat. Res.* **18**, 32–51.

Hobbs, R. J. & Gimmingham, C. H. 1987 Vegetation, fire and herbivore interactions in heathland. *Adv. ecol. Res.* **16**, 87–173.

Horn, H. 1975 Forest succession. *Scient. Am.* **232**, 90–98.

Hubbell, S. P. & Foster, R. B. 1990 Short-term population dynamics of trees and shrubs in a neotropical forest: El Nino effects and successional change. *Ecology* **71**. (In the press.)

Hutchings, M. J. 1986 The structure of plant populations. In *Plant ecology* (ed. M. J. Crawley), pp. 97–136. Oxford: Blackwell Scientific Publications.

Hutchings, M. J. 1987 The population biology of the early spider orchid, *Ophrys sphegodes* Mill. *J. Ecol.* **75**, 711–727, 729–742.

Inghe, O. & Tamm, C. O. 1985 Survival and flowering in perennial herbs. IV. The behaviour of *Hepatica nobilis* and *Sanicula europaea* on permanent plots during 1943–1981. *Oikos* **45**, 400–420.

Jerling, L. 1988 Population dynamics of *Glaux maritima* (L.) along a distributional cline. *Vegetatio* **74**, 161–170.

Kadmon, R. & Shmida, A. 1990 Spatiotemporal demographic processes in plant populations: an approach and a case study. *Am. Nat.* **135**, 382–397.

Keddy, P. A. 1981 Experimental demography of the sand-dune annual, *Cakile edentula*, growing along an environmental gradient in Nova Scotia. *J. Ecol.* **69**, 515–630.

Kershaw, K. A. 1973 *Quantitative and dynamic plant ecology* (2nd edn). London: Arnold.

Law, R. 1981 The dynamics of a colonizing population of *Poa annua*. *Ecology* **62**, 1267–1277.

Leck, M. A., Parker, V. T. & Simpson, R. L. 1989 (eds) *Ecology of seed banks.* San Diego: Academic Press.

Louda, S. M. 1982 Limitation of the recruitment of the shrub *Haplopappus squarrosus* (Asteraceae) by flower- and seed-feeding insects. *J. Ecol.* **70**, 43–53.

Mack, R. N. & Pyke, D. A. 1983 The demography of *Bromus tectorum*: variation in time and space. *J. Ecol.* **71**, 69–93.

Mueller-Dombois, D. 1981 Vegetation dynamics in a coastal grassland of Hawaii. *Vegetatio* **46**, 131–140.

Nakashizuka, T. 1987 Regeneration dynamics of beech forests in Japan. *Vegetatio* **69**, 169–175.

Noble, J. C. 1989 Fire studies in mallee (*Eucalyptus* spp.) communities of western New South Wales: the effects of fires applied in different seasons on herbage productivity and their implications for management. *Aust. J. Ecol.* **14**, 169–187.

Noy Meir 1975 Stability of grazing systems: an application of predator–prey graphs. *J. Ecol.* **63**, 459–481.

Ogden, J. 1985 An introduction to plant demography with species reference to New Zealand trees. *N.Z. Jl Bot.* **23**, 751–772.

Owens, M. K. & Norton, B. E. 1989 The impact of 'available area' on Artemisia tridentata seedling dynamics. *Vegetatio* **82**, 155–162.

Pacala, S. & Silander, J. A. 1985 Neighborhood models of plant population dynamics. I. Single-species models of annuals. *Am. Nat.* **125**, 385–411.

Pacala, S. & Silander, J. A. 1990 Field tests of neighborhood population dynamics models of two annual weed species. *Ecol. Monogr.* **60**, 113–134.

Parker, M. A. & Root, R. B. 1981 Insect herbivores limit habitat distribution of a native composite, *Machaeranthera canescens*. *Ecology* **62**, 1390–1392.

Pielou, E. C. 1977 *Mathematical ecology.* New York: John Wiley.

Prentice, I. C. 1986 Vegetation response to past climatic variations. *Vegetatio* **67**, 131–141.

Rabonowitz, D., Rapp, J. K., Cairns, S. & Mayer, M. 1989 The persistence of rare prairie grasses in Missouri: environmental variation buffered by reproductive output of sparse species. *Am. Nat.* **134**, 525–544.

Rees, M. & Crawley, M. J. 1989 Growth, reproduction and population dynamics. *Functional Ecology* **3**, 645–653.

Rice, E. L. 1984 *Allelopathy* (2nd edn). New York: Academic Press.

Roberts, H. A. 1981 Seed banks in soil. *Adv. appl. Biol.* 1–55.

Sheppard, A. W. 1987 Insect herbivore competition and the population dynamics of *Heracleum sphondylium* L. (Umbelliferae). Ph.D. thesis, University of London.

Sprugel, D. G. 1976 Dynamic structure of wave-generated *Abies balsamea* forests in the north-eastern United States. *J. Ecol.* **64**, 889–911.

Svensson, B. M. & Callaghan, T. V. 1988 Small-scale vegetation pattern related to the growth of *Lycopodium annotinum* and variations in its micro-environment. *Vegetatio* **76**, 167–177.

Symonides, E. 1984 Population size regulation as a result of intra-population interactions. III. Effect of *Erophila verna* (L.) C.A.M. population density on the abundance of the new generation of seedlings. *Ekologia Polska* **32**, 557–580.

Symonides, E., Silvertown, J. W. & Andreasen, V. 1986 Population cycles caused by overcompensating density-dependence in an annual plant. *Oecologia* **71**, 156–158.

Watkinson, A. R. 1986 The dynamics of plant populations. In *Plant ecology* (ed. M. J. Crawley), pp. 137–184. Oxford: Blackwell Scientific Publications.

Watkinson, A. R. 1990 The population dynamics of *Vulpia fasciculata*: a nine-year study. *J. Ecol.* **78**, 196–209.

Watt, A. S. 1964 The community and the individual. *J. Ecol.* **52**, 203–211.

Wells, T. C. E. 1981 Population ecology of terrestrial orchids. In *The biological aspects of rare plant conservation* (ed. H. Synge), pp. 281–295. Chichester: Wiley.

Whitmore, T. C. 1990 Changes over 21 years in the Kolombangara rain forests. *J. Ecol.* **78**. (In the press.)

Williams, E. D. 1978 *Botanical composition of the Park Grass Plots at Rothamsted 1856–1976*. Rothamsted, England.

Discussion

G. J. S. Ross (*Rothamsted Experimental Station, Harpenden, Herts. U.K.*). In Dr Crawley's historical survey of plant population changes he mentions the Rothamsted Park Grass experiment. The relative abundance of different species on each plot has been estimated for several occasions from 1862 onwards, and the divergence from the original composition at the start of the experiment is well recorded.

Unfortunately the difficulty and expense of maintaining this tradition has meant that there are no comparable recent estimates. I hope that the interest of those at Silwood Park and elsewhere will be fruitful.

Density dependence, regulation and variability in animal populations

ILKKA HANSKI

Department of Zoology, University of Helsinki, P. Rautatiekatu 13, SF-00100, Helsinki, Finland

SUMMARY

This paper reviews a series of approaches to the study of density dependence, regulation and variability in terrestrial animals, by using single-species, multispecies and life table time series data. Special emphasis is given to the degree of density dependence in the level of variability, which is seldom discussed in this context, but which is conceptually related to population regulation. Broad patterns in density dependence, regulation and variability in vertebrates and arthropods are described, with some more specific results for moths and aphids. Vertebrates have generally less variable populations than arthropods, which is the only well documented, consistent pattern in population variability. The degree of density dependence of variability is negatively correlated with the average level of variability, suggesting that generally the more regulated populations are less variable. Most population studies, especially on insects, have involved outbreak species with complex dynamics, which may explain the common failures to detect density dependence in natural populations. In British moths, density dependence is less obvious in the more abundant species. The study of uncommon and rare species remains a major challenge for population ecology.

1. INTRODUCTION

Density dependence, population regulation and variability in population size are three related and recurrently debated concepts in population ecology. Low variability implies the operation of some regulatory processes, which by definition involve one or more density dependent components. But the reverse is not true: not all density dependent processes regulate a population towards a stable equilibrium point or reduce variability.

The purpose of this paper is to present a selective review of studies of density dependence, regulation and variability in terrestrial animals, supplemented with original analyses on moths and aphids. By density dependence I mean 'some dependence of average growth rate on present and/or past population densities' (Murdoch & Walde 1989). Variability refers to the level of variation in population size or density from one generation to another. Population regulation has been given different definitions even in the most recent literature. To some ecologists, regulation is 'the process whereby a population returns to its equilibrium' (Varley *et al.* 1973; Dempster 1983; Sinclair 1989). If such a definition is adopted, it would be preferable to include stable non-point attractors among the equilibria. To many others, however, regulation simply means 'long-term persistence and fluctuations within limits, with the lower limit > 0' (see, for example, Mountford (1988); Murdoch & Walde (1989)), a definition that essentially equates regulation with persistence. In this paper, I will discuss and employ a concept which in a sense unites these two facets of regulation.

This paper is limited to observational data, and no experimental studies or techniques are cited, even if they are often superior to observational studies in answering questions about density dependence and population regulation (Murdoch 1970; Murdoch & Walde 1989; Gaston & Lawton 1987). Table 1 outlines the kinds of data and the types of analyses that have been used to study density dependence and variability in terrestrial animals. The following three sections focus in turn on single-species, multi-species and life table time series data. Most studies are concerned with what happens in local populations, but it is possible that a metapopulation perspective would provide a better understanding of population regulation in some species (§5).

Much of the debate about population dynamics is framed in sharp dichotomies: is there density dependence or not? Are local populations regulated or not? Posing such dichotomies is likely to be misleading, because the answers depend on the spatial and temporal scales under consideration, and on the kind of data available. This paper has been written in the spirit that observational data are better suited for comparative analyses of many species rather than for population dynamic studies of single species.

2. SINGLE-SPECIES TIME SERIES

(*a*) *Conceptual and methodological issues*

The most commonly used measure of variability is

Table 1. *Types of data and analyses that have been used to examine density dependence, regulation and variability in terrestrial animals*

type of data and analysis	object of analysis
single-species time series	
autoregressive and other statistical techniques	density dependence of population change
standard deviation of log-transformed census data	level of variability, population regulation
slope of the temporal variance-mean regression	density dependence of variability (= population regulation?)
multi-species time series	
regression techniques	causal explanation of population regulation
constancy in species composition	mechanisms of population regulation
life table time series	
k factor analysis	density dependence of particular mechanisms

the standard deviation of the logarithms of population sizes, s, measured over generations. Unfortunately, there are two major difficulties in using s or any other single parameter as a measure of variability of natural populations. First, although s is often assumed to be a density independent measure of variability, and hence suitable for comparisons of species varying in average abundance (Connell & Sousa 1983), generally it is not (below). A particular problem is created by time series for rare species with many zeros, in which variability is necessarily underestimated (now using the standard deviation of $\log[N+1]$). Second, the value of s depends on the spatial and temporal scales of sampling, which greatly complicates most analyses. In the real world, different processes affecting variability operate at different timescales, and there is no guarantee that variability will reach an asymptotic value with time before the population goes extinct (Pimm & Redfearn 1988). Figure 1 gives an example, for five species of forest insects, in which variability continues to increase

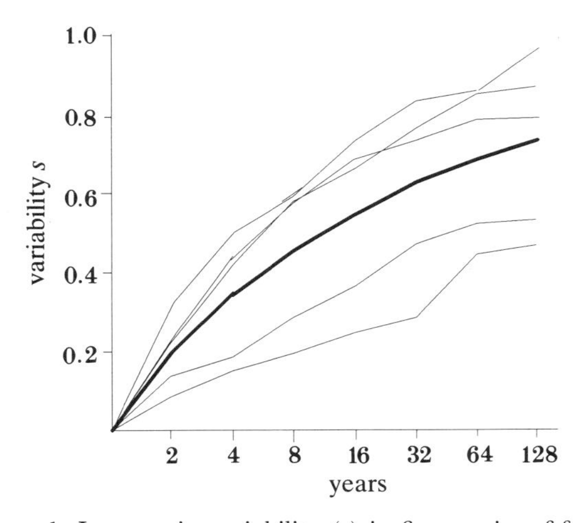

Figure 1. Increase in variability (s) in five species of forest insects with increasing length of the census period (the species are *Bupalus piniarius*, *Panolis flammea*, *Lymantria monácha*, *Dendrolimus pini* and *Diprion pini*; thin lines are the averages for four localities, the thick line is the average for the five species; data measure the extent of local and regional population outbreaks, from Klimetzek (1979)). Calculations were done on nested data as explained by Pimm & Redfearn (1988). Analysis of covariance indicated that species, locality and census period all had a highly significant ($p < 0.0001$) effect on s.

after 100 years of data-collecting! In brief, simple as the concept of variability is in theory, complex are the problems in using it in practice.

There has been much debate in the literature in recent years about how to use time series data to show density dependence. Various statistical techniques have been described, discussed and 'tested' by many authors (Morris 1959; Williamson 1972; Bulmer 1975*a*; Slade 1977; Vickery & Nudds 1984; Gaston & Lawton 1987; Pollard *et al.* 1987; Turchin 1990). None the less, many difficulties remain: (i) most techniques are suitable for detecting only linear density dependence, whereas various nonlinearities are common in nature; (ii) most techniques are unsuitable for detecting delayed density dependence, and may underestimate the overall level of density dependence (Turchin 1990); (iii) our chances of detecting density dependence increase with the length of the time series (Hassell *et al.* 1989), hence many failures to detect density dependence may simply be because of short runs of data, and (iv) density dependence may occur only infrequently, and only at some spatial scales. In view of these difficulties, I suggest that observational data are better suited for comparative studies of density dependence rather than for the elusive search of whether density dependence 'exists' or not. Three useful undertakings are to ascertain at which spatial and temporal scales density dependence is strongest (§5), which are the agents of density dependence (the focus of life table studies; §4), and how patterns of density dependence vary between taxa (Fowler 1981) and environments (Stubbs 1977).

Turning to population regulation, the most convincing demonstrations of regulation are expected to come from experimental studies (Murdoch 1970; Sinclair 1989), regardless of which definition of regulation is used (§1). When only observational data are available, one may attempt to show either that the underlying (deterministic) dynamics involve stabilizing density dependence (Taylor & Turchin, in preparation), or that population variability approaches an asymptotic value with time (Murdoch & Walde 1989; regulation used in the sense of persistence). Two problems with the latter approach are that, in fact, variability often increases with time without reaching an asymptotic value (figure 1), and variability often depends on density.

The degree to which the level of variability is density

dependent is an interesting question which has not been discussed much in this context. Strongly density dependent variability (s, measured on a logarithmic scale) means that the amplitude of population fluctuations (on an arithmetic scale) is relatively independent of average density, which changes when, for example, different limiting (but density independent) factors become more or less important. Density dependence of variability combines, in a sense, the two notions about population regulation, namely stabilizing density dependence (of population change) and constrained variability. The degree of density dependence of variability is conveniently measured by the slope of the temporal variance-mean regression (Taylor 1961). Note that variability (s) is density independent only if the slope of the variance-mean regression equals two (Hanski 1982).

(*b*) *Patterns of density dependence, variability and population regulation*

(i) *Density dependence of population change*

Several recent analyses of published population studies have concluded that the frequency of significant density dependence in animal populations is surprisingly low, often less than 50% (Dempster 1983; Strong *et al.* 1984; Stiling 1987, 1988; Gaston & Lawton 1987). In the previous section I enumerated several technical reasons for the many failures to detect density dependence. One biological reason, to be further discussed in §2*d*, is the high frequency of outbreak (pest) species among the species that have been studied.

In insects, the type of density dependence has been found to vary with the degree of temporariness of the habitat, from over-compensating density dependence in species living in more-temporary habitats to less severe density dependence in species living in more-permanent habitats (Stubbs 1977). Fowler (1981) suggested that while in large mammals most density dependent change occurs at high densities, in insects most density dependent change occurs at densities far below the environmental carrying capacity (not consistent with Stubbs' findings for species living in temporary habitats). Sinclair (1989) reviews in some detail between-taxon differences in the pattern of density dependence. It would be satisfying to conclude something general about the strength of density dependence in vertebrates versus arthropods, but we have too little comparable data to draw such conclusions.

(ii) *Variability*

A series of recent reviews has examined the level of variability in animal populations. A pioneering study by Connell & Sousa (1983) found no difference between terrestrial vertebrates and arthropods, but their database was limited and taxonomically biased. In the pooled results of many studies, terrestrial vertebrates have significantly less variable populations, on average, than arthropods (figure 2). Previous studies have reported that lizards have especially constant populations among vertebrates (Schoener 1985); that birds have more constant populations than mammals

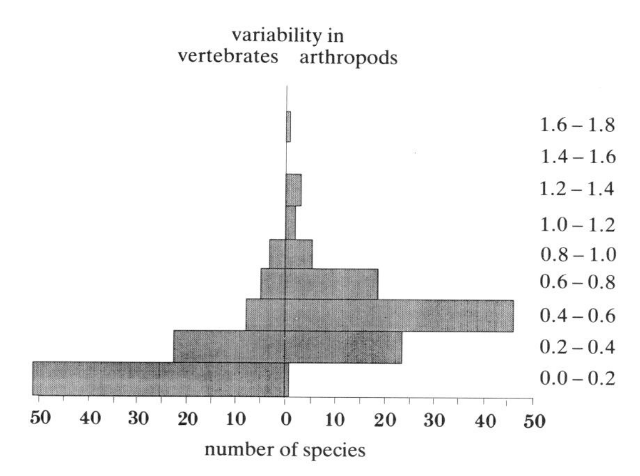

Figure 2. Population variability (s) in 91 species of terrestrial vertebrates (mammals, birds and lizards, from Connell & Sousa (1983); Schoener (1985); Ostfeld (1988); T. Solonen, unpublished data), and in 99 species of terrestrial arthropods (moths, aphids, hoverflies, grasshoppers, etc., from Connell & Sousa (1983); Owen & Gilbert (1989); Joern & Pruess (1986); I. Woiwod, unpublished data). Variability has been measured for generations where possible (most studies). The distributions for vertebrates and arthropods are highly significantly different (two-tailed Kolmogorov–Smirnov statistic 0.58, $p < 0.0001$). Note that average variability in the five forest insects in figure 1 remains between 0.4 and 0.6, the modal class for arthropods in this figure, for the time periods between 5 and 30 years, which bracket the typical lengths of time series data available for most species. The results in figures 1 and 2 are thus consistent with each other.

(Connell & Sousa 1983); and that aphids have more variable populations than hoverflies and moths (Owen & Gilbert 1989; figure 3). Some of these conclusions are probably correct, e.g. temperate small mammals tend to have more variable populations than temperate birds (figure 5), but others are based on small and possibly biased samples, and most results are hampered by the technical problems discussed in §2(*a*).

Theoretical arguments can be developed for both negative and positive correlations between variability and population growth rate (Pimm 1984), and between variability and the degree of polyphagy in herbivorous insects (Redfearn & Pimm 1988); and both negative and positive correlations have been found in empirical studies (table 2). One unexpected pattern to emerge is lack of correlation between variability and latitude in insects (Wolda 1983). Once again, however, it is not clear which of these studies are not compounding variability and density, and which results are not artefacts of the scale of sampling. I conclude that there are no well documented, consistent patterns in variability apart from the difference between vertebrates and arthropods (figure 2).

The broad difference between vertebrates and arthropods should not distract us from the fact that variability is primarily a property of populations living under particular environmental conditions. Recent demonstrations of conspicuously increasing variability in small mammals with latitude in northern Europe (Hansson & Henttonen 1985, 1988) is a case in point. Increasing variability is associated with increasingly regular multiannual cyclicity, with the cycle length

Table 2. *Ecological correlates of population variability in insects*

ecological factor	effect on variability	reference[a]
population growth rate	positive	1
	negative	8
body size	negative	2, 4
polyphagy	positive	2, 6, 5
	negative	7
geographical distribution	positive	2, 3, 4
number of competitors	positive	5
latitude	no correlation	9
	positive	10

[a] 1, Spitzer & Leps (1988); 2, Gaston (1988); 3, Glazier (1986); 4, Gaston & Lawton (1988); 5, Watt (1965); 6, Rejmanek & Spitzer (1982); 7, Redfearn & Pimm (1988); 8, Pimm (1984); 9, Wolda (1983); 10, Hansson & Henttonen (1985).

increasing from 3 years in southern Fennoscandia to 5 years in northern Lapland (Hanski *et al.* 1991). In this case large variability is not associated with lack of population regulation, but to a change from predominantly direct (not delayed) density dependence in the south to delayed density dependence in the north, possibly because of latitudinal shift in the type of predation (Hanski *et al.* 1991).

(iii) *Density dependence of variability*

Figure 3 summarizes the temporal variance–mean regression slopes for aphids, moths and birds sampled throughout the United Kingdom (Taylor & Woiwod 1980, 1982). There are highly significant differences between these taxa, birds having by far the smallest slopes, suggesting that bird populations are more strongly regulated than insect populations. Among birds, territorial species have significantly smaller slopes (average 1.08) than non-territorial species (1.28), which is consistent with the known regulatory function of territoriality (analysis in Hanski & Tiainen (1989)).

(*c*) *Relations between density dependence, regulation and variability*

The comparisons in the previous sections suggested that terrestrial vertebrates tend to have more regulated and less variable populations than arthropods. In this section I will examine the relations between density dependence, regulation and variability in 10 species of moths and 10 species of aphids, each species sampled for an average of 16 years at about 10 localities distributed throughout the United Kingdom, yielding a total of 190 time series. (The data are from the Rothamsted Insect Survey (Taylor 1986); a more comprehensive analysis will be published with I. Woiwod.)

Density dependence of population change was tested using two methods, an extension of the Ricker equation (Turchin 1990; delayed density dependence) and Bulmer's (1975*a*) autoregressive method (statistic R, direct density dependence). Turchin (1990) found a significant delayed effect in 8 out of 14 forest insects. In the 190 time series for British moths and aphids, there were 10 significant ($p < 0.05$) coefficients for the delayed effect, just the number expected by chance. The striking contrast between this result and Turchin's (1990) result is probably because many of the well-studied forest insects are outbreak species or have cyclic dynamics. In contrast, the evidence for direct density dependence in British moths and aphids is overwhelming: significant density dependence was found in 73 and 92% of the moth and aphid time series, respectively (figure 3). There was significant variation in the degree of density dependence between species but not between localities. These results show a higher incidence of density dependence than reported in most previous surveys. A possible explanation is discussed in the next section.

We are now ready to turn to the relations between density dependence, population regulation and variability. Theoretical considerations suggest that generally there is no simple relation between density dependence and variability, but assuming that populations have a stable equilibrium point, we would expect species with stronger density dependence to show less variability than species with weaker density dependence. Such a relation is found in moths but not in aphids (figure 4). This result suggests that aphid populations are not generally regulated towards a stable equilibrium point. Turning to the relation between the average level of variability and the degree of density dependence of variability, we find that they are negatively correlated in both moths and aphids (figure 4). This correlation parallels the result from the comparison between vertebrates and arthropods, and suggests that the more regulated populations tend to be less variable.

(*d*) *Density dependence of density dependence*

The frequent failures to detect significant density dependence in empirical studies may be explained by the methodological problems enumerated in §2*a*, but there is also an interesting biological possibility. Many population studies have been conducted on common species with tendency to outbreaks. Latto (1989) found that of the 63 insect life table studies quoted in Stiling (1988) and Hassell *et al.* (1989), 40 studies (63%) were of species that can clearly be called 'pests'. Turchin's (1990) observations on frequent delayed density dependence in forest insects suggest that many of these best-known population studies involve outbreak species, in which density dependence may occur less

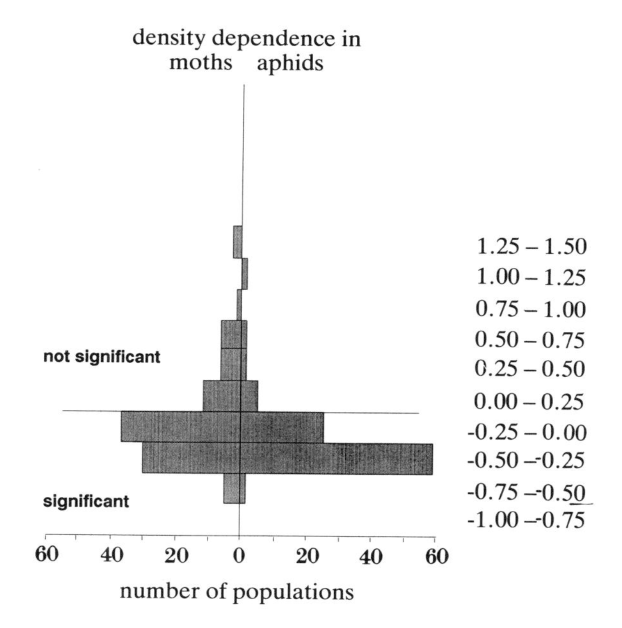

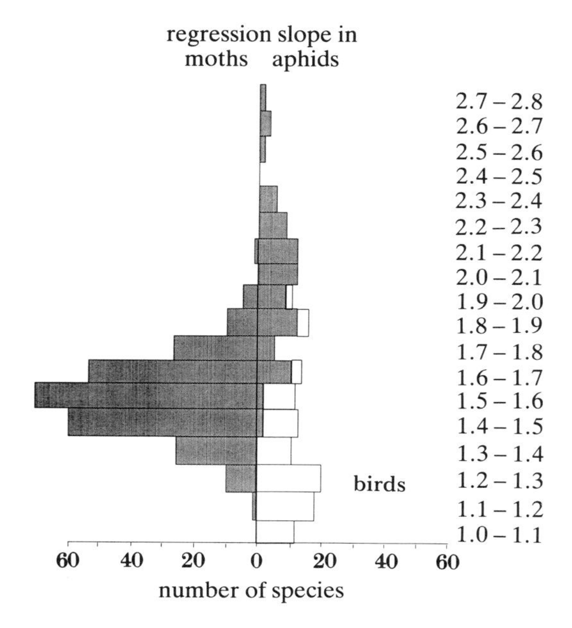

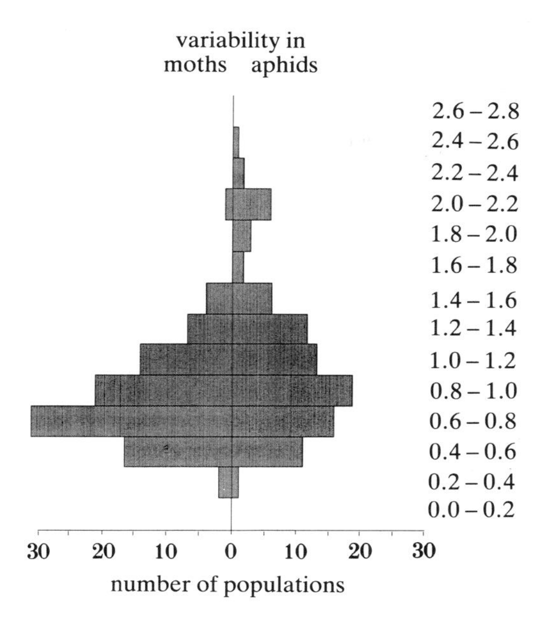

constantly than in many less common species, or which may have otherwise more complex dynamics than most uncommon species, making detection of density dependence more difficult. Ironically, detecting density dependence may be especially difficult in the kinds of species that are most frequently studied by population ecologists!

Additional support for this suggestion is given by the high incidence of density dependence in 'ordinary' (non-outbreak) British moths and aphids (figure 3), and by the observations that among these 'ordinary' species, density dependence of population change becomes less obvious, and density dependence of variability decreases, with increasing average density (table 3; the latter result has been previously reported by Hanski (1982) and Taylor & Woiwod (1982)). In summary, our fascination with outbreak and cyclic species may seriously bias our understanding of population dynamics in the vast majority of species.

3. MULTI-SPECIES TIME SERIES

(*a*) *Constancy of species composition*

A simple classification of communities may be based on the average level of variability (s) in the species, and on the concordance of their temporal abundance changes, measured by the average value of pair-wise rank correlations (r) in figure 5. The four 'ideal community types' of Strong *et al.* (1984) occupy the four corners in this scheme, with the caveat that their type (iii) may show high or low correlation. Ecologists have speculated on the mechanisms that are likely to dominate in the different kinds of communities. Correlated abundance changes have been assumed to imply the operation of some 'deterministic' factors, particularly interspecific competition (Grossman 1982; Strong *et al.* 1984); low correlation and low variability have been assumed to characterize species assemblages with independently regulated populations (Strong *et al.* 1984); while the combination of large variability but low correlation suggests the operation of strong environmental stochasticity (Strong *et al.* 1984). Although such inferences about population dynamic processes based on variability and interspecific correlation have only heuristic value, I shall advance some new suggestions below.

Figure 5 shows the position of 14 north temperate and boreal insect and vertebrate communities with

Figure 3. Comparison of British moths, aphids and birds in density dependence of population change, density dependence of variability and in the level of variability. Density dependence was measured by using Bulmer's (1975*a*) R (the values shown are R–R_c, where R_c is the lower 5% point of R; negative values thus show significant density dependence at 5% level). Density dependence of variability was measured by the slope of the variance–mean regression as explained in the text. Variability was measured by s, the standard deviation of log-transformed abundances. The regression slopes are from Taylor & Woiwod (1980, 1982). Calculations of density dependence and variability were done on 190 populations of 20 species of moths and aphids (data from the Rothamsted Insect Survey).

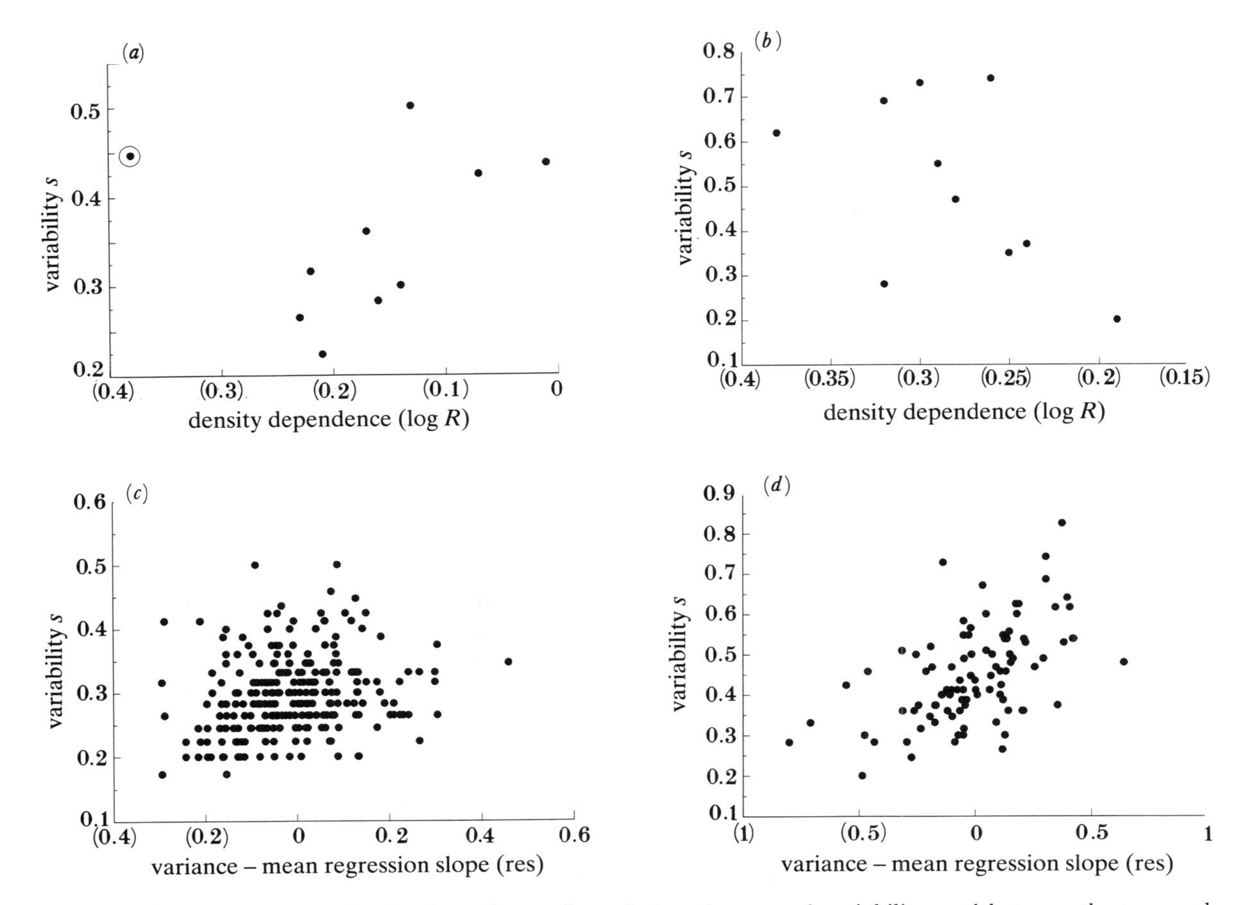

Figure 4. Relations between density dependence of population change and variability, and between the temporal variance–mean regression slope and variability, in British moths (*a*), (*c*) and aphids (*b*), (*d*). Density dependence was measured by using Bulmer's R (note that density dependence increases with decreasing log R). To remove the effect of mean density on the regression slope (cf. table 3), I first regressed the slope against the mean, and used the deviation from this regression line instead of the value of the slope. Statistics: (*a*), $r = 0.71$, $p < 0.05$ (excluding the circled migrant species, *Plusia gamma*); (*b*), $r = -0.52$, $p = 0.10$; (*c*), $r = 0.39$, $p < 0.001$; and (*d*), $r = 0.51$, $p < 0.001$. Data: variability and the regression slope from Taylor & Woiwod (1980); density dependence is the average value for 10 species sampled at about 10 localities throughout the United Kingdom (data from the Rothamsted Insect Survey).

Table 3. *The relations between density dependence of population change (Bulmer's R) and abundance, and between density dependence of variability (temporal variance-mean regression slope b) and abundance, in British moths*

(Results on density dependence are based on 98 populations of 10 species (data from the Rothamsted Insect Survey). Results on variance-mean regressions are based on the 263 species reported in Taylor & Woiwod (1980). t is the t-test value of the regression slope.)

	dependent variable: abundance			
independent variable	t	sign.	n	R^2
Bulmer's R (density dependence)	3.13	0.002	98	0.10
slope b (regulation)	11.74	<0.001	263	0.35

respect to interspecific correlation and variability. The bird assemblages are characterized by high correlation but low variability; the insect communities have high variability and high correlation; while small mammals display the highest level of variability and the greatest spread of correlations (figure 5). The combination of low correlation and high variability has been suggested to show strong environmental stochasticity, but the temperate small mammals present a counter-example: their cyclic populations show strong but delayed density dependence, and the low average correlation is due to different species participating to different degrees in different peaks of the multiannual cycle (Henttonen 1986). Many temperate small mammals comprise an exception to the generally low level of variability in vertebrates (figure 2).

I suggest three tentative conclusions on the basis of figure 5. First, there are no communities with low variability and low correlation, which would suggest strong but independent population regulation in the species. Second, species in the communities with high variability and high correlation (insects in figure 5) are

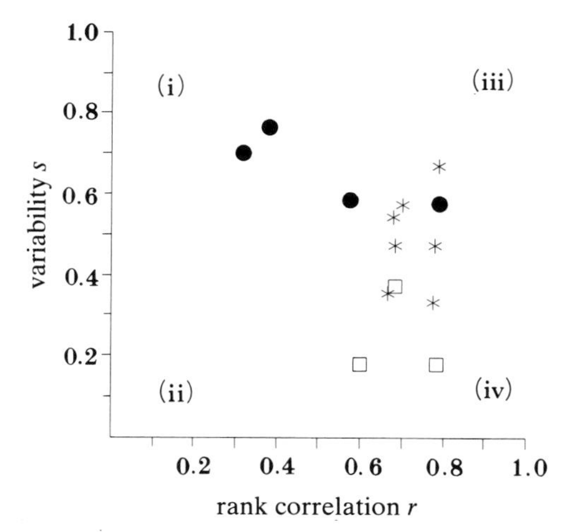

Figure 5. Position of 14 communities of north temperate and boreal insects (crosses), birds (squares) and small mammals (circles) with respect to the average of the Spearman's rank correlation between yearly abundances of pairs of species, and the average level of variability in the species. Variability was measured using annual census data, but in all cases shown here this measure is a good approximation of variability measured over generations. Data from Henttonen *et al.* (1987); T. Solonen (unpublished data); Skarén (1972); Vlasak (1987); Takeda (1987); den Boer (1977); Sheftel (1989); Doube *et al.* (1991); Ruszkowski *et al.* (1981); Sellin (1988); Enemar *et al.* (1984); Gilbert & Owen (1990); I. Woiwod (unpublished data).

affected by strong environmental stochasticity, with correlated effects on all or most species. And third, communities with low variability but large correlation (birds) are structured primarily by interspecific competition, while communities with high variability but low correlation (small mammals) are more affected by predation. It will be interesting to see what modifications to these conclusions are needed when more data of this sort are available.

(*b*) *Statistical modelling techniques*

Autoregressive analyses of single-species data aim at revealing whether abundance changes are affected by density dependent processes, for instance intraspecific competition. If time series data are available for several interacting species, more complicated regression and other statistical models may be used to examine predator–prey and competitive interactions between two or more species (Bulmer 1974, 1975*b*; Teräsvirta 1982; Owen & Gilbert 1989). It is however debatable how strong inferences one may draw about interspecific interactions from such observational data. This is an area where experimental techniques are much preferable to analyses of observational data.

4. LIFE TABLE TIME SERIES

I suggested in a previous section that the interesting task for a population ecologist is not to show what is strongly expected, the occurrence of some density dependence, but to uncover the temporal and spatial scales, and the stages in species' life cycles, where density dependence occurs (Williamson 1972; Sinclair 1989). Following Varley & Gradwell (1960), hundreds of ecologists have regressed *k*-values against density (N_t) to detect the operation of specific density dependent mortality factors (the *k*-value is $\log[N_t/N_s]$, where N_s is the number or density of individuals which survived during the focal life stage).

There are several recent reviews of the incidence of density dependence in life table studies (Dempster 1983; Strong *et al.* 1984; Stiling 1987, 1988). These reviews purport to show that, contrary to what most ecologists would expect, 30–50 % of the original studies failed to find any density dependent factors at all (Dempster 1983; Stiling 1988), and where density dependence was detected, it was more often because of an upper 'ceiling' set by limiting resources rather than to natural enemies operating at all densities. By using the same data that were analysed by Stiling (1988), Hassell *et al.* (1989) showed how the percentage of studies in which density dependence was detected increases with the duration of the study, suggesting that many studies have been conducted over such short periods of time that conclusions about density dependence should be drawn with great caution.

Another particular issue which has been much debated recently is the relation between spatial and temporal density dependence. Hassell (1985) suggests that many populations may be primarily regulated by spatial ('within-generation') density dependence, and he argues that it may be difficult, given spatial heterogeneity and omnipresent stochasticity, to detect temporal, year-to-year density dependence by using conventional life table analysis. More recently, Stewart-Oaten & Murdoch (1990) have shown that spatial heterogeneity can indeed translate spatial density dependence into temporal density dependence, and thereby increase stability, but their model suggests that spatial heterogeneity is generally more likely to have a destabilizing than stabilizing effect.

The emphasis on spatial density dependence is welcome as it brings the issues to the level of individuals that actually experience and are affected by the lower or higher density. There is an increasing interest in ecology to replace phenomenological, population level theories and approaches by more mechanistic alternatives with clearly measurable quantities at the individual level (Tilman 1988; Lomnicki 1988). None the less, as far as population regulation is concerned, what matters is temporal density dependence (Mountford 1988). Population regulation by definition means changes in the overall population size within a finite range of values. What is meant by 'overall population size' is not self-evident, however, and is discussed in the next section, population regulation may occur at different spatial scales.

5. POPULATION, OR METAPOPULATION, REGULATION?

The previous sections have followed the prevailing paradigm in population ecology in discussing questions about population regulation and variability at the level of local populations. There is an alternative perspective in the literature, which emphasizes how 'a natural population occupying any considerable area will be made up of a number of ... local populations' (Andrewartha & Birch 1954), how 'the risk of wide fluctuation in animal numbers is spread unequally over a number of subpopulations', and how 'the consequences of this spreading of the risk in space will be a relative reduction in the amplitude of fluctuations of animal numbers in the entire population' (den Boer 1968). Following Levins (1969), the 'entire population', or the ensemble of local populations, is frequently called a metapopulation.

Much confusion has been created by the claim (den Boer 1968, 1987) that metapopulation regulation or persistence requires no density dependence at the level of local populations. The claim is incorrect. Without any density dependence, populations would grow indefinitely large, or they would go rapidly extinct, and in the latter case no metapopulation would persist for long. Density dependence is required for metapopulation persistence; what is not needed is regulation of local populations.

Two other requirements for metapopulation regulation, when local regulation fails, are dispersal between local populations, and asynchronous dynamics in these populations. Asynchronous local dynamics may be due to population dynamic factors, such as predator–prey interaction (Taylor 1988), or to stochastic environmental factors. Metapopulation regulation is likely to play the greatest role in species with large local variability but little synchrony among populations. Figure 6 shows that in 20 species of moths and aphids sampled throughout the United Kingdom, the species with more variable populations tended to have more synchronous dynamics. A positive relation between variability and regional synchrony suggests that high variability in these insects is largely because of regionally correlated environmental stochasticity, and not to local dynamics. Alternatively, the result in figure 6 could be because of a high level of dispersal among unstable (highly variable) local populations, but this explanation seems unlikely in the scale of the U.K. In any case, positive correlation between variability and regional synchrony is not favourable for metapopulation regulation in the face of weakly regulated local populations with high variability. It remains to be shown by future studies whether the pattern in figure 6 is the rule or the exception. More generally, it is yet a very open question how frequently species persist as metapopulations with unstable local populations (Taylor 1988; Harrison 1991).

In spite of the negative result for metapopulation regulation in figure 6, I cannot resist the temptation to speculate on a special group of species, but a group that may actually comprise the majority of species on Earth. I have in mind the arthropods which are living in the canopy of tropical forests. Given the tremendous diversity and patchy distribution of tree species in tropical forests, and the probably quite high (but unknown) frequency of monophagous and oligophagous insect species (Erwin & Scott 1980; May 1988), the canopy of tropical forests may well appear to many species like the classical metapopulation scenario as originally envisaged by Levins (1969): small and like patches of suitable habitat placed randomly in the matrix of inhospitable surroundings. To what extent are the immense numbers of canopy-living arthropods in tropical forests (Erwin & Scott 1980; Stork 1987; May 1988) regulated by metapopulation dynamics? To what extent is the coexistence of these species based on metapopulation dynamics? These are fascinating questions without even a hint of an answer.

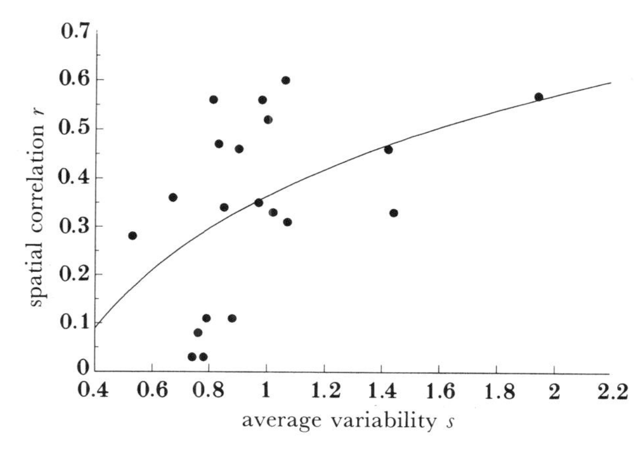

Figure 6. Relation between average local variability and average spatial correlation of population fluctuations (regional synchrony) in 10 species of moths and 10 species of aphids, each sampled at about 10 localities distributed throughout the United Kingdom.

6. CONCLUSION

The majority of population dynamic studies deal with common species, which have large local populations and are widely distributed, or species that have outbreaks at regular or irregular intervals. Our perception of density dependence, regulation and variability in animal populations is biased towards such species with perhaps especially complex dynamics. Directing more attention towards uncommon and rare species would be most welcome, because they represent the majority of species, but there remain formidable practical problems in estimating variability of rare species, and in showing at which spatial scale their populations are regulated.

I am most indebted to T. Solonen and I. Woiwod for making unpublished data available to me. Comments and suggestions by K. J. Gaston, J. Latto, W. W. Murdoch, A. D. Taylor, P. Turchin, I. Woiwod and X. Xia helped me greatly.

REFERENCES

Andrewartha, H. G. & Birch, L. C. 1954 *The Distribution and Abundance of Animals*. The University of Chicago Press.

Bulmer, M. G. 1974 A statistical analysis of the 10-year cycle in Canada. *J. Anim. Ecol.* **43**, 701–718.

Bulmer, M. G. 1975*a* The statistical analysis of density dependence. *Biometrics* **31**, 901–911.

Bulmer, M. G. 1975*b* Phase relations in the ten-year cycle. *J. Anim. Ecol.* **44**, 609–622.

Connell, J. H. & Sousa, W. P. 1983 On the evidence needed to judge ecological stability or persistence. *Am. Nat.* **121**, 789–824.

Dempster, J. P. 1983 The natural control of populations of butterflies and moths. *Biol. Rev.* **58**, 461–481.

den Boer, P. J. 1968 Spreading of risk and stabilization of animal numbers. *Acta Biotheor.* **18**, 165–194.

den Boer, P. J. 1977 Dispersal power and survival. Carabids in a cultivated countryside. *Miscellaneous Papers 14, Landbouwhogeschool, Wageningen.*

den Boer, P. J. 1987 Detecting density dependence. *Trends Ecol. Evol.* **2**, 77–78.

Doube, B. M., Macqueen, A., Ridsdill-Smith, T. J. & Weir, T. A. 1991 Native and introduced dung beetles in Australia. In *Dung Beetle Ecology* (ed. I. Hanski & Y. Cambefort). Princeton University Press.

Enemar, A., Nilsson, L. & Sjöstrand, B. 1984 The composition and dynamics of the passerine bird community in a subalpine birch forest, Swedish Lapland. A 20-year study. *Ann. Zool. Fennici* **21**, 321–338.

Erwin, T. L. & Scott, J. C. 1980 Seasonal and size patterns, trophic structure, and richness of Coleoptera in the tropical arboreal ecosystem: the fauna of the tree *Luehea seemannii* Triana and Planch in the Canal Zone of Panama. *Coleopt. Bull.* **34**, 305–322.

Fowler, C. W. 1981 Density dependence as related to life history strategy. *Ecology* **62**, 602–610.

Gaston, K. J. 1988 Patterns in local and regional dynamics of moth populations. *Oikos* **53**, 49–57.

Gaston, K. J. & Lawton, J. H. 1987 A test of statistical techniques for detecting density dependence in sequential censuses of animal populations. *Oecologia* **74**, 404–410.

Gaston, K. J. & Lawton, J. H. 1988 Patterns in body size, population dynamics, and regional distribution of bracken herbivores. *Am. Nat.* **132**, 662–680.

Gilbert, F. & Owen, J. 1990 Size, shape, competition, and community structure in hoverflies (Diptera: Syrphidae). *J. Anim. Ecol.* **59**, 21–40.

Glazier, D. S. 1986 Temporal variability of abundance and the distribution of species. *Oikos* **47**, 309–314.

Grossman, G. D. 1982 Dynamics and organization of a rocky intertidal fish assemblage: the persistence and resilience of taxocene structure. *Am. Nat.* **119**, 611–637.

Hanski, I. 1982 On patterns of temporal and spatial variation in animal populations. *Ann. Zool. Fennici* **19**, 21–37.

Hanski, I. & Tiainen, J. 1989 Bird ecology and Taylor's variance-mean regression. *Ann. Zool. Fennici* **26**, 213–217.

Hanski, I., Hansson, L. & Henttonen, H. 1990 Specialist predators, generalist predators, and the microtine rodent cycle. *J. Anim. Ecol.* (In the press.)

Hansson, L. & Henttonen, H. 1985 Gradients in density variations of small rodents: the importance of latitude and snow cover. *Oecologia* **67**, 394–402.

Hansson, L. & Henttonen, H. 1988 Rodent dynamics as community processes. *Trends Ecol. Evol.* **3**, 195–200.

Harrison, S. 1991 Local extinction in a metapopulation context: an empirical evaluation. In *Metapopulation dynamics* (ed. M. Gilpin & I. Hanski). London: Academic Press. (In the press.)

Hassell, M. P. 1985 Insect natural enemies as regulating factors. *J. Anim. Ecol.* **54**, 323–334.

Hassell, M. P., Latto, J. & May, R. M. 1989 Seeing the wood for the trees: detecting density dependence from existing life-table studies. *J. Anim. Ecol.* **58**, 883–892.

Henttonen, H. 1986 Causes and geographic patterns of microtine cycles. Ph.D. thesis, University of Helsinki, Finland.

Henttonen, H., Oksanen, T., Jortikka, A. & Haukisalmi, V. 1987 How much do weasels shape microtine cycles in the northern Fennoscandian taiga? *Oikos* **50**, 353–365.

Joern, A. & Pruess, K. P. 1986 Temporal constancy in grasshopper assemblages (Orthoptera: Acrididae). *Ecol. Entomol.* **11**, 379–385.

Klimetzek, D. 1979 Insekten-Grossschädlinge an Kiefer in Nordbayern und der Pfalz: Analyse und Vergleich 1810–1970. *Freiburger Waldschutz-Abhandlungen, Freiburg.*

Latto, J. 1989 The analysis of density dependence in insect populations. Ph.D. thesis, Imperial College, London.

Levins, R. 1969 Some demographic and genetic consequences of environmental heterogeneity for biological control. *Bull. Entomol. Soc. Am.* **15**, 237–240.

Lomnicki, A. 1988 *Population Ecology of Individuals.* Princeton University Press.

May, R. M. 1988 How many species are there on Earth? *Science, Wash.* **241**, 1441–1449.

Morris, R. F. 1959 Single-factor analysis in population dynamics. *Ecology* **40**, 580–588.

Mountford, M. D. 1988 Population regulation, density dependence, and heterogeneity. *J. Anim. Ecol.* **57**, 845–858.

Murdoch, W. W. 1970 Population regulation and population inertia. *Ecology* **51**, 497–502.

Murdoch, W. W. & Walde, S. J. 1989 Analysis of insect population dynamics. In *Toward a more exact ecology* (ed. P. J. Grubb & J. B. Whittaker). Oxford: Blackwell.

Ostfeld, R. S. 1988 Fluctuations and constancy in populations of small rodents. *Am. Nat.* **131**, 445–452.

Owen, J. & Gilbert, F. S. 1989 On the abundance of hoverflies (Syrphidae). *Oikos* **55**, 183–193.

Pimm, S. L. 1984 Food chains and return times. In *Ecological communities. Conceptual issues and the evidence* (ed. D. R. Strong, D. Simberloff, L. G. Abele & A. B. Thistle). Princeton: Princeton University Press.

Pimm, S. L. & Redfearn, A. 1988 The variability of population densities. *Nature, Lond.* **334**, 613–614.

Pollard, E., Lakhani, K. L. & Rothery, P. 1987 The detection of density dependence from a series of annual censuses. *Ecology* **68**, 2046–2055.

Redfearn, A. & Pimm, S. L. 1988 Population variability and polyphagy in herbivorous insect communities. *Ecol. Monogr.* **58**, 39–55.

Rejmanek, M. & Spitzer, K. 1982 Bionomic strategies and long-term fluctuations in abundance of Noctuidae (Lepidoptera). *Acta Ent. Bohemoslov.* **79**, 81–96.

Ruszkowski, A., Zadura, M. & Kaczmarska, K. 1981 The composition of the bumblebee species, *Bombus* Latr. (Apidae, Hymenoptera) at Pulawy, Poland, in 1961–1972. *Bull. Entomol. de Pologne* **51**, 307–322.

Schoener, T. W. 1985 Are lizard population sizes unusually constant through time? *Am. Nat.* **126**, 633–641.

Sellin von, D. 1988 Zur Dynamik des Sommervogelbestandes eines isolierten Birken-Stieleichen-Waldes während einer 12 jährigen Untersuchungsperiode. *Beitr. Vogelkd.* **34**, 157–176.

Sheftel, B. I. 1989 Long-term and seasonal dynamics of shrews in Central Siberia. *Ann. Zool. Fennici* **26**, 357–370.

Sinclair, A. R. E. 1989 Population regulation in animals. In *Ecological concepts* (ed. J. M. Cherrett). Oxford: Blackwell.

Skarén, U. A. P. 1972 Fluctuations in small mammal populations in mossy forests of Kuhmo, eastern Finland, during eleven years. *Ann. Zool. Fennici* **9**, 147–151.

Slade, N. A. 1977 Statistical detection of density dependence from a series of sequential censuses. *Ecology* **58**, 1094–1102.

Spitzer, K. & Leps, J. 1988 Determinants of temporal variation in moth abundance. *Oikos* **53**, 31–36.

Stewart-Oaten, A. & Murdoch, W. W. 1990 Temporal consequences of spatial density dependence. *J. Anim. Ecol.* (In the press.)

Stiling, P. D. 1987 The frequency of density dependence in insect host–parasitoid systems. *Ecology* **68**, 844–856.

Stiling, P. D. 1988 Density-dependent processes and key factors in insect populations. *J. Anim. Ecol.* **57**, 581–594.
Stork, N. E. 1987 Arthropod faunal similarity of Bornean rain forest trees. *Ecol. Entomol.* **12**, 219–226.
Strong, D. R., Lawton, J. H. & Southwood, R. 1984 *Insects on plants. Community patterns and mechanisms.* Oxford: Blackwell.
Stubbs, M. 1977 Density dependence in the life-cycles of animals and its importance in K- and r-strategies. *J. Anim. Ecol.* **46**, 677–688.
Takeda, H. 1987 Dynamics and maintenance of collembolan community structure in a forest soils system. *Res. Popul. Ecol.* **29**, 291–346.
Taylor, A. D. 1988 Large-scale spatial structure and population dynamics in arthropod predator–prey systems. *Ann. Zool. Fennici* **25**, 63–74.
Taylor, L. R. 1961 Aggregation, variance and the mean. *Nature, Lond.* **189**, 732–735.
Taylor, L. R. 1986 Synoptic dynamics, migration and the Rothamsted Insect Survey. *J. Anim. Ecol.* **55**, 1–38.
Taylor, L. R. & Woiwod, I. P. 1980 Temporal stability as a density-dependent species characteristic. *J. Anim. Ecol.* **49**, 209–224.
Taylor, L. R. & Woiwod, I. P. 1982 Comparative synoptic dynamics. I. Relationships between inter- and intraspecific spatial and temporal variance/mean population parmeters. *J. Anim. Ecol.* **51**, 879–906.
Teräsvirta, T. 1982 Mink and muskrat interaction: a structural analysis. *Research Report, Department of Statistics, University of Helsinki, Finland.*
Tilman, D. 1988 *Dynamics and structure of plant communities.* Princeton University Press.
Turchin, P. 1990 Rarity of density dependence or population regulation with lags? *Nature, Lond.* **344**, 660–663.
Varley, G. C. & Gradwell, G. R. 1960 Key factors in population studies. *J. Anim. Ecol.* **29**, 399–401.
Varley, G. C., Gradwell, G. R. & Hassell, M. P. 1973 *Insect population ecology.* Oxford: Blackwell.
Vickery, W. L. & Nudds, T. D. 1984 Detection of density-dependent effects in annual duck censuses. *Ecology* **65**, 96–104.
Vlasak, P. 1987 Small mammals, their population cycle and production in stands of the Alliance Molinion Koch, 1926. *Acta Univ. Car.-Biol.* **31**, 313–348.
Watt, K. E. F. 1965 Community stability and the strategy of biological control. *Can. Entomol.* **97**, 887–895.
Williamson, M. 1972 *The analysis of biological populations.* Edward Arnold, Unwin Brothers Ltd.
Wolda, H. 1983 Spatial and temporal variation in abundance in tropical animals. In *Tropical rain forest: ecology and management* (ed. S. L. Sutton, T. C. Whitmore & A. A. Chadwick). Oxford: Blackwell.

Regulation in fish populations: myth or mirage?

J. G. SHEPHERD[1] AND D. H. CUSHING[2]

[1] *Ministry of Agriculture, Fisheries & Food, Directorate of Fisheries Research, Pakefield Road, Lowestoft, Suffolk, U.K.*
[2] *198 Yarmouth Road, Lowestoft, Suffolk, U.K.*

SUMMARY

There is abundant evidence of long-term changes in the abundance of fish populations, but the causes are not known. It is almost certain that climatic changes are responsible in part, but the role of population regulatory mechanisms is unclear. The evidence is conflicting. The ability of fish populations to sustain levels of fishing mortality several times the level of natural mortality suggests strong regulatory mechanisms. The persistence of stocks for centuries, with few extinctions or explosions, also implies some regulation, but not necessarily strong regulation. The high levels of fluctuation in recruitment suggest weak regulation except in the earliest stages of the life history. Under weak regulation the time taken for effective explosions or extinctions is long, maybe a century for 1000-fold changes in abundance. There are few historical records that imply greater stability (persistence) of stocks than this.

Analysis of stock-recruitment diagrams (the fisheries biologists' version of k-factor analysis) rarely yields clear evidence for or against regulation, because of high levels of fluctuation, which cannot therefore be because of single-species deterministic chaos (though multispecies chaos remains a possibility). Even the exceptions to this rule (North Sea herring, Georges Bank haddock) are not wholly convincing. Conversely, it is credible that these and other long-term declines of recruitment (Northeast Arctic cod, North Sea haddock) could be due to regulation, since stock sizes also fell. Regrettably we cannot distinguish the chicken and the egg.

It is indeed quite plausible that the only regulatory process operating for fish populations is a stochastic one: increased (and non-normal) variability at low stock sizes. This would give strong regulation in the mean, because of the increasing excess of the mean over the median at low stock sizes, but only because of increasingly large, but increasingly infrequent, outstanding year-classes. This sounds such an accurate description of heavily fished stocks that further exploration of this mechanism seems warranted.

1. INTRODUCTION

Recruitment to marine fish stocks often varies by one or two orders of magnitude, for reasons that are not well understood. This causes considerable difficulty when one attempts to study regulatory processes, since the variations obscure any relation that may be present. The variations seem to occur on a very wide range of timescales, with major fluctuations from year to year, and over decades and centuries.

The fishery biologist's standard approach to the study of regulation is the examination of stock-recruitment diagrams. For most fish species the breeding population consists of several age groups, and the most useful measure of stock is usually the spawning stock biomass. Total egg production is an even better measure, but the necessary data on fecundity are rarely available on a year-to-year basis. The major interest of fish populations when considering regulation is that many have been exploited very heavily. Total mortality rates exceeding five times the natural mortality rate are not uncommon, and may be sustained for decades. This led Shepherd & Cushing (1980) to believe that fish stocks must be strongly regulated. It also means that the populations have been 'exercised' over a wide range of spawning stock sizes, and the data for such stocks are potentially of great interest.

These aspects are discussed in more detail below, and a brief account is given of the results of attempts to find and quantify the regulatory process. This provides the motivation for the 'myth' in the title of this paper, since it is fair to say that most fish biologists believe in regulation even though they have been able to find remarkably little direct evidence for it. The latter part of the paper addresses the 'mirage' aspect, through a discussion of the possibility that deterministic density-dependent processes may be very weak, while stochastic processes may provide a rather effective mechanism for regulation of population size in practice. The mirage metaphor seems appropriate since the process becomes less clear as one tries to study it more closely. If this idea has any validity, it may resolve the classic conflict between the pro-regulation school of Nicholson (1933), and the anti-regulation school of Andrewartha & Birch (1954), since they would both be right if the variability is in fact an essential part of the regulatory mechanism.

2. THE EFFECT OF FISHING, AND EVIDENCE FOR DETERMINISTIC REGULATION

In the absence of regulatory processes, the expected relation between recruitment and parent stock size would be strict proportionality, a straight line through the origin on a stock-recruitment plot. This would be the result of constant fecundity and mortality rates, and this should in principle be the most appropriate null hypothesis, to be adopted in the absence of evidence to contradict it.

The consequence of such an assumption is, however, that the stock would tend exponentially to either zero or infinite size, with a characteristic timescale of a few years, depending on the level of total mortality on the stock (including fishing mortality), unless this happened to be just sufficient to keep the stock in a state of neutral equilibrium. Simple graphical simulations (see, for example, Beverton & Holt (1957) figure 6.1) can easily be used to show such behaviour, and the results are quite unlike anything normally observed in practice. It is for this reason that almost all fishery biologists believe that some sort of regulatory processes are at work, even though they cannot be described in any detail.

If the relation between recruitment and parent stock is nonlinear, then the possibility of regulation towards equilibrium states exists. These may, however, be either stable or unstable points, and are unlikely to be clearly observable in the presence of large variability. For single-cohort spawning stocks, these occur wherever the stock-recruit curve intersects a line of unit slope (through the origin) on a plot of recruits against numbers. The generalization of this to multiple-age spawning populations is straightforward, but seems not to be very well-known outside the world of fisheries. One simply calculates the steady-state level of spawning stock biomass per recruit for any desired regime of natural and fishing mortality, growth and maturity at age, and plots the line of the corresponding slope through the origin on the diagram of recruitment against spawning stock biomass. The intersections of these survival lines with the recruitment line again define potential equilibrium states. In fact the ratio of recruitment (R) to spawning stock biomass (SSB) must equal the reciprocal of the spawning biomass per recruit (BPR) so that:

$$\mathrm{R}/\mathrm{SSB} = 1/\mathrm{BPR}. \qquad (1)$$

Now, biomass-per-recruit is a measure of survival and decreases as total mortality increases because of fishing (see figure 1). Thus the slope of the survival lines increases (because we conventionally plot SSB on the x-axis) as fishing mortality (F) increases. The equilibrium stock size therefore decreases as F increases, as shown in figure 2. It is easy to see that the stock will collapse when fishing mortality is so high that 1/BPR exceeds the slope of the recruitment curve at the origin. This slope at the origin is therefore a quantity of great concern to fishery biologists. Regrettably it is not guaranteed to be positive definite. It could be zero, or effectively infinite, and techniques for determining this slope for real data sets are in their infancy.

In practice there is a tendency among fishery biologists to take constant recruitment as the null hypothesis. This is dangerous, because a stock capable of generating constant recruitment at any stock level cannot be collapsed, however great a fishing mortality is applied. It is also remarkable, because this assumption is often maintained with great tenacity, and without any appreciation that it corresponds to the

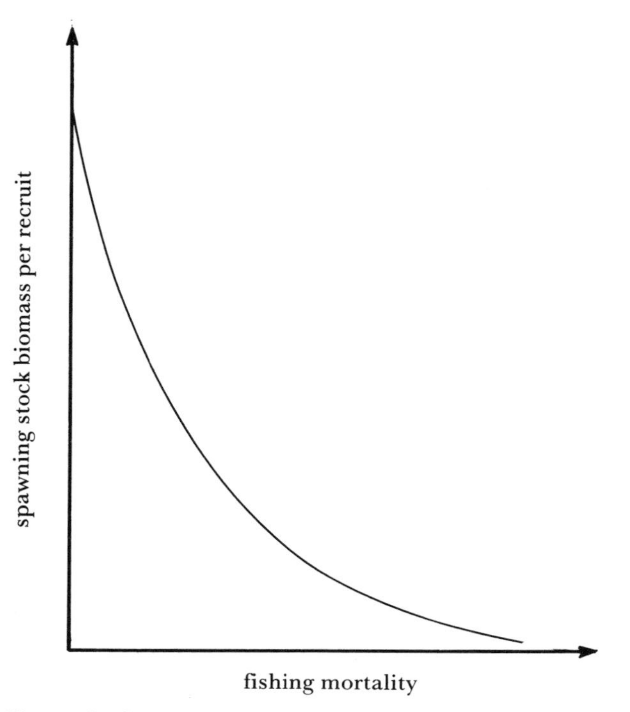

Figure 1. Survival in exploited stocks, as measured by spawning stock biomass per recruit.

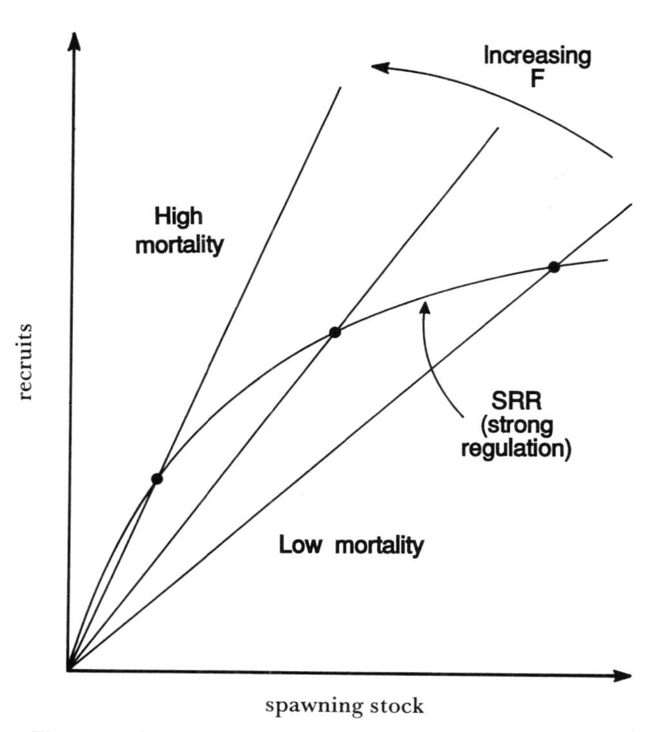

Figure 2. Stabilization of a regulated stock under various levels of exploitation.

operation of extremely strong regulatory processes. Such a powerful and persistent 'myth' is unlikely to have arisen by pure change, and it is presumably founded on the observation mentioned above, that fish populations do not in practice explode or collapse exponentially on decadal timescales, even where they have been subjected to large modulations of total mortality rates through fishing.

The evidence from detailed investigations of stock and recruitment data for marine fish is however distinctly equivocal. It is hard to distinguish on statistical grounds between constant recruitment (corresponding to strong regulation) and a constant R/SSB ratio (corresponding to no regulation). This is because of the high level of variability in the recruitment time series, and partly to the shortness of the time series available (rarely more than 30 years). Nevertheless, adding appropriate levels of noise to simple simulations of unregulated stocks, as described below, shows that simple variability does not alter the conclusion that an unregulated stock should collapse or explode in a quite characteristic manner, on a decadal timescale, and be very dependent on the level of fishing mortality applied.

Furthermore, it is clear that an unregulated stock must, to have survived, have evolved reproductive processes just capable of generating an R/SSB ratio which is sufficient to balance the effects of natural mortality. It is easily seen, by sketching or simulation, that such a stock would collapse rapidly under any significant fishing mortality (such as, for example, $F = M$, a very modest rate in the real world), as was pointed out many years ago by Beverton & Holt (1957).

It is therefore reasonable to conclude that there is some basis for the belief that fish populations are regulated in some way. In fact, the evidence is twofold: first, that they do not explode when subjected only to low (natural) levels of mortality: secondly, that they do not collapse at all quickly, when subjected to high levels of mortality. We regard these pieces of evidence as separate, as their implications are in fact rather different. It has been recognized for many years (see, for example, Reddingius (1971)) that only rather weak regulation is required to prevent a stock exploding at low mortality rates. This is easily verified, again, by a little sketching or simulation. In this context 'weak regulation' means a modest modulation of the R/SSB ratio. Such a weak regulation could be described for example by a Cushing-type non-asymptotic power-law relation with any power marginally less than one (Cushing 1973), or by any straight-line relation with a small positive intercept (though it should be remarked that both these relationships imply very strong regulation, large modulation of the R/SSB ratio, at low stock levels). There is no necessity to postulate a strongly regulatory process, such as a finite asymptotic value of recruitment at high stock level, to prevent unrealistic behaviour at low mortality. In passing, we note that there is *a fortiori* no need whatever to postulate an overcompensatory relation, where recruitment ultimately reduces as stock size increases, such as that inherent in the Ricker formulation (Ricker 1954). This is not to say that overcompensation may not occur, only that it is a much stronger assumption than is required, and should not be regarded as typical.

Conversely, at the low stock sizes that may be (and are) generated under heavy fishing, strong regulation is required. As a first approximation, spawning biomass per recruit (BPR) is approximately inversely proportional to total mortality (see figure 1). Fishing mortality rates up to and beyond five times the natural mortality rate are by no means uncommon for heavily fished stocks (e.g. for the North Sea gadoids in recent years). If a stock is to survive for more than a few years under such a mortality regime, the R/SSB ratio must be capable of increasing more than fivefold so as to compensate.

Thus the qualitative evidence is that some possibly rather weak regulatory process is required to prevent stocks increasing indefinitely when fishing mortality is small ($F < M$, say), but that very strong regulation is needed to prevent stocks collapsing rapidly under heavy fishing ($F > 2M$, say). There is no particular reason to suppose that the same biological mechanisms operate in both cases (though there is no reason why they should not either). It may be, for example, that processes such as density-dependent growth or fecundity, which can hardly be envisaged as capable of generating the high modulation of R/SSB required at low stock size, could be entirely adequate to provide the necessary modest modulation at high stock size. The strength of regulation is of course important, because it determines the maximum fishing mortality which a stock can sustain for very long, a crucial piece of information in developing rational stock management.

Table 1 summarizes the strength of regulation

Table 1. *Strength of regulation implied by various stock-recruitment relations at low and high stock size*

mortality... stock level...	high low	low high
constant R/SSB	none	none
constant *R*	very strong	strong
Cushing-type	very strong	moderate/weak
linear (positive intercept)	very strong	weak
Beverton–Holt	moderate[a]	strong
Shepherd	moderate[a]	variable
Ricker	moderate[a]	very strong

[a] Depends on parameter values, but R/SSB has a finite maximum at zero stock size.

implied by various well-known stock-recruitment relations, at both low and high stock sizes, just to illuminate the situation. The qualitative evidence suggests that one needs strong regulation at low stock size, but only weak regulation at high stock size. Apart from the constant R/SSB (no regulation) case, which is clearly inadequate, it is evident that all of these provide more than enough regulation at both high and low stock size. Nothing quite so powerful is required to explain the observed behaviour of real fish stocks. There is no problem in postulating mechanisms capable of producing weak regulation at high stock size, i.e. capable of modulating R/SSB by a factor of, say, two either way around the value appropriate to the natural (unfished) population. The problem is rather to identify the processes capable of generating very high R/SSB levels at low stock sizes. This remains an unsolved problem for most stocks. The possibility that this is because there is no deterministic regulatory mechanism at low stock size, only a stochastic one, is discussed below.

At first sight there seems to be a paradox here: qualitative evidence suggesting strong regulation at low stock size, but great difficulty in detecting any relation from the detailed data. However, we observe that a powerful regulatory mechanism at low stock size (i.e. one capable of generating high R/SSB ratios) may look quite unremarkable on a stock-recruitment diagram, and could be extremely difficult to detect by statistical analysis. For example, a straight-line relation with any small positive intercept would suffice; this relation generates R/SSB ratios which tend to infinity as stock size reduces. The same is true of a Cushing-type power law relationship with a slope even fractionally less than one. Such relations would be adequate to explain the qualitative observations, but would be extremely difficult to distinguish from a straight line through the origin, either visually or statistically. This could perhaps explain why it has proven so difficult to determine stock-recruitment relations from real data. The paradox may therefore be more apparent than real. Nevertheless, one should beware of adopting such relations for practical use, because they imply infinitely strong regulation at zero stock size, and exclude the possibility of stock collapse even under massive fishing mortality. This is a much more extreme assumption than is warranted by the evidence, and it would be imprudent to make it.

3. THE DEPENDENCE OF RECRUITMENT ON PARENT STOCK

In the light of the above discussion, we now consider the data for a number of stocks for which long time series of data (more than 30 years) are available. In each case we first show the time series of recruitment, and then the plot of these data against spawning stock biomass. To these data we have simply fitted the functional form proposed by Shepherd (1982), that is,

$$R = aB/\{1+(B/K)^{\beta}\}, \tag{2}$$

where R is recruitment in numbers; B is spawning stock biomass in tonnes; a, the slope at the origin, is maximal survival (recruits per unit biomass); K is the threshold biomass (at which recruitment is reduced to half aB), and above which density dependent effects predominate; β expresses the degree of compensation. These fitted lines are simply to guide the eye.

Five stocks were examined; North Sea sole, North Sea plaice, Iceland cod, Northeast Arctic cod and North Sea herring. They have been chosen because each is supported by about 30 or 40 years of observations; in all cases, recruitment and stock were estimated by virtual population analysis (VPA). The data have been taken from the reports of relevant International Council for the Exploration of the Sea (ICES) working groups, augmented by recent data provided by T. Jakobsen, J. Jakobsson, and staff at the Fisheries Laboratory, Lowestoft. The equation was fitted with the Marquardt algorithm by using a program called Fishparm (C. Walters, personal communication).

The five stocks are described in two figures for each: the time series of recruitment, with mean and standard deviation, both logarithmic, and the stock recruitment relation (see figures 3–7). In the North Sea sole there are three large year-classes, 1958, 1963 and 1987. There was a sharp decline in stock as the 1963 year-class was finished out. During a cold winter (as in 1963), with very cold easterly winds across the southern North sea, sole move into deeper water where the fishermen go to catch them, for example, in the Silver Pit (so named from the money made in a cold winter

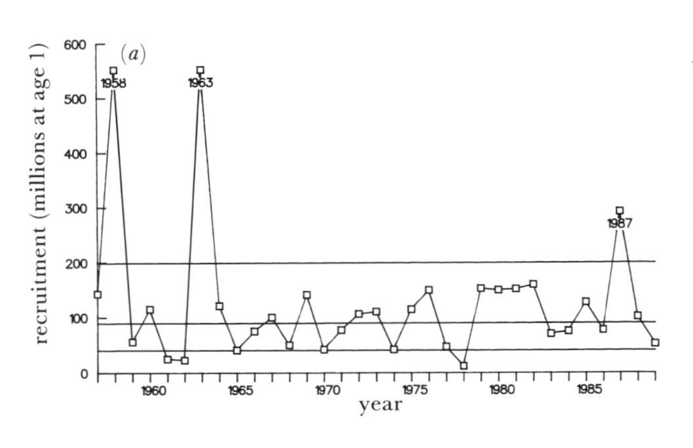

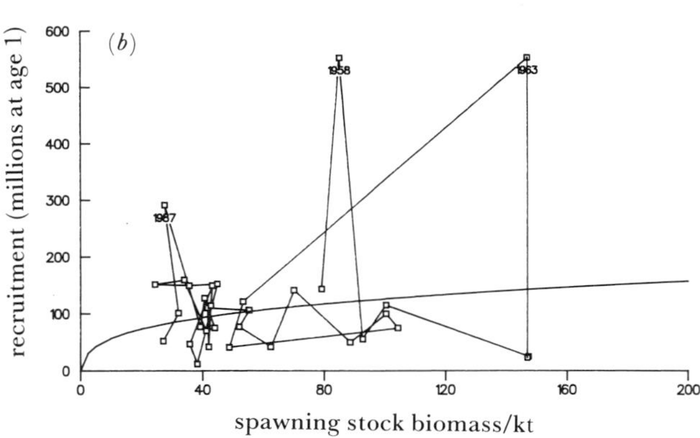

Figure 3. North Sea sole. (*a*) Time series of recruitment. (*b*) Stock-recruitment diagram.

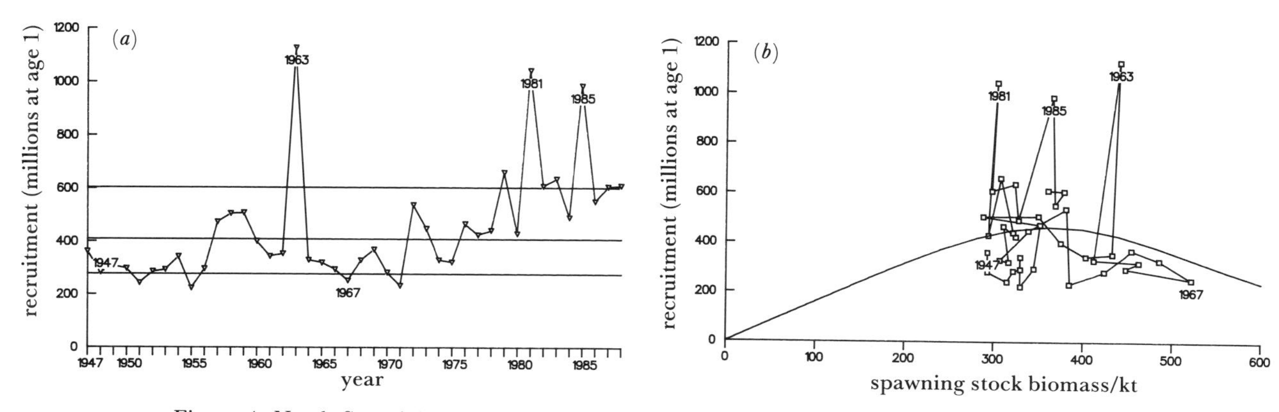

Figure 4. North Sea plaice. (*a*) Time series of recruitment. (*b*) Stock recruitment diagram.

in 1838). In contrast, after the 1958 year-class, stock built up steadily to 1962. The time series shows quite low variability except for the three high year-classes, but in 1979–1982 there were four quite good year-classes in succession.

In the plaice of the southern North Sea, the coefficient of compensation (β) appears to be high because of the low recruitment from 1964 to 1971 when stock size was quite high. There are three high year-classes, 1963, 1981 and 1985, the first and last of which were associated with cold winters. The cold winters are believed to have produced high recruitment because the shrimps (which generate density dependent mortality), move away into deeper water (van der Veer 1986), but there was no high recruitment from the cold winter of 1947. The time series of recruitment shows a slight decline to 1971 and a steady rise subsequently. The reason for this increase is unknown, but the increment is responsible for the anomalously high coefficient of compensation.

In the Iceland cod we see the quick exploitation of year-classes as they appear (1964, 1973 and 1983) because the fish are first caught at three years of age, and do not spawn until seven or eight, and so suffer a long period of exploitation as juveniles. The time series shows the high year-classes 1964, 1976 and 1986. During earlier years the spawning stock at Iceland received immigrant spawners from West Greenland. The West Greenland stock collapsed in the late sixties, and the last immigrations occurred in 1961 and 1963. An interesting point is that there was no reduction in recruitment during the passage of the Great Salinity Anomaly of the 1970s described by Dickson *et al.* (1988) through north Icelandic waters in 1965–71, presumably because the cod spawn off southern Iceland.

In the Northeast Arctic cod, there was an increase in recruitment, even though the stock declined after the second world war, culminating in a strong year-class in 1950. But there were also three strong year-classes at low stock in 1963, 1964 and 1970. There seems to be a fairly systematic seven-year quasi-periodicity in these data, but the very low year-classes of 1977–80 may also be associated with the passage of the Great Salinity Anomaly (Cushing 1988). It is worrying that the last peak in recruitment in 1983 was not very high, and the data are quite suggestive of a reduction of recruitment at low stock size, especially if the seven-year cycle is filtered out.

The North Sea herring shows a real collapse in the early 1970s but also strong year-classes in 1956, 1960 and 1985. The time series shows the collapse and recovery; catches were banned between 1977 and 1982 when the very poor year classes, 1974 to 1977, appeared. The ban on catches was very timely and assisted recovery. Here again the reduction of recruitment at low stock size looks quite convincing.

These data show the difficulty of determining the relation between recruitment and parent stock from historic data. They are atypical only in that they are some of the longest time series available and include two of the most convincing examples of possible reduction of recruitment at low stock (for the North Sea herring and the Northeast Arctic cod).

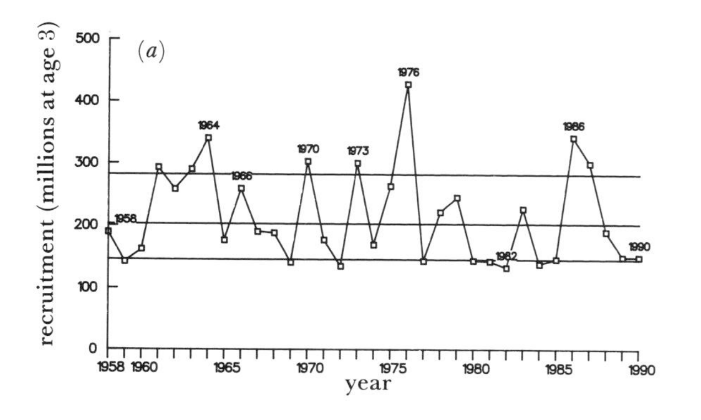

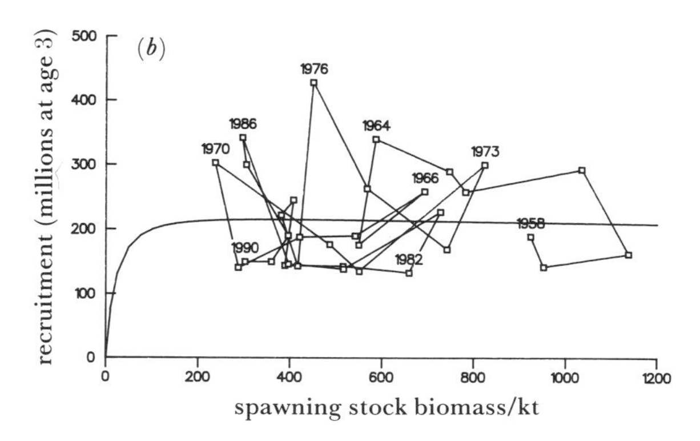

Figure 5. Iceland cod. (*a*) Time series of recruitment. (*b*) Stock recruitment diagram.

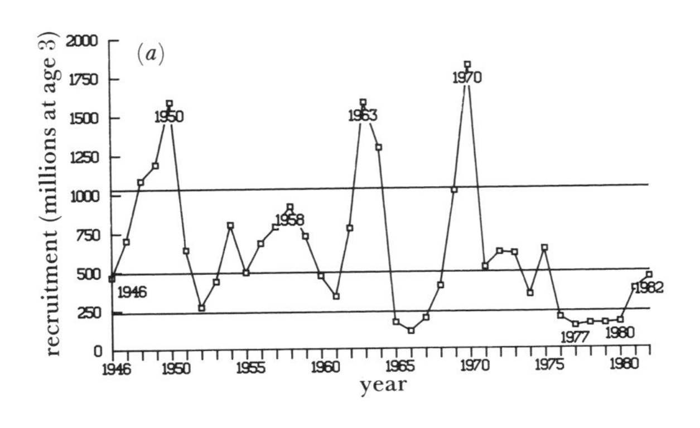

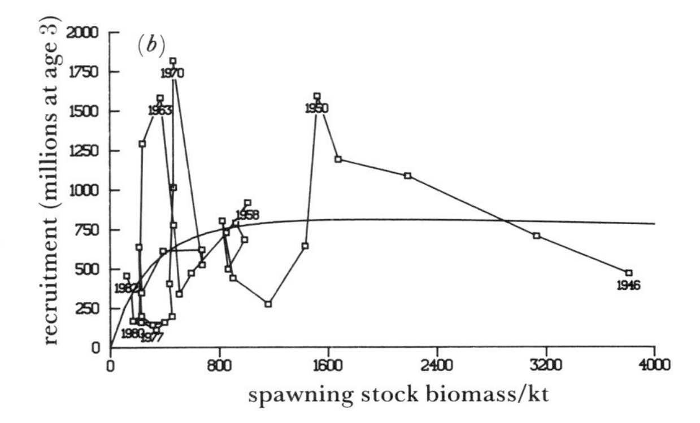

Figure 6. North east Arctic cod. (*a*) Time series of recruitment. (*b*) Stock recruitment diagram.

The relations we have fitted are only illustrative, and not to be taken seriously as attempts to determine the relation. We have ignored serious technical difficulties associated with autocorrelation in the recruitment time series (for which the reasons are incidentally far from clear), and the lagged cross-correlation due to the known causal relation between recruitment and the resultant stock size. It is still possible that purely empirical attempts to estimate such relations may yet be successful, if appropriate statistical techniques are used. It is more likely that it will be possible to obtain reasonable estimates of the slope at the origin by using suitable weighted averages that emphasize values obtained at low stock size (J. Pope, personal communication) and this may be sufficient for pragmatic fishery management. Nevertheless, attempts to analyse such data in similar ways have been conducted for several decades, without much success. The stock-recruitment relation remains elusive.

Is this because we are dealing with a chaotic system? On the face of it, this seems unlikely. First, to enter a chaotic regime, it is necessary that the recruitment curve should cut the survival line while descending rather steeply (Beverton & Holt 1957). This can only occur with a strongly domed over-compensatory relation, and there is no evidence that these are common. Secondly, and more conclusively, the stock-recruitment relation ought to be the underlying deterministic map for the system. If we were just dealing with deterministic chaos, the stock-recruitment plot would be clean, even though the time-series would be chaotic (Holden 1986). Clearly, this is not so. However, before dismissing chaotic dynamics entirely, we should recall that most fish stocks are components of multispecies systems, so that a single-species stock-recruitment diagram is just a projection of a more complicated system. For the moment the possibility of multispecies chaos must therefore be left open. However, attempts by the ICES Multispecies Working Group (Anon 1986) to clean up the stock-recruitment picture by allowing for predation by other species have not so far been successful, and most investigators remain convinced that much of the variation is due to environmental factors, or predation during the egg and larval stages.

4. MECHANISMS FOR DENSITY-DEPENDENT REGULATION

Despite the difficulty of showing density-dependence of recruitment in fish populations, there is no shortage of proposed mechanisms for generating the effect. One obvious possibility is the modulation of fecundity or maturation rates. The other main possibility is, of course, density-dependent juvenile mortality.

The processes concerned are presumably predation, starvation and cannibalism. The predator may be considered as an adventitious one, and so, as the larva or juvenile grows through fields of successively larger and less abundant predators, mortality tends to decrease with age. A predator could aggregate on its prey as Ricker (1957) suggested, but fish larvae tend to be rather thinly distributed (one per cubic metre or less). Starvation must increase mortality, but a more realistic thesis is that of Ricker & Foerster (1948), which combines the effects of predation and lack of food; if the specific growth rate is less than the maximum, so the animals are exposed longer to

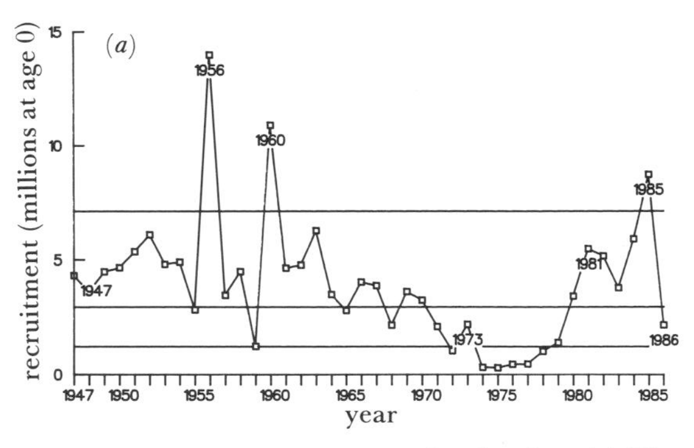

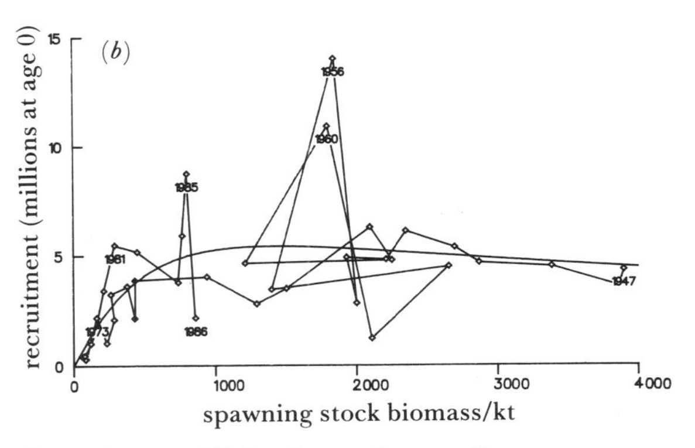

Figure 7. North Sea herring. (*a*) Time series of recruitment. (*b*) Stock recruitment diagram.

mortality generated by the adventitious predator. This idea was developed by Shepherd & Cushing (1980). Cannibalism is potentially a source of stock-dependent mortality, as big cod eat codling, but decisive evidence is lacking perhaps because the big cod could play the part of any adventitious predator. The only process that determines compensation so far discovered is the establishment of territories by the migratory trout in Black Brows Beck (Elliott 1990). In a small area of Black Brows Beck in Cumbria, Elliott (1984*a*) sampled eggs, alevins and parr of migratory trout for 22 years. Eggs were counted, and alevins were sampled by quadrat or cylinder sampling in February, and the parr were caught by repeated electrofishing in May–June and in August–September. The parr were thus sampled in the first and second year of their lives. The mortalities of eggs and alevins were very low, as were losses due to migration as shown by low surface net catches downstream.

Elliott (1984*b*, 1990) found that the number of parr surviving depends on the parent stock according to a Ricker curve, with remarkably little scatter of the data. This study is a very special one in that the population is very small, and the scale of observation is really that of the parr territories. The density-dependent mortality occurs during the earliest free feeding stage of the life history, and endures for about four or five weeks. The most remarkable point is the very low standard error about the Ricker curve, in sharp contrast to the observations on the larger stocks at sea. The conclusion is that the effect seems to be well-established, though it is still not clear how or why the mortality is proportional to initial cohort size, nor whether similar processes occur in marine populations.

In general, density-dependent growth is not very prominant, although it can be seen quite clearly in a large and long-lived fish like the Pacific halibut (Deriso 1985). After the predatory pike had been fished out in Windermere during the second world war, the stock of perch increased by a factor of 33, and then growth and fecundity were density-dependent (Le Cren 1958; le Cren *et al.* 1977). Horwood *et al.* (1986) found that the fecundity of plaice increased by 60% for a small increase in stock of 16%, which was not a compensatory change. Bailey & Almatar (1989) came to a similar conclusion for herring, but suggest that density-dependent fecundity might be more prominent in unexploited stocks.

Density-dependent mortality has been equally difficult to establish. Long ago, Hjort (1914) suggested that there might be a critical period of high mortality when the yolk had been exhausted, and the larvae started to feed. Sette (1935), Marr (1956) and May (1974) denied this thesis but in the past decade or so such an increment in mortality has been detected in single years (Lasker & Smith 1977; Kondo 1980). Daan (1981) and Ware & Lambert (1985) have found density-dependent mortality in the egg and larval stages; van der Veer (1986) established that the mortality of plaice at settlement was density-dependent due to predation by shrimp. Cook & Armstrong (1986) showed that the mortality of young haddock between the pelagic and demersal phases was density-dependent. The death rate of the Northeast Arctic cod between pelagic and demersal phases is also density-dependent (Sundby *et al.* 1989). Density-dependent mortality can occur at any early stage in the life history, but there is not yet enough information to relate it to the coefficient of density-dependence in the stock recruitment relation equation, or to establish that the mechanisms considered are in fact the right ones.

An interesting point is revealed by the observations on plaice eggs and larvae in the Southern Bight of the North Sea. Of eleven observations on eggs, and of four on larvae, there was no relation of either instantaneous or cumulative mortality upon initial numbers (Bannister *et al.* 1974). Estimates of the mortalities of eggs and larvae are available for four year-classes; at the larval stage the 1969 year-class was 100 times more abundant than the other three. It became an average year-class, the excess being presumably eaten by the shrimps at settlement. Yet, with another six year-classes, van der Veer (1986) found a link between the numbers of late larvae and an index of recruitment. The essential processes may occur between late larval stages and settlement.

Sinclair (1988), Sinclair & Iles (1989) have proposed the member-vagrant hypothesis: members of a population return to their native spawning grounds (as do the Pacific salmon) and vagrants do not. Then 'there is no necessary requirement for food availability or predators... losses due to spatial processes can be sufficient for the regulation of absolute abundance and temporal variability'. There must be some loss of animals on the migration circuit because the water movements are variable. The proportion of vagrants could be density-dependent, more leaving as stock increases, like the locusts, but the precise mechanism for this is not known. This would be best established by the appearance of tagged fish at a spawning ground distant from the native one. The stray of the Pacific salmon is very low indeed, and this is part of the evidence of return to the native stream. The west Greenland stock of cod was probably supported by emigration of eggs, larvae and juveniles from Iceland in the Irminger current across the Denmark Strait. For a period a proportion of mature fish returned from west Greenland to Iceland to spawn (see Cushing (1982), for an account of these migrations). These are, however, probably exceptional events, and such losses are in general unlikely to be density-dependent.

5. THE EFFECT OF ENVIRONMENTAL FACTORS

This paper is concerned with regulation in fish populations, and therefore deals primarily with endogenous (stock dependent) effects. However, exogenous effects such as weather and currents almost certainly play a large part in determining recruitment, and are believed to be the cause of the large variability characteristic of most marine fish stocks. We may use the term 'crucial period' to denote that stage in the life-history when density-dependent processes are largest, to avoid confusion with the 'critical period' of Hjort (1914) which has been re-interpreted in various ways, but is often taken to be the period when cumulative mortality is greatest, or when year-class strength is determined. Then exogenous variability occurring before the crucial period will be suppressed to some extent by the density-dependent processes, while that occurring during or after the crucial period is likely to be transmitted in full to the final year-class size. The consequences of this have been explored recently by an ICES Study Group on Models of Recruitment Processes (Anon 1990).

There is a large literature on the effects of the environment on recruitment to marine fish stocks (see, for example, Cushing (1982); Koslow *et al.* (1987); Shepherd *et al.* (1984); Garrod & Colebrook (1978)). There is once again no shortage of candidate mechanisms, but severe difficulty in separating correlations due to cause and effect from those due to a common cause (such as coincident trends). Given that controlled experiments are hardly possible for marine systems, this situation is likely to endure.

6. STOCHASTIC REGULATION

In view of the difficulties encountered in establishing the existence of density-dependent regulation, we now explore an intriguing possibility, that the deterministic regulatory processes may indeed be weak, weaker than anything envisaged within conventional stock-recruitment relations, and that regulation at low stock size may be due almost entirely to stochastic processes.

The genesis of this idea is quite simple. Stock-recruitment data invariably display high variability of recruitment. Being highly variable, but nevertheless non-negative, the probability distribution of recruitment is inevitably skewed, and better described by, for example, a log-normal distribution than by a symmetric (e.g. normal) distribution. It has also often been remarked that it seems that the variability of recruitment increases as stock size declines, although we are not aware of any very convincing demonstrations of this, nor is it probable that one could be prepared with less than 40 or 50 years of data. Garrod (1983) sought but failed to find the effect. Nevertheless, if it does occur, it could lead to a stochastic regulatory process, in the following way.

Assume that the variability of recruitment, expressed as the logarithmic standard deviation (σ) of a log-normal distribution, increases as stock size declines. It is well-known that the arithmetic mean of a log-normal distribution exceeds its geometric mean (or median) by an amount which is approximately $\exp(\sigma^2/2)$. This excess would therefore also increase as the stock size decreased. The stock size is of course some weighted arithmetic moving average of previous recruitment values. Dynamically, therefore, the stock size will depend on something like the recent arithmetic mean of recruitment. If, however, it is the median or geometric mean of the distribution of recruitment which is determined by a weakly regulatory (or even non-regulatory) deterministic process, the effect will be to generate larger arithmetic average R/SSB ratios at lower stock sizes.

The same mechanism would of course work equally well for any skewed distribution whose skewness increased as stock size declined: the assumption of log-normality is merely convenient. The essential ingredients are: (i) skewness of the distribution; (ii) increasing variability as stock size decreases; (iii) median recruitment controlled by deterministic process. Of these, (i) is inevitable, (ii) is plausible, and (iii) is perhaps more controversial. These aspects are discussed later. For the moment it is interesting to explore the consequences in more detail.

First, there need be little or no deterministic regulation in operation: this means that the ratio R/SSB may be distributed around some level which is effectively constant over the range of stock sizes observed. The points on a stock-recruitment diagram may be scattered widely around a straight line through the origin, with no apparent pattern which would correspond to a conventional stock-recruitment relation. A plot of ln(R/SSB) against SSB would in fact show more-or-less symmetrical scatter about a horizontal line, with (if it were detectable) a rather greater scatter at low stock sizes. Thus conventional methods for seeking stock-recruitment relations would almost certainly fail.

Secondly, the mechanism provides only regulation in the mean, and does so because of the increasing contribution made, as the stock declines, by increasingly rare, increasingly extreme recruitment events, as the distribution of year-class strengths becomes increasingly skewed. This sounds so much like a photofit description of recruitment to heavily fished stocks that we feel that this possibility warrants serious consideration.

7. SIMULATION STUDY

We have therefore carried out a simulation study of this process, to see whether it is indeed capable of maintaining non-zero stocks over a wide range of fishing mortality rates. We set up a standard age-disaggregated catch forecast model, in which the recruitment is controlled by a stock-recruitment relation with pseudo-random noise superimposed, and where the variance of the noise is (if required) determined by the stock size. The vital parameters were based on the North Sea plaice stock, for which the observed standard deviation of ln(R/SSB) is about 0.5, but the details are almost certainly of no consequence.

By using various stock-recruitment relations (i.e. none (constant R/SSB), constant recruitment, Beverton–Holt, and Cushing) we first verified that the system behaved as expected in the absence of recruitment variability. For these tests we simulated the effects of various fishing mortality rates in the range 0–1.0 per year. The stock recruitment relations were set up to pass through the observed mean recruitment and SSB values, any other parameters being chosen by intelligent guesswork. The behaviour, in terms of expansion, collapse and stabilization was qualitatively as expected from the simple graphical approach to the problem.

Next, we verified that adding a constant level of variability (as a normally distributed perturbation of log recruitment) did not alter the results in any essential respect. In fact, multiplying by a log-normal perturbation in this way increases the average level of recruitment by the constant factor $\exp(\sigma^2/2)$ mentioned above, and this effect is visible in the simulation results, as will be seen. It does not, however, alter the expansion/decline/stabilization behaviour in any significant way, as expected.

Finally, we implemented the variable variance stochastic mechanism discussed above, as an embellishment of several of the stock-recruitment models. In principle, and indeed in practice to some extent, the mechanism can be superimposed on a purely unregulated deterministic relationship (constant R/SSB). However, such a system is very sensitive to the parameters selected, the population has a tendency to grow very large at low mortality levels, and it is difficult to control the simulations so as to produce plottable results. Wishing to introduce a little regulation, we superimposed the mechanism on both the Beverton–Holt and Cushing (power law) relations. This achieved rather more than the desired result, however, because (as shown in table 1) the Beverton–Holt relation provides strong regulation at high stock levels, and the Cushing relation at low levels.

To achieve the desired minimal effect, we therefore again used instead the functional form of Shepherd (1982), given in equation (2). Selecting $a = 2R_0/B_0$, $K = B_0$ and $\beta = 0.5$, (where R_0 and B_0 are the mean observed recruitment and SSB) this interpolates between a Cushing curve (with a power of 0.5) at high stock levels, and an unregulated relation at low stock levels. The maximal deterministic value of R/B at zero stock size is in fact only double that at the reference level K (taken to be the average stock size).

The variability was applied as a normally distributed perturbation $N(0, \sigma^2)$ with:

$$\sigma^2/\sigma_d^2 = 21.2\,B_0/(21.2\,B + B_0), \qquad (3)$$

where σ_d^2 is the observed variance of ln(R/SSB), equal to 0.24 for the North Sea plaice stock. This strange expression effectively makes the variance inversely proportional to biomass, which is just a plausible way of making it increase as biomass decreases, chosen by analogy with the variance of a mean (which is inversely proportional to the size of the sample). The extra terms involving 21.2 and B_0 are just to limit the ratio σ/σ_d to 4.6 when the stock gets very small. Without this modification, one can get computer overflow when the occasional large random number occurs, which can crash the calculation. This effectively limits the arithmetic mean slope at the origin of the effective stock-recruitment curve, where otherwise it would be infinite.

The arithmetic mean recruitment is obtained by multiplying the deterministic (median) value given by equation (2) by the inflation factor $\exp(\sigma^2/2)$. Both the median and the arithmetic mean values are plotted in figure 8. The weakness of the regulation in the deterministic relationship, and the effectiveness of the stochastic mechanism in increasing R/SSB at low stock size, are both clearly apparent.

We first carried out a series of 100 year simulations

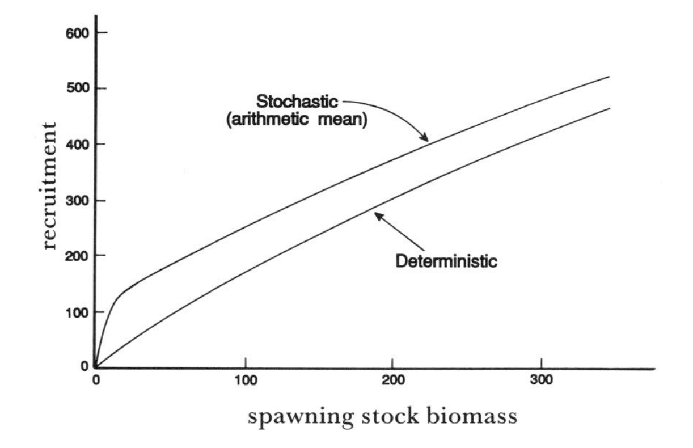

Figure 8. Deterministic and stochastic stock-recruitment relations used for simulation.

by using these relations, starting from their recent average levels of recruitment (488 million) and SSB (372 kt). The results, for a range of fishing mortalities from zero to 0.8 are summarized in figures 9–12. These show the evolution of recruitment for three cases: (i) deterministic; (ii) with constant noise level; (iii) with stock-dependent noise. When there is no fishing (figure 9) recruitment rises to about 2000 million in the deterministic case. With constant noise, the average level is (of course) a little higher, and the variability is very apparent. With the stock-dependent stochastic relationship, the variability is reduced at the high stock levels, and the result is almost indistinguishable from the deterministic case.

At a fishing mortality of 0.30 (figure 10) deterministic recruitment stays more or less constant at about 500 million, and there is little difference between the constant and variable noise cases (the variance for the latter is now only a little reduced).

At a fishing mortality of 0.5 (figure 11), recruitment declines to about 100 million in the deterministic case. The recruitment for the variable variance case now exceeds that for the others by a substantial margin, and recruitment appears to have stabilised at around 500 million after 100 years.

At a fishing mortality of 0.8 (figure 12) the stock collapses to zero for both the deterministic and the constant variance cases, but with variable variance it persists, with recruitment still fluctuating in the low hundreds after 100 years, following a massive year-class around year 58.

All of this is pretty much as expected. It merely shows that the mechanism postulated does in fact work as predicted. However, when we attempted to find the level of F at which the stock would collapse, even with the variable variance mechanism, we got a surprise. This is shown in figure 13, for a fishing mortality of 2.0, now with a 1000-year time horizon.

The stock survives its first 100 years, probably because of the massive year-class in year 58, now even more pronounced. Thereafter it 'collapses' to a very low level, and remains there for 150 years. At around year 250, another extreme year-class occurs, leading to a century or so of modest but variable recruitment. The stock then disappears for almost 200 years, before reappearing for three centuries or so. After year 850 it

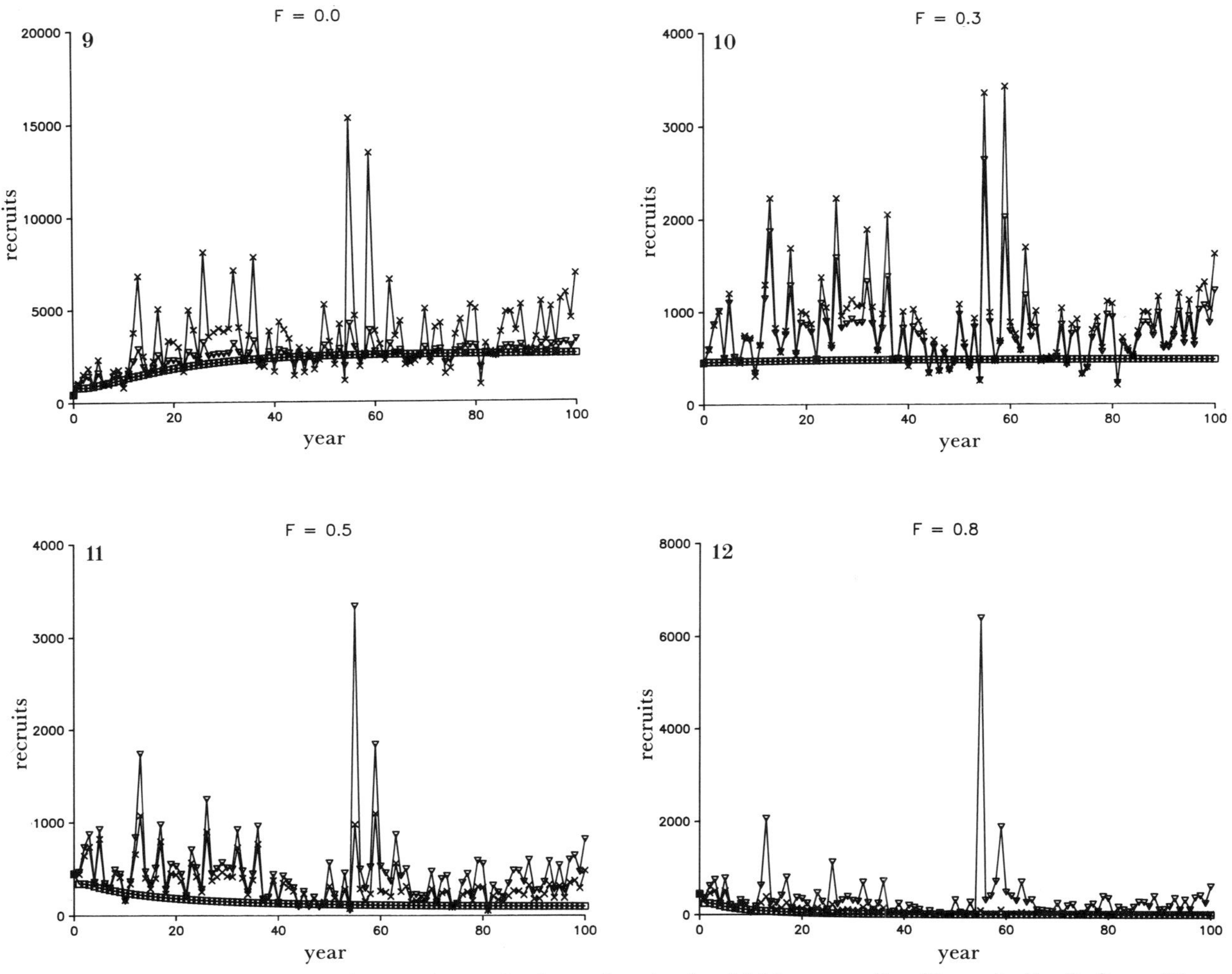

Figures 9–12. Simulated recruitment time series for various levels of fishing mortality. Figure 9, $F = 0$; figure 10, $F = 0.3$; figure 11, $F = 0.5$; figure 12, $F = 0.8$. (□, deterministic (no variability); ×, constant variability; △, stock-dependent variability.)

collapses again, but we clearly cannot rule out the possibility of further recoveries; indeed, they seem to be rather probable.

At a fishing mortality of 2.5 (figure 14) the last episode of good recruitment is restricted to only about one century starting at year 700. Only with $F = 2.8$ (figure 15) does the stock collapse, and stay extinct for the remaining 900 years of the simulation.

This behaviour is of course explicable with hindsight, but was not expected by us. It should be stressed that this episodic behaviour on a timescale of centuries is obtained without any time-variation of the parameters of the model. There is no climatic variation, no change of fishing mortality, predator abundance or anything. It is entirely self-generated: the stock bootstraps itself back up from near-extinction every now and then, purely by chance events.

We have in fact, without deliberately setting out to do so, created a highly non-stationary stochastic process. Neither the mean nor the variance of this process is constant, but each depends on the other, and they fluctuate together over time, without any systematic external control at all. Results generated by such a process violate the assumptions of stationarity implicit in conventional methods of time series analysis. They will defy understanding using these tools. If such a mechanism operates, it is perhaps not surprising that attempts to analyse such data have met with little success.

The ability of the stock to bounce back after six centuries in the wilderness (figure 14) even when held down by a level of fishing mortality which is extremely high even by North Sea standards, is a little uncanny. It is, nevertheless, distinctly reminiscent of the long-term behaviour of quite a few real stocks (as described earlier in this paper). These long-term changes are usually attributed to climatic changes, or interaction between stocks (for example, the Californian sardine–anchovy switching). Our simulations now suggest an alternative hypothesis: that they are due to pure chance. There may, in fact, be no deterministic regulatory mechanism to be found, beyond that which would lead to increasing variability at low stock size. So far as we are aware, this is a new and rather exciting result. It removes the apparent conflict between the ideas of regulation and variation of recruitment as dominant mechanisms. Here they are inextricably linked aspects of the same stochastic process.

For completeness, we plot the 'observed' stock-recruitment plot from the simulations (figure 16): this is a composite diagram built up from the results for fishing mortalities of 0.3, 0.5 and 0.8. Whether or not

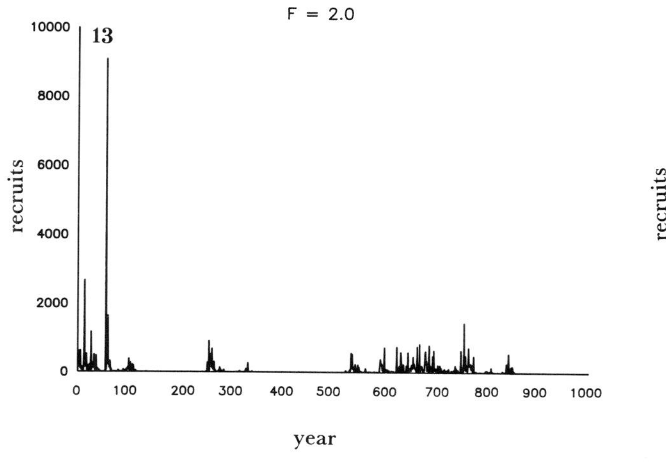

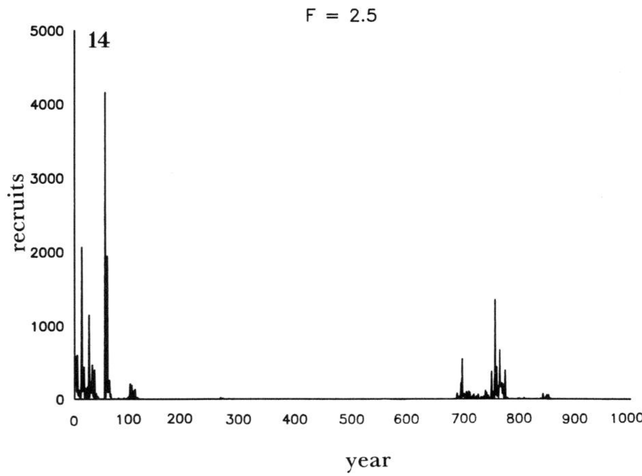

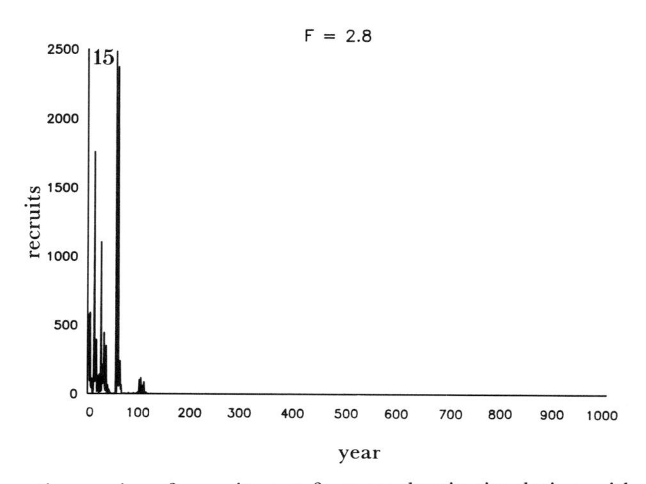

Figures 13–15. 1000-year time series of recruitment from stochastic simulation with stock-dependent variability. Figure 13, $F = 2.0$; figure 14, $F = 2.5$; figure 15, $F = 2.8$.

it resembles the real data sets is perhaps a matter of opinion! Finally, figure 17 shows the time series of ln(R/SSB), the crucial parameter describing reproductive success, for $F = 0.8$. This shows no systematic trend, as expected, and the changing variability is not immediately apparent, and would be difficult to detect statistically.

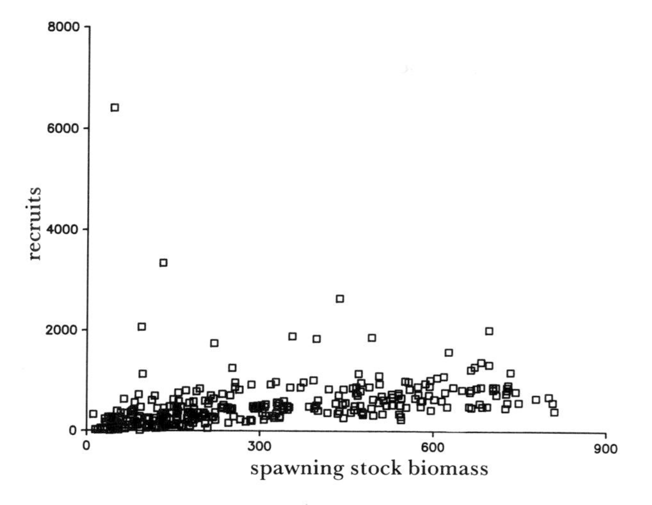

Figure 16. Simulated stock-recruitment scatter plot (composite diagram for $F = 0.3$, 0.5 and 0.8).

8. MECHANISMS FOR STOCHASTIC REGULATION

These results are rather suggestive, and would have considerable implications for stock-recruitment research; these are discussed later. First, it is necessary to ask whether the mechanism proposed bears any relation to reality, or whether it is just a figment of our

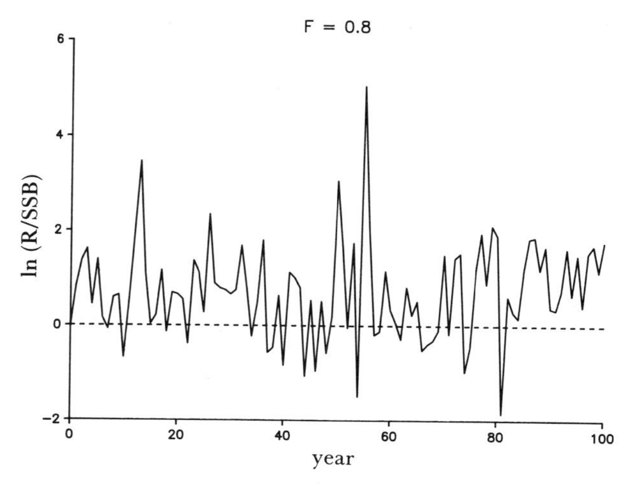

Figure 17. Simulated time series of ln(R/SSB) for $F = 0.8$.

imagination. The three essential ingredients are skewness, increasing variability as the stock size decreases, and control of the median (rather than the arithmetic mean) by whatever deterministic processes operate. The first of these is almost inevitable for a highly variable non-negative quantity, is confirmed by observation, and need not detain us.

The second is plausible. There are doubtless various mechanisms that could lead to such behaviour. The most obvious is patchiness, in either space or time. There is considerable literature on this subject (see Steele (1976) for an account in relation to marine systems), and there is no doubt that it occurs, probably for most populations and certainly for marine fish. The recruitment to the stock (a rather large unit) is the result of a summation over space and time of all the components of the stock. We know that there is extreme variability of abundance in the plankton at small space and timescales, and almost certainly of mortality too. By the central limit theorem, we should expect the variability (expressed by the coefficient of variation, perhaps) of the resultant total to be less than that of any of the component parts. Just how much less depends on the relative space and timescales, and the distribution of variability itself. However, will modulation of the size of the stock lead to any variation of the total variability? Clearly, if the effect is simply to reduce the abundance everywhere by the same factor, the answer would be no. If, on the other hand, the reduction were expressed as a reduction in the range, in space or time, of the population, the answer is most likely yes, because there would be fewer components contributing to the total. Beyond that, if as the stock declines the areas of high abundance are affected more than those of low abundance (or vice versa), then greater or lesser modulation of variability could be expected. Observations on fish stocks suggest that such effects do in fact occur (Myers & Stokes 1989).

Effects of this sort would of course also be generated by density-dependent processes acting locally, at high abundance the variability would be reduced by regulation. It is, however, an intriguing possibility that the effect could be generated just by chance. The coefficient of variation of a compound probability distribution describing the patchy spatial distribution of animals (e.g. a negative binomial distribution) seems quite likely to behave as hypothesized, for example. The answer to this question must therefore be: 'quite possibly'. It is a concrete question which can be studied by field observations: the hypothesis is testable. The strength of the effect will depend on the spectrum of space and timescales concerned. To understand the process one would need to study the structure of the patchiness, and such work is closely related to modern ecological studies carried out purely for their intrinsic interest. Here perhaps, is yet another example of a practical question (of considerable economic importance) suddenly focussing attention on the results obtained from decades of basic research driven only by curiosity.

The third question is quite tricky. The answer depends on how the variability arises, on what is determined by what, and how the perturbations enter the process. The arithmetic mean is the classic measure of location of a statistical distribution, and statisticians even use the term 'expectation' for it, which lends it an air of universality it does not deserve. The key question is, as the variability reduces, onto which measure of location does the distribution collapse? A simple real example suffices to show that this may be the geometric mean or median rather than the arithmetic mean, in at least one relevant case. Consider, in fact, the abundance of fish eggs just before hatching. Suppose that they have been subjected to mortality by predators since they were laid, and that the abundance of the predators varies according to a normal distribution. Then, if the mortality rate is proportional to the predator abundance, the number of eggs which survive will be log-normally distributed. If the variation of the predator abundance is suppressed, so that it stabilizes at its arithmetic mean, then the egg abundance will stabilise at its previous geometric mean (the same as the median in fact). If the variation of predator abundance is not normally distributed, something more complicated will happen, but it seems plausible that, for perturbation processes of this sort, one should think of things being distributed about their medians, rather than their arithmetic means. This is not true for pure sampling (counting) variability but we have not so far thought of any other counter-examples. These are, however, deep probabilistic waters, and for the present purpose it suffices that deterministic control of medians rather than arithmetic means seems to be entirely plausible. Laurec *et al.* (1980) considered a somewhat similar model where the arithmetic mean rather than the median was controlled, and found that this increased the frequency of stock collapse. This is to be expected, because in this case the variability acts to produce depensatory rather than compensatory density-dependence.

9. DISCUSSION

We conclude that the stochastic mechanism proposed is plausible, and perhaps even likely to occur in practice. Whether it is indeed powerful enough to produce effects of the magnitude we have simulated is another question, that can only be answered by analysis of real data sets. However, our results do provide a new perspective for stock-recruitment studies, and suggest that we could be missing the point in seeking processes that would deterministically regulate total recruitment. We should instead perhaps study the causes of the variability itself, since that might be an equally or even more important part of the problem of regulation. The role of density-dependent processes may be primarily local, and they may act mainly to modulate variability, and therefore only indirectly affect total recruitment to the stock.

Studies of variability in space and time of abundance, at as wide a range of scales as possible, for a wide range of stock sizes, are clearly indicated. We may need not patch studies, but patchiness studies. These will be expensive. Before committing a lot of resources to such work, it would be wise to seek, in the few long-term series available, for any evidence of variation of variability as a function of stock sizes. This will be difficult to detect (Garrod 1983); it takes 10 years of data to determine one variance with a precision of 30% and several estimates of variance for each stock will be required to establish any changes. Nevertheless, something may show up, and the search should be made as soon as possible. Meanwhile, those responsible for operational stock assessments should bear in mind that not only is the casual assumption of constant recruitment dubious and dangerous, but also that the ability of fish stocks to withstand heavy exploitation may have been bought at the price of extreme variability at low stock sizes.

The simulation calculations for the evaluation of the stochastic mechanism were carried out by Dr M. Sun.

REFERENCES

Andrewartha, H. G. & Birch, L. C. 1954 *The distribution and abundance of animals.* (781 pages.) Chicago University Press.

Anon 1986 *Report of the ICES ad hoc Multispecies Assessment Working Group 1985.* ICES CM 1986/Assess: 9.

Anon 1990 *Report of the ICES-IOC Study Group meeting on Models for Recruitment Processes, Paris, May 1990.* ICES CM 1990/A: 5.

Bailey, R. S. & Almatar, S. 1989 Variations in fecundity and egg weight of herring (*Clupea harengus* L) Part II. Implications for hypotheses on the stability of marine fish populations. *J. Cons. perm. int. Explor. Mer.* **45**, 125–130.

Bannister, R. C. A., Harding, D. & Lockwood, S. J. 1974 Larval mortality and subsequent yearclass strength in plaice (*Pleuronectes platessa* L). In *Early life history of fish* (ed. J. H. S. Blaxter). Berlin: Springer.

Beverton, R. J. H. & Holt, S. J. 1957 On the dynamics of exploited fish populations. *Fishery Invest. Ser.* **2**, no. 19.

Cook, R. M. & Armstrong, D. W. 1986 Stock-related effects in the recruitment of North Sea haddock and whiting. *J. Cons. perm. int. Explor. Mer.* **42**, 272–280.

Cushing, D. H. 1973 The dependence of recruitment on parent stock. *J. Fish Res. Bd Canada* **30**, part 2, 1965–1976.

Cushing, D. H. 1982 *Climate and fisheries.* (373 pages.) London: Academic Press.

Cushing, D. H. 1988 The northerly wind. In *Toward a theory on biological–physical interactions in the World Ocean* (ed. B. J. Rothschild), pp. 235–244. Kluwer.

Daan, N. 1981 Comparison of estimates of egg production from the Southern Bight cod stock from plankton surveys and from market statistics. *Rapp. P.-v. Réun. Cons. perm. int. Explor. Mer.* **178**, 242–243.

Deriso, R. B. 1985 Stock assessment and new evidence of density dependence. In *Fisheries dynamics; harvest management and sampling* (ed. P. Mundy, T. S. Quinn & R. B. Deriso), pp. 49–60. Univ. Washington, Seattle.

Dickson, R. R., Meincke, J., Malmberg, S.-A. & Lee, A. J. 1988 The great salinity anomaly in the northern North Atlantic, 1968–1982. *Prog. Oceanogr.* **20**, 103–151.

Elliott, J. M. 1984*a* Numerical changes and population regulation in young migratory trout *Salmo trutta* in a Lake District stream, 1966–83. *J. Anim. Ecol.* **53**, 327–350.

Elliott, J. M. 1984*b* Growth size, biomass and production of young migratory trout *Salmo trutta* in a Lake District stream, 1966–83. *J. Anim. Ecol.* **53**, 979–994.

Elliott, J. M. 1990 Mechanisms responsible for population regulations in young migratory trout, *Salmo trutta* III the role of territorial behaviour. *J. Anim. Ecol* (In the press.)

Garrod, D. J. 1983 On the variability of yearclass strength. *J. Cons Perm. int. Explor. Mer.* **41**, 63–66.

Garrod, D. J. & Colebrook, J. M. 1978 Biological effects of variability in the North Atlantic Ocean. *Rapp. P.-v. Cons. int. Explor. Mer.* **173**, 128–144.

Hjort, J. 1914 Fluctuations in the great fisheries of northern Europe viewed in the light of biological research. *Rapp. P.-v. Réun. Cons. perm. int. Explor. Mer.* **20**, 1–228.

Holden, A. V. (ed.) 1986 *Chaos.* (324 pages) Manchester University Press.

Horwood, J. W., Bannister, R. C. A. & Howlett, G. J. 1986 Comparative fecundity of North Sea plaice (*Pleuronectes platessa* L). *Proc. R. Soc. Lond.* B **228**, 401–431.

Kondo, K. 1980 The recovery of the Japanese sardine – the biological basis of stock fluctuations. *R. P.-v. Réun. Cons. perm. int. Explor. Mer.* **177**, 322–354.

Koslow, J. A., Thompson, K. R. & Silvert W. 1987 Recruitment to north west Atlantic cod (*Gadus morhua* L) and haddock (*Melanogrammus aeglefinus* L) stocks: influence of stock size and climate. *Can. J. Fish Aquat. Sci.* **44**, 26–39.

Lasker, R. & Smith, P. E. 1977 Estimation of the effects of environmental variations on the eggs and larvae of the northern anchovy. *Calcofi Rep.* **19**, 128–137.

Laurec, A., Fonteneau, A. & Champagnat, C. 1980 A study of the stability of some stocks described by self-regenerating stochastic models. *Rapp. P.-v. Réun. Cons. perm. int. Explor. Mer.* **177**, 423–438.

Le Cren, E. D. 1958 Observations on the growth of perch (*Perca fluviatilis* L) over twenty-two years with special reference to the effects of temperature and changes in population density. *J. Anim. Ecol.* **27**, 287–334.

Le Cren, E. D., Kipling, C. & McCormack, J. 1977 A study of the numbers, biomass and yearclass strengths of perch (*Perca fluviatilis* L) in Windermere from 1941 to 1966. *J. Anim. Ecol.* **46**, 281–307.

Marr, J. C. 1956 The 'critical period' in the early life history of marine fishes. *J. Cons. perm. int. Explor. Mer.* **21**, 160–170.

May, R. C. 1974 Larval mortality in marine fishes. In *The early life history of fish* (ed. J. H. S. Blaxter), pp. 3–20. Heidelberg: Springer.

Myers, R. A. & Stokes, T. K. 1989 Density-dependent habitat utilisation of groundfish and the improvement of research surveys *ICES CM 1989/D: 15.*

Nicholson, A. J. 1933 The balance of animal populations. *J. Anim. Ecol.* **2**, 132–178.

Reddingius, J. 1971 Gambling for existence. *Acta Biotheor.* **20** (Suppl. 1), 1–208.

Ricker, W. E. & Foerster, R. E. 1948 Computation of fish production. In *A symposium on fish populations* (*Bull. Bingh. Oceanogr. Collection* **11**), pp. 173–211.

Ricker, W. E. 1954 Stock and recruitment. *J. Fish Res. Bd. Canada* **11**, 559–623.

Ricker, W. E. 1958 Handbook of computations for biological statistics of fish populations. *Bull. Fish Res. Bd. Canada* **119**, 1–300.

Sette, O. F. 1935 Biology of the Atlantic mackerel (*Scomber scombrus*) of North America: its early life history including the growth, drift and mortality of egg and larval populations. *Fish Bull. U.S. Fish Wildlife Serv.* **50**, 149–234.

Shepherd, J. G. 1982 A versatile new stock-recruitment relationship for fisheries and the construction of sustainable yield curves. *J. Cons. perm. int. Explor. Mer* **40**, 67–75.

Shepherd, J. G., Pope, J. G. & Cousins, R. D. 1984 Variations in fish stocks and hypotheses concerning their links with climate. *Rapp. P.-v. Réun. Cons. perm. int. Explor. Mer.* **185**, 255–267.

Shepherd, J. G. & Cushing, D. H. 1980 A mechanism for density-dependent survival of larval fish as the basis of a stock-recruitment relationship. *J. Cons. perm. int. Explor. Mer.* **39**, 160–167.

Sinclair, M. J. 1988 *Marine populations: an essay on population regulation and speciation.* (252 pages), Seattle and London: University of Washington Press.

Sinclair, M. J. & Iles, T. D. 1989 Population richness of marine fish species. *Aquat. Living Res.* **1**; 71–83.

Steele, J. H. 1976 Patchiness. In *The ecology of the seas* (ed. D. H. Cushing & J. J. Walsh), pp. 98–115. Philadelphia & Toronto: W. B. Saunders.

Sundby, S., Bjorke, H. Soldal, A. V. & Olsen, S. 1989 Mortality rates during the early life stages and yearclass strength of north east Arctic cod (*Gadus morhua* L). *Rapp. P.-v. Réun. Cons. perm. int. Explor. Mer.* **191**, 351–358.

van der Veer, H. W. 1986 Immigration, settlement and density dependent mortality of a larval and post-larval O group plaice (*Pleuronectes platessa* L) population in the western Wadden See. *Mar. Ecol. Progr. Ser.* **29**, 223–236.

Ware, D. M. & Lambert, T. C. 1985 Early life history of Atlantic mackerel (*Scomber scombrus* L) in the southern Gulf of St Lawrence. *Can. J. Fish Aquat. Sci.* **42**, 577–592.

Discussion

R. J. H. Beverton (*Montana, Old Roman Road, Langstone, Gwent, U.K.*). Dr Shepherd has shown a most intriguing way in which extreme episodic fluctuations, apparently density-independent, can nevertheless have a regulating influence on the long-term dynamics of the population. I would like to make two comments.

One is that I believe that a further search for evidence of higher variability of recruitment at low population sizes may prove rewarding, perhaps by using the statistics of extreme events as well as the conventional measures of variance.

The other is that an exceptionally large recruitment may set in train events other than the purely structural changes in the size and age composition of the adult population when the yearclass becomes mature.

The North Sea haddock is a good example. The three largest yearclasses ever recorded, those of 1962, 1967 and 1974, were each followed by year-classes significantly smaller than average; indeed, the inhibiting effect of the very large year-class was detectable, though muted, in two or even three of the year-classes that immediately followed it. Whatever the mechanism (not intra-specific predation in the case of haddock, which is primarily a benthic feeder), the result is that the haddock system evidently does not permit an extremely large year-class more often than once every five years or so (at best), however favourable the other circumstances may be. This puts a form of density-dependent 'capping' on the upper end of the range of populations size, but at the cost of higher short-term variability. Perhaps these dynamics would be worth investigating as a follow-up of Dr Shepherds paper.

J. G. Shepherd. I agree with Professor Beverton that we need to look harder at levels of variability at various stock sizes, and that we may need to use non-standard statistical tools to do so. His point about the apparent effect of large year-classes is also interesting. Armstrong & Cook (loc cit) have carried out an investigation for the North Sea haddock, but in general serial autocorrelation, including the negative correlation he mentions, appears to be quite common. I am sure that this aspect also warrants further investigation. For example, is there a similar effect after small year-classes?

Geometry, heterogeneity and competition in variable environments

PETER L. CHESSON†

Department of Zoology, Ohio State University, 1735 Neil Avenue, Columbus, Ohio 43210, U.S.A.

SUMMARY

The effects of environmental fluctuations on coexistence of competing species can be understood by a new geometric analysis. This analysis shows how a species at low density gains an average growth rate advantage when the environment fluctuates and all species have growth rates of the particular geometric form called subadditive. This low density advantage opposes competitive exclusion. Additive growth rates confer no such low density advantage, while superadditive growth rates promote competitive exclusion.

Growth-rate geometry can be understood in terms of heterogeneity within populations. Total population growth is divided into different components, such as may be contributed by different life-history stages, phenotypes, or subpopulations in different microhabitats. The relevant aspects of such within-population heterogeneity can be displayed as a scatter plot of sensitivities of different components of population growth to environmental and competitive factors, and can be measured quantitatively as a covariance. A three-factor model aids the conceptual division of population growth into suitable components.

INTRODUCTION

Do environmental fluctuations explain the high diversity of some animal and plant communities (Abrams 1984; Armstrong & McGehee 1976; Chesson & Huntly 1989; Comins & Noble 1985; Connell 1978; Grubb 1977; Sale 1977; Shmida & Ellner 1985; Woodin & Yorke 1975)? Hutchinson (1961) suggested that environmental fluctuations could favour different species at different times, and that this would permit coexistence of many species. Similar ideas form part of Grubb's (1977, 1986) conception of the regeneration niche. These ideas are now supported by mathematical models, with some provisos, as can be seen below.

It has also been suggested that environmental events affecting all species similarly may promote coexistence when they occur in the form of disturbance, causing pulses of mortality (Connell 1978). This idea is best developed for models of successional systems in patchy environments (Hastings 1980).

Models argue against the common conception that environmental variability promotes diversity by keeping species' densities at levels where there is little competition (Chesson & Huntly, in preparation). Diversity maintenance usually involves different species being favoured at different times by environmental or competitive conditions (Chesson & Huntly 1989). Thus while the initial effects of disturbance may be similar mortality for all species, diversity maintenance involves differential responses to the ensuing sequence of competitive conditions.

† Present address: Ecosystem Dynamics Group, Research School of Biological Sciences, Australian National University, P.O. Box 475, Canberra, ACT 2601, Australia.

It is also important to keep in mind that environmental variability need not have the same effect in different systems. Models show a strong dependence on the biological details of the component species, including life-history properties, physiology and behaviour (Chesson 1988). Depending on these details, environmental variability may promote diversity, promote competitive exclusion, or have no effect at all on diversity.

At least for the case of temporal environmental variability, there is a general understanding of the sorts of situations leading to these different effects of environmental variability (Chesson & Huntly 1989). It is the purpose of this article to explore this in two specific ways. First we consider a geometric approach to stochastic competition that shows how geometric properties of population growth rates can oppose competitive exclusion in some circumstances, but promote it in others. We will then explore the origin of the important geometric relations in the biology of a population. Heterogeneity within a single species population is of special importance to the geometry of a species' population growth rate and therefore of special importance in the outcome of interactions with other species.

A GENERAL COMPETITION MODEL

Environmental effects in competition models are usually represented through some population parameter that is sensitive to fluctuating environmental factors. The choice of an environmentally dependent parameter depends on the actual ecological community that one is trying to represent, but in general, it is something like a density-independent birth rate, mortality rate, seed germination rate, or resource uptake rate. A population can be expected to have several environmentally dependent parameters (Chesson & Warner 1981), but for simplicity it is here assumed that each species has just one key environmentally dependent parameter, denoted by the symbol $E_i(t)$, where i is the species and t is time. Often time, t, will be suppressed for notational simplicity.

A second important ingredient in stochastic competition models is the competition parameter. This is taken as a single number $C_i(t)$ representing for the time period t to $t+1$, the effect of competition both with and between species on the growth rate of the given species i. This competition parameter is presumed to depend on the densities of the species in the system and also on their environmentally dependent parameters. Thus, it can be represent as

$$C_i(t) = c_i(E_1, X_1, E_2, X_2 \ldots, E_n, X_n), \tag{1}$$

where $X_1, X_2, \ldots, X_n$ are the population densities of the n competing species in the system, and c_i is some function.

Environmentally dependent parameters can be expected to affect the amount of competition occurring through their effects on the abundances of competing forms. For example, if the environmentally dependent parameter is a germination rate, it will affect the abundance of seedlings, which potentially strongly affects the intensity of competition as these seedlings grow. This indirect effect of E_i through its effect on competition depends on the density of the species, and vanishes when the species' density is 0.

With these definitions, population dynamics in discrete time can be represented as:

$$X_i(t+1) = G_i(E_i, C_i)\, X_i(t), \tag{2}$$

where the function G_i combines E_i and C_i to give the finite rate of increase of species i from time t to time $t+1$. A particular example of a model within this general class of models is shown in Table 1.

To analyse the model, we express population growth on the log scale, writing

$$\ln X_i(t+1) - \ln X_i(t) = g_i(E_i, C_i), \tag{3}$$

where g_i is simply $\ln G_i$. Changes in ln population size over any period of time can be found as the sum of g_i over that period. It follows that the observed trend in population growth over any period of time is the simple arithmetic average of g_i for the period. Thus, to predict future trends in population growth, we average $g_i(E_i, C_i)$ over the theoretical probability distribution of E_i and C_i.

According to the standard invasibility analysis (Turelli 1981; Chesson & Ellner 1989), a species will persist in the presence of its competitors if its population recovers after fluctuations to low density. Such recovery is determined by the mean low density growth rate, which is defined as the average value, Δ_i, of $g_i(E_i, C_i)$, with $X_i(t)$ set equal to 0. This is expressed mathematically as

$$\Delta_i = E[g_i(E_i, \hat{C}_i)], \tag{4}$$

where the E outside the square brackets is the mathematical symbol for averaging the quantity inside over its theoretical probability distribution. The notation $\hat{C}_j$ means C_j evaluated with X_i set equal to 0.

If Δ_i is positive, we know that species i will tend to increase from low density, and thus avoid competitive exclusion. If it is negative, species i will tend to decrease from low density and may become extinct.

For two-species communities, under certain conditions, Chesson & Ellner (1989) showed that positive values for the Δs of both species imply coexistence in the sense that both species show steady fluctuations, and always recover from excursions to lower densities (figure 1). Negative values for both species mean that they have positive probability of converging to extinction and that one of them must eventually do so, a situation that can be described as random competitive exclusion.

Finally, if the Δs of the two species are of opposite sign, the species with the negative Δ is driven extinct by the other species. No such rigorous analysis has been completed for the n species situation, but Turelli (1980) and Ellner (1985) have found the invasibility approach to be generally supported by computer simulations.

Invasibility analysis requires knowledge of the probability distribution of $\hat{C}_i$. To get this, one assumes

Table 1. *Seed-bank model*

$X_i(t+1) = [(1-E_i(t))\, s_i + E_i(t)\, Y_i/C_i(t)]\, X_i(t)$.	
$X_i(t)$:	size of seed bank of species i at the beginning of year t before germination.
$E_i(t)$:	germination fraction of species i in year t.
s_i:	yearly survival rate for ungerminated seeds.
Y_i:	seed yield per germinated seed in the absence of competition.
$C_i(t)$:	Reduction in seed yield due to competition. A function of the number of seeds germinating, e.g. $C_i(t) = 1+\Sigma_{j=1}^{n} \alpha_{ij} E_j(t)\, X_j(t)$, for constants α_{ij}.
$G_i(E_i, C_i)$:	$[(1-E_i(t))\, s_i + E_i(t)\, Y_i/C_i(t)]$, the finite rate of increase for species i.

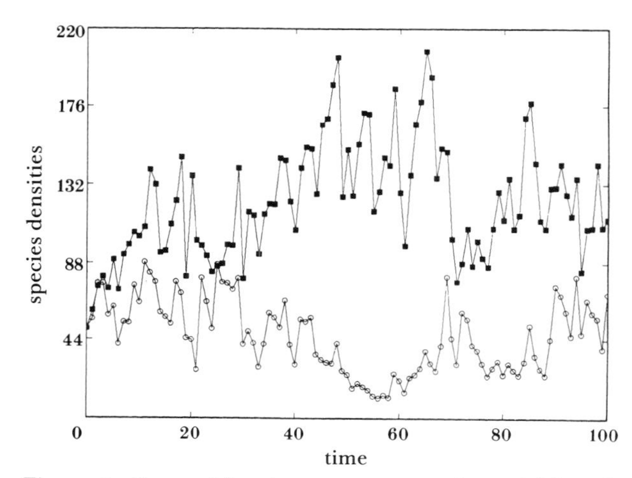

Figure 1. Competition between two species satisfying the seed-bank model of table 1. The species with open circles is competitive subordinate with Y value 80% of that of the other species. The E_i have independent rectangular distributions, $s_1 = s_2 = 0.9$, and $\alpha_{ij} = 1$ for all i, j.

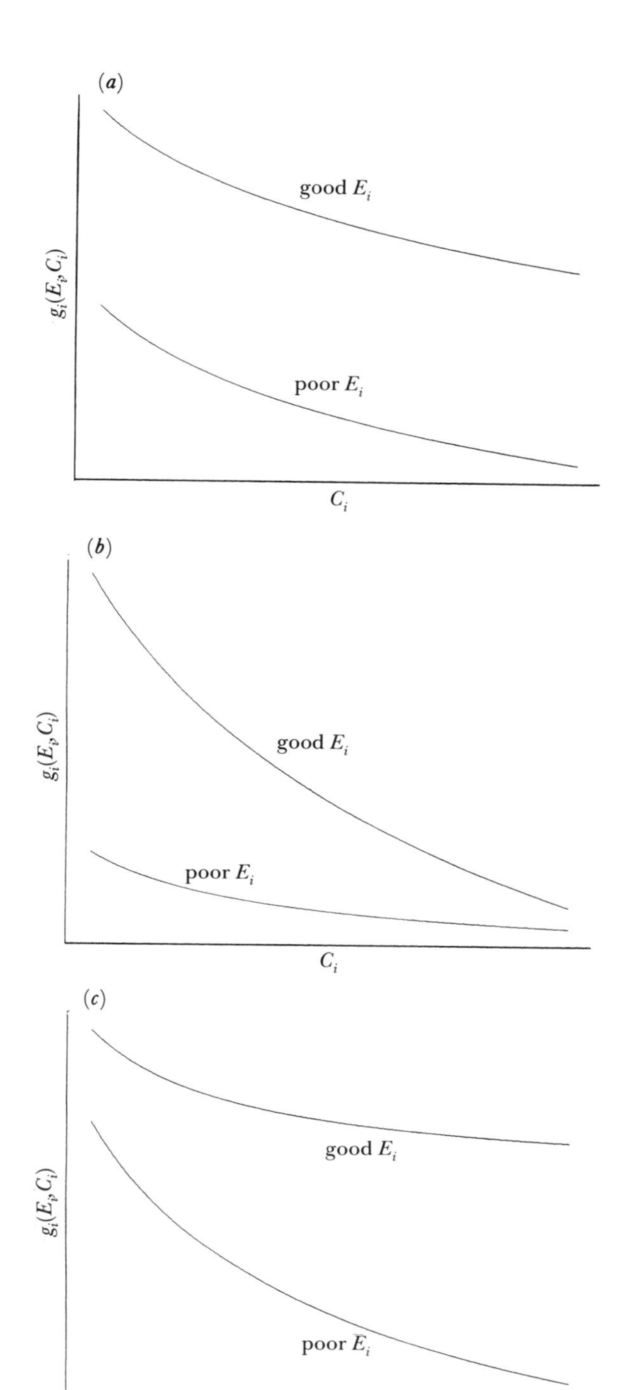

Figure 2. Change in log population size ('growth rate'), as a function of C_i for different values of the environmentally dependent parameter. (*a*) additive; (*b*) subadditive; (*c*) superadditive.

that the other species in the system have settled down to the sort of fluctuations that they achieve in the long-run after the removal of species i. This means in particular that the probability distribution of any species j, say, has no further changes over time, i.e. is described by a stationary probability distribution (Turelli 1981).

Stationary distributions are difficult to calculate in general, which means that we must resort to various tricks to get enough information about them. One such trick is recognizing that if species j is at its stationary distribution, then $E[\ln X_j(t+1)]$ and $E[\ln X_j(t)]$ are the same because the probability distribution of X_j used to calculate these averages is the same at both times t and $t+1$. It follows that the difference between them is 0, i.e.

$$E[g_j(E_j, \hat{C}_j)] = 0. \tag{5}$$

By using these ideas it is possible to show that the shapes of the functions g_i are critical in determining the effect of environmental fluctuations on coexistence. Figure 2 shows the three possible shapes that growth rates can have. These different general types of growth rates are called additive, subadditive or superadditive, depending on whether the plot of $g_i(E_i, C_i)$ against C_i for different values of E_i gives parallel curves, converging curves or diverging curves. These shapes are generally associated with diversity neutral, diversity promoting and diversity demoting effects of environmental fluctuations (Chesson 1988), respectively. How these different effects come about is best understood by the geometric approach presented here.

COEXISTENCE AND THE GEOMETRY OF POPULATION GROWTH RATES

To introduce this geometric approach, I make a number of simplifying assumptions. I assume that the species differ only in their responses to environmental factors. Thus I shall assume that the g_is are all the same, and so we can drop the subscript. Similarly, I assume that the C_i are all the same. Thus all species are affected by the same, possibly composite, competitive factor.

With these assumptions we can set about calculating Δ_i for any species, arbitrarily designated as i, within this n species system. We use the technique of conditional expectations, which allows us to evaluate Δ_i in stages. We compare the growth rate of species i at low density with the growth rate of some other species j, assuming that j and all other species are at their stationary distribution. Thus species j has a mean growth rate equal to 0 in accordance with equation (5).

The environmentally dependent parameters of species i and j, (E_i, E_j), have some bivariate probability distribution. Let (ϵ_+, ϵ_-) be one point chosen from this distribution with $\epsilon_+ > \epsilon_-$. We consider the effects of (E_i, E_j) fluctuating between the two values (ϵ_+, ϵ_-) and (ϵ_-, ϵ_+). This is equivalent to conditioning on the event that (E_i, E_j) takes one or the other of these two values. We also condition on the environmentally dependent parameters of all other species, and all species population densities. Thus, we consider all these things as fixed for the time being. The only uncertainty left is whether (E_i, E_j) takes the value (ϵ_+, ϵ_-) or (ϵ_-, ϵ_+). These restrictions let us focus on the most important aspects of population dynamics geometrically (figure 3).

Under the imposed conditions, $\hat{C}$ can only vary with E_j as $\hat{C}$ does not depend on E_i when $X_i = 0$, and everything else is being conditioned on (is fixed for now). Thus $\hat{C}$ can take just the values

$$c_- = c(E_1, X_1, E_2, X_2, \ldots, \epsilon_-, X_j, \ldots, E_n, X_n), \tag{6}$$

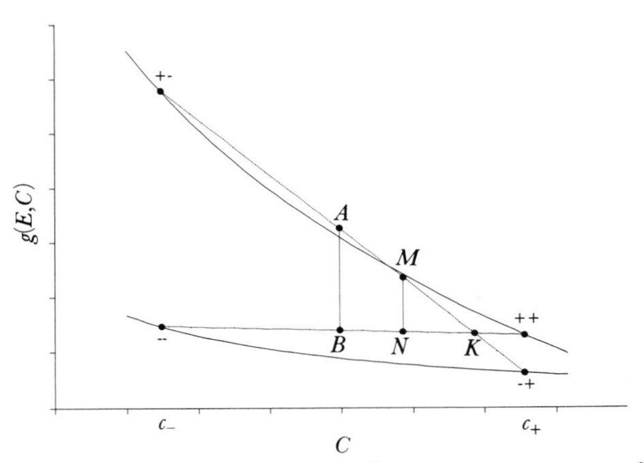

Figure 3. The growth rates $g(\epsilon_+, \hat{C})$ (top curve) and $g(\epsilon_-, \hat{C})$ (bottom curve), as functions of $\hat{C}$ for the subadditive case.

and

$$c_+ = c(E_1, X_1, E_2, X_2, \ldots, \epsilon_+, X_j, \ldots, E_n, X_n). \tag{7}$$

It follows that $g(E_i, \hat{C})$ fluctuates between the two values $g(\epsilon_+, c_-)$ and $g(\epsilon_-, c_+)$, while $g(E_j, \hat{C})$ fluctuates between the values $g(\epsilon_-, c_-)$ and $g(\epsilon_+, c_+)$. Labelling these values of g as g_{+-}, g_{-+}, etc. these fluctuations can be located on the figure 3 as fluctuations between the point $+-$ and $-+$ for species i, and $--$ and $++$ for species j. These points have the respective coordinates (c_-, g_{+-}), (c_+, g_{-+}), (c_-, g_{--}), (c_+, g_{++}).

Their probabilities will generally not be equal. If p is the probability that (E_i, E_j) takes the value (ϵ_+, ϵ_-), then the points $+-$ and $--$, have probability p and the points $++$ and $-+$ have probability $q = 1-p$.

The first stage in calculating Δ_i is averaging g_i for the restricted conditions of figure 3. This average is found at the point $M = (\bar{c}, \bar{g}_i) = (pc_- + qc_+, pg_{+-} + qg_{-+})$, which lies on the line joining $+-$ and $-+$, dividing it in the ratio $q:p$. Depicted here is a situation where $q > p$. If q were equal to p, M would lie at the point A dividing the line segment $+-$ to $-+$ in equal portions. The corresponding point $(\bar{c}, \bar{g}_j)$ for species j lies at the point N, shown here below M.

Figure 3 shows that M lies above N, so that $\bar{g}_i > \bar{g}_j$, whenever these points are to the left of the point K. In this situation species i has a higher average growth rate than species j for fluctuations between the two chosen environmental states. Thus species i has an advantage in average growth rate. This advantage comes from the fact that its growth rate fluctuates between $+-$ and $-+$ while species j has fluctuations between $--$ and $++$, which in turn comes from the fact that species i is at low density (an invader) while species j is a resident.

This low-density advantage means that species i can experience a sizeable disadvantage in frequency of favourable environmental states ($p \ll q$), and yet still have a higher average growth rate than other species. For example, in the specific geometry of the figure, K divides the lines $--++$ and $+--+$ roughly in the ratio 6:1, which means that species i would have to have a greater than 6:1 disadvantage to species j in frequency of favourable conditions, before it had a lower average growth rate.

Under what conditions does this low-density advantage mean that a species will be able to coexist with its competitors? The low density average growth rate, Δ_i, is the average value of $\bar{g}_i$ when all the variables that were fixed to construct figure 3, are now varied over their distribution of possible values. Because species j is assumed at its stationary distribution, the average of $\bar{g}_j$ is 0. It follows that

$$\Delta_i = E[\bar{g}_i - \bar{g}_j], \tag{8}$$

which is the average of the difference between points M and N in figure 3, when the environmentally dependent parameters and species densities vary over their joint probability distribution of possible values. Thus if species i is never at such a disadvantage to its competitors that M and N lie to the right of K, then species i must persist in the system. If this is true for each species considered individually in the low-density state, then all species coexist together in spite of potentially significant average disadvantages that some species may have.

The situation just considered, however, is not very realistic. It is more likely, when there are inequalities between species, that the points M and N will lie to the right of the point K for some values of ϵ_+, ϵ_-, for some population densities and some values of the environmentally dependent parameters of other species, which are held fixed to construct figure 3. The final average value, Δ_i, of $\bar{g}_i - \bar{g}_j$ is then an average of situations in which MN is to the left of K and situations where it is to the right of K. The geometry of the figure biases the outcome such that positive values of $\bar{g}_i - \bar{g}_j$ will tend to predominate, leading to a positive result for Δ_i unless species i is too greatly inferior to other species. The situation is far from straightforward, however, as is best understood through some specific examples.

Consider a two-species model of strong competition for space, such as the lottery model (Chesson & Warner 1981). If reproduction is always sufficient to fill the available space in spite of fluctuations in the environmentally dependent parameters, then the population density of the resident species j does not fluctuate with time. The line $--++$ is horizontal and $\bar{g}_j = 0$. If the environment does only fluctuate between two states, figure 3 tells the complete story, for then $\Delta_i = \bar{g}_i$. If the inferior species is not so inferior that NM is at or to the right of K, the species coexist.

How do we proceed to the more useful situation where the environmentally dependent parameters take on a continuum of values? This is surprisingly simple for the lottery model if the birth rates are the environmentally dependent parameters and the logs of these are normally distributed, possibly correlated between species, with equal variances but different means. For low adult death rates, and any given set of model parameters, it can be shown that MN is always to the left of K or always to the right. This is also true if one makes the assumption that the mean differences between species and the variances are small, but that the adult death rates have any value. In other cases, it can be expected that some chosen values of (ϵ_+, ϵ_-) will have MN to the left of K while for others it will be to the right, leading to some positive values of $\bar{g}_i$ and some negative values. The final sign of Δ_i in such situations necessarily involves averaging the $\bar{g}_i$ values over the distribution of (ϵ_+, ϵ_-).

As a further illustration, consider symmetric competition between species, as discussed by Chesson (1988). This means in particular that the competition function depends on all species in the same way, and the probability distributions of environmentally dependent parameters are the same for all species. Until now, we have not made the usual 'white noise' assumption that the environment process is independent over time. Indeed, the forgoing results apply for any colour of the noise process. However, it is now important to make the white noise assumption that the environment process takes values in the next time period that do not depend on values it has taken in previous time periods. This assumption means that conditioning on population densities, whatever they may be, does not upset the symmetric relation between the environmentally dependent parameters of the invading species i and the chosen test species resident j. (The analysis can be done with any of the $n-1$ resident species without affecting the results.)

In this setting, the symmetry of the relations among the species means that $p = \frac{1}{2}$, MN coincides with AB and divides the lines $+--+$ and $--++$ exactly in half. In the subadditive situation of figure 3, K lies right of AB, unless $\epsilon_+ = \epsilon_-$, in which case the figure collapses to a single point. Thus whenever ϵ_+ differs from ϵ_-, $\bar{g}_i - \bar{g}_j$ is positive. It follows that if the species can have different values for their environmentally dependent parameters, i.e. show some differences in their responses to the environment, Δ_i is positive for all species. Thus these n species coexist.

Symmetry is not realistic, but the power of the symmetric example is that very few assumptions are needed along with symmetry to prove coexistence. Thus, setting symmetry aside, it is very general indeed. Moreover, the final result, coexistence of all n species, can be expected to be robust to small departures from symmetry, because the Δ_i will usually change continuously with continuous deformation of the model. Thus small departures from symmetry are unlikely to make any of the Δ_i negative or zero, and so coexistence is likely to be preserved. Sufficient departures from symmetry, however, will lead to competitive exclusion of some species.

Our analysis suggests that the stronger the convergence of the growth curves in figure 3, the more asymmetry can be tolerated before exclusion occurs, and this expectation is supported by approximate formulae for the Δ_i (Chesson 1989). The reason for this is that K will then be a long way to the right of AB, and so quite small p-values, i.e. large disadvantages to species i, can be tolerated before $\bar{g}_i - \bar{g}_j$ becomes negative.

In some cases the growth rates do not merely converge, but cross. This is the case with the seed bank model of table 1, depicted in figure 4. Then it is possible for fluctuations to occur between environmental states that always lead to a greater growth rate

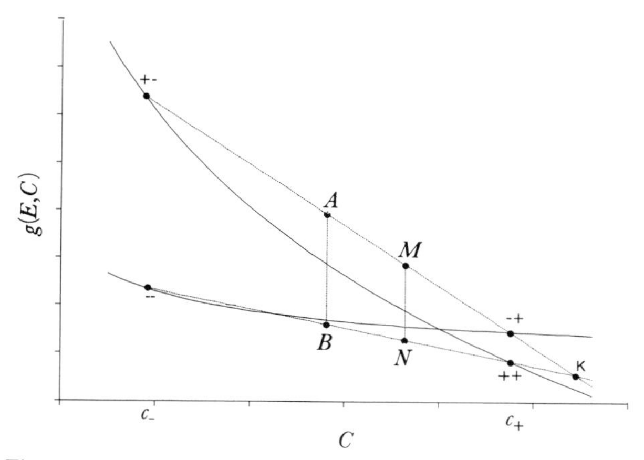

Figure 4. Intersecting subadditive growth rates of the seed-bank model.

for a species at low density, and $\bar{g}_i - \bar{g}_j$ is positive even when $p = 0$ (figure 4).

COMPETITIVE EXCLUSION

Figure 2 shows three possibilities for the growth rate $g_i(E_i, C_i)$. In addition to the subadditive case (figure 2*b*) discussed above, the growth rate could be additive (figure 2*a*) or superadditive (figure 2*c*). The geometry of the additive case means the sign of $\bar{g}_i - \bar{g}_j$ is simply the sign of $p-q$. Thus additivity confers no advantage on a species at low density that might offset other disadvantages, and we cannot expect environmental fluctuations to promote coexistence or competitive exclusion.

The superadditive case is easily seen to confer a disadvantage on a species at low density. This can offset advantages that a species may have, allowing dominance by other species, or promoting a situation where all species have negative low-density growth rates. The latter situation leads to random competitive exclusion (Chesson & Ellner 1989).

GENERATION OF NON-ADDITIVITY BY WITHIN POPULATION HETEROGENEITY

The simplest sorts of population models tend to be additive (Chesson 1988). Additivity comes about when the action of competition does not affect the action of the environment and vice versa, given their values. This means that the finite rate of increase, $G(E, C)$, takes the general form

$$G(E, C) = A(E)/B(C), \tag{9}$$

where now I have dropped the species subscript for simplicity of notation. Note that this assumption is quite independent of the idea above that the actual value of C depends on the value of E. Taking logs of this gives

$$g(E, C) = \ln A(E) - \ln B(C), \tag{10}$$

which is the additive form giving the parallel growth rates of figure 2*a*. Departures from the additive situation can be measured by the quantity γ, defined by

$$\gamma = \frac{\partial^2 g}{\partial E \partial C}, \tag{11}$$

which is zero in the additive case.

The independence of action of E and C expressed in (9) can also be thought of as meaning that competitive and environmental factors operate uniformly across all individuals in the population. Thus the fact that an individual has been affected a certain way by the environment does not alter the effect that competition will have. The seed-bank model of table 1 gives an obvious departure from this situation. The effect of environment is to bring on seed germination or not. If an individual does not germinate, it is assumed to experience no competition. On the other hand, an individual that germinates is exposed to competition from other individuals as they grow. It follows that the environmental factor and the competitive factor are far from independent in action. If the environment increases the germination fraction, it necessarily increases the fraction of the population that is exposed to competition.

Dependence of action comes about because the population is subdivided into groups of individuals that are doing different things at the same time, and these different groups are affected differently by environmental and competitive factors. Thus we can represent the growth rate of the entire population, subdivided into different contributions to population growth, as follows:

$$G(E, C) = \sum_{l=1}^{k} G_l(E, C), \tag{12}$$

where $G_l(E, C)$ simply represents a contribution to the growth rate from the lth of k possible sources. In the seed bank model, for example, we have $G_1(E, C) = (1-E)s$, and $G_2(E, C) = EY/C$.

It is important to note that the components of the growth rate in equation (12) could each be additive; that is within components, the action of environment and competition could be independent, but the overall effect when they are combined is interactive, not independent (Chesson & Huntly 1988). To see how this comes about, we define sensitivity to environment, competition, and their interaction for each of the components of population growth as follows:

$$\alpha_l = \frac{\partial g_l}{\partial E}, \quad \beta_l = -\frac{\partial g_l}{\partial C}, \quad \gamma_l = \frac{\partial^2 g_l}{\partial E \partial C}, \tag{13}$$

where $g_l = \ln G_l$.

We can now express the deviation of the total population growth rate, $g(E, C)$, from additivity by the formula

$$\gamma = \bar{\gamma} - \sum_{l=1}^{k} (\alpha_l - \bar{\alpha})(\beta_l - \bar{\beta})\, G_l / \sum_{l=1}^{k} G_l \tag{14}$$

where G_l is just a shorthand for $G_l(E, C)$, and $\bar{\alpha}$, $\bar{\beta}$, $\bar{\gamma}$, are the weighted averages

$$\bar{\alpha} = \sum_{l=1}^{k} \alpha_l G_l / \sum_{l=1}^{k} G_l, \tag{15}$$

etc.

Equation (14) gives the measure of nonadditivity, γ, as an average of the nonadditivity that may occur in each component by itself minus the weighted covariance between α_l and β_l. The weights in these calculations, G_l, are each component's contribution to population growth.

This representation of non-additivity, γ, is most useful when the γ_l are zero, meaning that the components have been chosen such that within a component, environment and competition act uniformly across all individuals. If the γ_l are not zero this may mean that the components are themselves heterogeneous with respect to the operation of environment and competition, and can be further subdivided. In some cases, however, non-additivity arises within a component as a behavioural or physiological response that is not easily reducible (Chesson 1988).

In the case when the γ_l are zero, the population-level nonadditivity is simply minus the weighted covariance of the α_l and the β_l, which can best be understood by means of a scatter plot of the α_l and β_l. For example, the seed bank model of table 1 is shown as the two-point scatter plot, figure 5. Point A is for the seeds that remain in the soil. The fraction of seeds in this component is $1-E$, where E, the germination fraction, is the environmentally dependent parameter. The number of seeds in this component is therefore negatively sensitive to E. Dormant seeds do not compete, and so the sensitivity of this component to competition is zero. Thus the point A has a negative value of α_l and a zero value of β_l.

The point B in the figure corresponds to seeds that germinated, and therefore grow and compete with each other and plants of other species. This component is increased by germination, and experiences competition; thus it represents positive values for both α_l and β_l. The covariance measures common linear variation among the two variables α_l and β_l, which is perhaps best understood by appreciating that it equals the correlation between them times the product of their standard deviations. Thus it involves how closely they lie on a straight line, the sign of the line's slope, as well as the amount of variation shown by each variable.

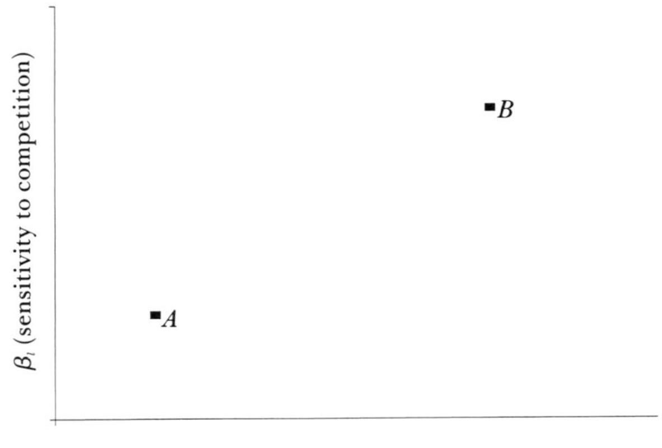

Figure 5. Two-point scatter plot for a population with two distinct components to its growth rate showing positive covariance of sensitivities to environmental and competitive factors.

The relation between the covariance and non-additivity can be seen intuitively, also. As the environmentally dependent parameter is increased, components with higher sensitivity to the environment contribute relatively more to population growth compared with other components. Positive covariance of sensitivity means that these same components are those with higher sensitivity to competition, and so the population as a whole has more average sensitivity to competition. Thus, the change from low slope to a more negative slope shown in figure 2*b*, which means that γ is negative.

The seed bank example is perhaps a further worthwhile illustration. As E increases, more of the seeds germinate, and are exposed to competition. Thus, average population sensitivity to competition increases. One can also view this from the perspective that as E increases, more of the population is at point B in figure 5 (where there is high sensitivity to competition) and less at point A (where there is low sensitivity to competition).

One can use scatter plots like this to assess nonadditivity whenever a population can be divided into components with differing sensitivities to environment and competition. A few more examples will serve to illustrate. In many organisms the juvenile phase is especially sensitive to environmental and competitive factors, compared with the adult organism. Thus point B in figure 6 could represent these juveniles and point A the adult phase. Thus, again there is positive covariance of sensitivities, and a negative value of γ. Previously this situation has been discussed under the heading of the storage effect (Warner & Chesson 1985), which now can be seen as a special case of the more general concept, subadditivity.

Another example involves subpopulations in different habitats. Spatial variation in the quality of living conditions occurs for all organisms, as has been documented extensively in plants (Harper 1977; Gross 1984; Silvertown 1987). Are there places, however, that are less sensitive to environmental factors than elsewhere, and also less sensitive to competition? It may well be that in plant populations, areas that are relatively and permanently harsh may qualify. Such harshness may limit individual growth making it unresponsive to other factors, including temporally fluctuating environmental factors and fluctuating competition.

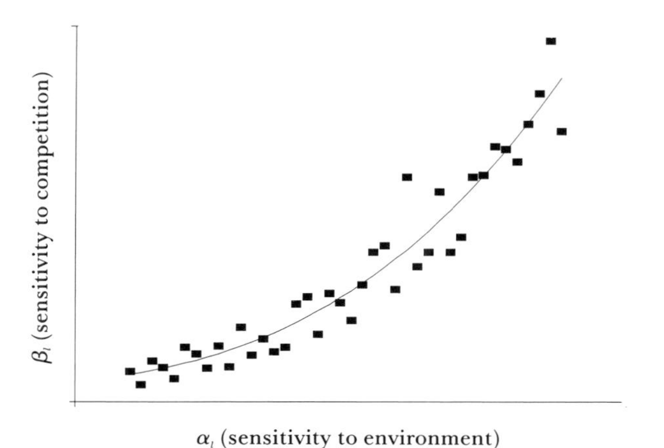

Figure 6. Scatter plot for a population whose growth rate has many components, showing positive covariance of sensitivities. The line represents an ideal l factor causing a strict relation between α_l and β_l.

An alternative example involves cases where recruitment into especially favourable spots is limited, thus keeping densities low. For example, considering species that compete for moisture, small naturally moist areas may never achieve densities high enough for much competition if they are so small that most dispersal is away from them (DeAngelis *et al.* 1979). Lowered sensitivity to the environment must go along with this lowered sensitivity to competition if γ is to be negative. In the situation just envisaged, lowered stress might make individuals less sensitive to the environment. Clearly, this need not always be so, as removal of moisture limitation might make individuals more responsive to other factors. If that occurred, the covariance of sensitivities would be negative, and the measure of non-additivity, γ, would be positive giving the superadditive situation of figure 2*c*.

In addition to spatial variation, variation with life-history stage or population growth process, one could also consider phenotypic variation. This is of particular interest in relation to the seed bank model, because the same individual plant may produce different phenotypes of seeds with the potential for different germination properties and different competitive abilities (Cavers & Steel 1984; McGinley *et al.* 1987; Fenner 1985; Venable & Brown 1988; Silvertown 1989). Covariation in these traits might well be another source of non-additivity.

THE THREE FACTOR SCHEME

There seems to be an endless variety of ways in which populations could be heterogeneous and through which non-additivities could occur. The purpose of this section is to provide some general way of viewing the possibilities to aid the investigation of the relevant sorts of heterogeneity. In general, it can be expected that there will be a host of factors contributing to individual survival, growth and reproduction, and ultimately population growth. Some factors are consumed, and as a result of consumption become less available. More generally, some factors create a negative feedback loop for the affected species in the community (Andrewartha 1970). We associate such factors with competition, and we refer to them as C factors.

Other factors may not lead to any kind of feedback for the relevant biological populations. I divide these into two sorts, E factors and l factors. The E factors are those that vary over time. The l factors are those that do not vary over time, but instead vary within a single species population (one member of a community of similar species). For simplicity, we shall think of a population as having just one of each of the three sorts of factors. These of course could be composite factors, but thinking of them as single factors will help sharpen intuition. These three factors, C, E, and l, lead to competitive parameters, environmentally dependent parameters, and population subdivision respectively, as discussed in the models above.

The importance of the factor l is its effects on the other two factors. For plant species it is known that some factors (essential or interactively essential resources) have strong effects on the sensitivities of the organisms to other factors (Tilman 1988). An example is perhaps shade as a factor in the dynamics of species in the understory of forests (Dahlem & Boerner 1985). Shade is an l factor if we assume that understory species are shaded by trees but do not shade each other, and that tree canopy density varies.

Plant growth is sensitive to temperature (Larcher 1983), but light limitation through shade restricts a plant's ability to respond to warmer weather. Perhaps more importantly, shade moderates temperatures. Thus shade should affect the sensitivity of plant growth to variation in the weather, which can be the E factor, and would be measured in some standardized way such as plant temperature in full sunlight.

Shade, through light limitation is also likely to affect competition for resources such as nutrients (Tilman 1988), which might be the C factor. Thus we see that shade, the l factor, leads to joint positive variation in sensitivity to the E and C factors. A scatter plot such as in figure 6 would be obtained. The different points in the figure represent different places on the forest floor. Ideally, if E, C, and l were all truly single factors, and there were no other factors present, these points would all lie on the single curve through the points. In reality, however, other factors varying from place to place are likely to make it a scatter. The positive covariance of sensitivities shown here means that growth rates for the population encompassing the points in the scatter plot, would be subadditive. Thus according to the community theory presented here, we should expect that a scatter plot like that in figure 6 should help maintain species diversity in a fluctuating environment.

The factor l might in some cases be qualitatively different from the other factors. In the seed bank model, for instance, the factor l simply distinguishes between the class of individuals growing above ground and those remaining as seeds. Thus it is simply a label for two somewhat incomparable components. In the case of seed phenotypes, the factor l might be seed size, and there is evidence that seed size affects response to both environmental and competitive conditions (McGinley *et al.* 1987; Fenner 1985; Venable & Brown 1988; Silvertown 1989).

While this theory involving E, C and l factors may have elements in common with Tilman's (1982) theory of plant competition, it nevertheless has important differences. For example, in this theory coexistence can occur if all species are limited by the same competitive factor. In other words C can be identical for different species. In contrast Tilman's theory requires C to be different for different species. In this regard, it is important to note that the l factor here is not a competitive factor and can affect all species similarly. The environmental factor, E, however, must distinguish between species at least quantitatively. For example, different species might have different abilities to benefit from higher temperatures.

CONCLUSION

The likely broad occurrence of non-additivity in nature implies that temporal environmental fluctuations quite generally modify the outcomes of competitive interactions between species. However, the actual effect of environmental fluctuations depends on whether non-additivity is in the subadditive or superadditive direction. How much environmental fluctuations modify the outcomes of competition depends on the strength of non-additivity as measured here by a covariance of within-population heterogeneity.

Discussions with Nancy Huntly have been of great value in developing the material in this article. This work was supported by a grant from the National Science Foundation (BSR-8615028).

REFERENCES

Abrams, P. 1984 Variability in resource consumption rates and the coexistence of competing species. *Theor. Popul. Biol.* **25**, 106–124.

Andrewartha, H. G. 1970 *Introduction to the study of animal populations* (283 pages). London: Methuen.

Armstrong, R. A. & McGehee, R. 1976 Coexistence of species competing for shared resources. *Theor. Popul. Biol.* **9**, 317–328.

Cavers, P. & Steel, M. G. 1984 Patterns of change in seed weight over time on individual plants. *Am. Nat.* **124**, 324–335.

Chesson, P. L. 1988 Interactions between environment and competition: how fluctuations mediate coexistence and competitive exclusion. In *Community ecology* (ed. A. Hastings) (*Lect. Notes Biomath.* **77**), pp. 51–77. New York: Springer–Verlag.

Chesson, P. L. 1989 A general model of this role of environmental variability in communities of competing species. *Lect. Math. Life Sci.* **20**, 97–123. American Mathematical Society.

Chesson, P. L. & Ellner, S. P. 1989 Invasibility and stochastic boundedness in two-dimensional competition models. *J. Math. Biol.* **27**, 117–138.

Chesson, P. & Huntly, N. 1988 Community consequences of life-history traits in a variable environment. *Ann. Zool. Fenn.* **25**, 5–16.

Chesson, P. & Huntly, N. 1989 Short-term instabilities and long-term community dynamics. *Trends Ecol. Evol.* **4**, 293–298.

Chesson, P. L. & Warner, R. R. 1981 Environmental variability promotes coexistence in lottery competitive systems. *Am. Nat.* **117**, 923–943.

Comins, H. N. & Noble, I. R. 1985 Dispersal, variability and transient niches: species coexistence in a uniformly variable environment. *Am. Nat.* **126**, 706–723.

Connell, J. H. 1978 Diversity in tropical rainforests and coral reefs. *Science, Wash.* **199**, 1302–1310.

Dahlem, T. S. & Boerner, R. 1985 Effects of light gap and early emergence on the growth and reproduction of *Gernium maaculatum*. *Can. J. Bot.* **65**, 242–245.

DeAngelis, D. L., Travis, C. C. & Post, W. M. 1979 Persistence and stability of seed-dispersed species in a patchy environment. *Theor. Popul. Biol.* **16**, 107–125.

Ellner, S. 1985 ESS germination strategies in randomly varying environments. I. Logistic-type models. *Theor. Popul. Biol.* **28**, 50–79.

Fenner, M. 1985 *Seed ecology* (151 pages.) London: Chapman and Hall.

Gross, K. L. 1984 Effects of seed size and growth form on seedling establishment of six monocarpic perennials. *J. Ecol.* **72**, 369–387.

Grubb, P. J. 1977 The maintenance of species richness in plant communities: the regeneration niche. *Biol. Rev.* **52**, 107–145.

Grubb, P. J. 1986 Problems posed by sparse and patchily distributed species in species-rich plant communities. In *Community ecology* (ed. J. Diamond & T. Case), pp. 207–225. Harper and Row.

Harper, J. L. 1977 *Population biology of plants* (892 pages.) London: Academic Press.

Hastings, A. 1980 Disturbance, coexistence, history and competition for space. *Theor. Popul. Biol.* **18**, 361–373.

Hutchinson, G. E. 1961 The paradox of the plankton. *Am. Nat.* **95**, 137–145.

Larcher, W. 1983 *Physiological plant ecology* (303 pages.) New York: Springer–Verlag.

McGinley, M. A., Tenne, D. H. & Geber, M. A. 1987 Parental investment in offspring in variable environments: theoretical and empirical considerations. *Am. Nat.* **130**, 370–398.

Sale, P. F. 1977 Maintenance of high diversity in coral reef fish communities *Am. Nat.* **111**, 337–359.

Shmida, A. & Ellner, S. P. 1985 Coexistence of plant species with similar niches. *Vegetatio* **58**, 29–55.

Silvertown, J. W. 1987 *Introduction to plant population ecology* (229 pages). London: Longman.

Silvertown, J. 1989 The paradox of seed size and adaptation. TREE **4**, 24–26.

Tilman, D. 1982 Resource competition and community structure. Princeton, New Jersey: Princeton University Press.

Tilman, D. 1988 *Plant strategies and the dynamics and structure of plant communities* (360 pages.) Princeton, New Jersey: Princeton University Press.

Turelli, M. 1980 Niche overlap and invasion of competitors in random environments II. The effects of demographic stochasticity. In *Biological growth and spread, mathematical theories and applications* (ed. W. Jäger, H. Rost & P. Tautu), pp. 119–129, Lecture Notes in Biomathematics no. 38. Berlin: Springer–Verlag.

Turelli, M. 1981 Niche overlap and invasion of competitors in random environments I. Models without demographic stochasticity. *Theor. Popul. Biol.* **20**, 1–56.

Venable, D. L. & Brown, J. S. 1988 The selective interactions of dispersal, dormancy, and seed size as adaptations for reducing risk in variable environments. *Am. Nat.* **131**, 360–384.

Warner, R. R. & Chesson, P. L. 1985 Coexistence mediated by recruitment fluctuations: a field guide to the storage effect. *Am. Nat.* **125**, 769–787.

Woodin, S. A. & Yorke, J. A. 1975 Disturbance, fluctuating rates of resource recruitment, and increased diversity. In *Ecosystem analysis and prediction* (ed. S. A. Levin), pp. 38–41. *Proceedings of the SIAM-SIMS Conference*. Alta, Utah 1974.

Discussion

P. J. Grubb (*Botany School, University of Cambridge, U.K.*). Would Professor Chesson please show how his analyses may be used to determine the limiting similarity between species?

P. Chesson. This paper explores just one of many different ways that species may differ from one another. Within this restricted context it shows how dissimilarity of responses to environmental factors gives an advantage to a species at low density, opposing competitive exclusion. This effect depends on growth rates being subadditive (figure 2*b*), and its strength depends on the magnitude of subadditivity, or the rate of convergence of growth-rate curves for different environmental conditions. How increased dissimilarity increases the low-density advantage can be seen from figure 3. More dissimilarity means that the points labeled $+-$ and $--$ will be further separated from $++$ and $-+$. It is easiest to imagine these points as shifted to more extreme positions on the depicted growth rate curves, but in fact these curves also change making the contrast in slope greater. However this is viewed, the important outcome is the same: more dissimilarity increases the contrast between the distances $--$ to $+-$ and $-+$ to $++$, with the effect that the low density advantage given by the distance MN is increased.

This geometric approach helps in understanding how a limit to similarity arises in populations in a fluctuating environment. Quantitative measurement of the limit to similarity is best pursued by approximate formulae for the mean low density growth rate that I have presented elsewhere (Chesson 1989). These allow classical niche dissimilarity involving resource use to be considered in concert with dissimilarity arising through differences of responses to fluctuating environmental factors.

Population dynamics in spatially complex environments: theory and data

PETER KAREIVA

Department of Zoology, University of Washington, Seattle, Washington 98195, U.K.

SUMMARY

Population dynamics and species interactions are spread out in space. This might seem like a trivial observation, but it has potentially important consequences. In particular, mathematical models show that the dynamics of populations can be altered fundamentally simply because organisms interact and disperse rather than being confined to one position for their entire lives.

Models that deal with dispersal and spatially distributed populations are extraordinarily varied, partly because they employ three distinct characterizations of space: as 'islands' (or 'metapopulations'), as 'stepping-stones', or as a continuum. Moreover, there are several different ways of representing dispersal in spatially structured environments, as well as several possibilities for allowing environmental variation to come into play. In spite of this variety, a few common themes emerge from spatial models. First, island and stepping-stone models emphasize that little can be concluded from simply recording patterns of occupancy, instead a metapopulation's fate will be determined by the balance between local extinction and recolonization and how that balance interacts with random catastrophes. Island and stepping-stone models also make it clear that the spatial dimension, in particular spatial subdivision, can alter the stability of species interactions and opportunities for coexistence in both predator–prey and competitive systems. Continuum models, which usually take the form of reaction-diffusion equations, address slightly different questions. Reaction-diffusion theory suggests that in uniform environments, certain combinations of local dynamics and dispersal can produce persistent irregularities in the dispersion of species. These striking spatial patterns, which are called diffusive instabilities, can arise from predator–prey interactions, Lotka–Volterra competitive interactions, and from density-dependent population growth in an age-structured population. Moreover, although they differ fundamentally in their structure, the three major classes of spatial models share the common generalization that spatial effects should be expected only for: (i) selected spatial scales; (ii) specific dispersal rates, and (iii) particular patterns of environmental variation relative to the frequency and range of dispersal. The theoretical possibilities are thus contingent on spatial scale and dispersal rates.

Although explicit experimental tests of spatial models are non-existent, a handful of studies report general changes in species interactions associated with manipulations of habitat subdivision. Observational studies with adequate data concerning dispersal and scale are also scarce; but those few observational studies with the appropriate supporting information consistently show profound spatial effects, especially effects due to habitat subdivision.

The challenge for empiricists is to investigate more rigorously the roles of spatial subdivision and dispersal in natural communities. The challenge for theoreticians is to make the empiricist's job easier; this can best be done by delineating when spatial effects are most likely to be influential, and by offering guidance on how to design appropriate experiments. Simply saying that the spatial environment is important is to mouth a platitude: what we need to know is whether this presumed importance amounts to much in natural systems.

INTRODUCTION

Spatial heterogeneity is one of the most obvious features of the natural world. It may also be one of the most important factors influencing population dynamics. Yet until recently, the spatial dimension of species interactions has been neglected or glossed over in both experimental and theoretical investigations. Fortunately, within the past two decades mathematical ecologists have had great success in showing how specific aspects of the spatial environment can alter population and community dynamics. Meanwhile, in stark contrast, experimental ecology has offered only half-hearted and disappointingly primitive investigations into the effects of heterogeneous environments on population dynamics (Doak *et al.* 1991).

The mismatch between theory and experiment in the area of spatial effects may be because of the fact that spatial models have become so numerous, diverse and complex, that it is difficult to identify which theoretical results (if any) are pertinent to particular systems. The goal of this paper is to narrow the gap between theory and experiments regarding the implications of the spatial dimension for population

Table 1. *A categorization of spatial models according to the manner in which the spatial dimension is represented. The table selects only the most recent models whenever there are multiple publications dealing with essentially similar formulations*

Island models

Andreasen & Christiansen (1989). The effect of population subdivision on the spread of a disease is examined.

Burkey (1989). A stimulation model is used to predict how the degree of habitat fragmentation and rate of migration interact to determine a specie's probability of extinction.

Chesson (1981). Stochastic versus deterministic models for spatially distributed single species models are contrasted.

Chesson (1985). The effect of environmental variability on the coexistence of competitors that occupy patchy environments is examined.

Chewning (1975). Conditions under which migration can stabilize predator-prey interactions are specified.

Comins & Noble (1985). Coexistence of competitors is shown to be promoted by random environmental variability spread over a spatially distributed system, as long as there is a modest rate of dispersal.

Diekmann, Metz & Sabelis (1988*a*). A detailed analysis of how within-patch dynamics and between-patch dispersal processes interact to determine the stability of a predator-prey-plant interaction.

Diekmann, Metz & Sabelis (1988*b*). The above detailed model is reduced to three ordinary differential equations that are meant to capture the key ingredients of the system without requiring any reference to spatial details.

Gurney & Nisbet (1978). Fluctuations in a predator-prey system are predicted on the basis of migration and extinction, total number of patches, and patterns of environmental variability.

Hanski (1983). Coexistence of competitors is analysed by using a metapopulation model, with special attention to the importance of the relative timescales of within-patch versus regional processes.

Harrison & Quinn (1989). The effect of correlated environments and the persistence of metapopulations is examined via stochastic simulations.

Hastings (1977). The stability of a predator-prey interaction in a subdivided environment is analysed with respect to dispersal rates and the timecourse of local within-patch dynamics.

Hastings (1978). The stability of one predator-two prey interaction in a subdivided environment is examined.

Hastings & Wolin (1989). A metapopulation model is analysed in which subpopulations face random disasters as a function of local population size (and hence patch 'age').

Hilborn (1975). A simulation is used to investigate how dispersal between 'cells' influences the stability of predator-prey interactions, modelled after Huffaker's laboratory system.

Kuno (1981). The effect of dispersal on rate of population growth in a temporally and spatially varying environment is examined.

Levin (1974). The effect of subdivision on pairwise predator-prey or competitive interactions is analysed.

Nakano (1981). The importance of dispersal for population regulation is assessed using connected systems of ordinary differential equations.

Pulliam (1989). The consequences for population regulation of consistence differences in habitat suitability ('sources' versus 'sinks') plus density-dependent dispersal are examined.

Quinn & Hastings (1987). Models that predict the effects of spatial subdivision on single-species dynamics are reviewed.

Reeve (1988). The effect of environmental variability and dispersal on the stability of subdivided host-parasitoid systems is examined.

Reddingius (1970). An age-structured simulation model for the dynamics of a subdivided population subject to random catastrophes suggests that subdivision enhances the stability of the population.

Sabelis & Laane (1986). A detailed simulation of a predator mite-prey mite interaction exhibits stable limit cycle behaviour on a regional scale (averaged over all patches), although locally all subpopulations are doomed to extinction.

Shorrocks, Atkinson & Charlesworth (1979). The effect of transient heterogeneous habitat patches on coexistence in a competitive system are examined.

Slatkin (1974). Competition in a subdivided environment with extinction and recolonization is analysed.

Takafuji, Tsuda & Miki (1983). Simulations are used to explore the role of migration in stabilizing a predator mite-prey mite-plant interaction that consists of numerous patches (plants), each of which can be overexploited by the prey mites.

Vance (1984). The contribution of dispersal to the stability of a metapopulation is analysed.

Stepping-stone models

Crowley (1977). The effects of spatial correlation in environmental disturbances are analysed with respect to how organisms are themselves distributed among different patches.

Crowley (1985). The effect of dispersal rates and the arrangement of patches (or cells) on predator-prey stability is examined.

Doak (1989). The effect of various degrees of habitat clustering on spotted owl persistence is examined.

Fujita (1983). The interplay of migration rates, number of patches, and within-patch dynamics in determining the persistence of a predator-prey-plant interaction is explored.

Nachman (1987*a*, *b*). A stochastic simulation is used to study the effects of patch number and dispersal rates on the temporal fluctuations exhibited by an acarine predator-prey system.

Ziegler (1977). By stimulating a predator-prey interaction with emigration from patches occurring only during specific stages in a patch's cycle of occupancy, the effect of predator and prey dispersal rates on the regional persistence is analysed.

Continuum models

Andow, Kareiva, Levin & Okubu (1991). A simple population growth and diffusion model is applied to three well-documented case studies of ecological invasions.

Table 1. (*cont.*)

Comins & Blatt (1978). The ability of a heterogeneous environment to stabilize predator–prey interactions is analysed, and interpreted in terms of a 'refuge'.

Diekmann (1978). Integral equations are used to model the spatial spread of epidemics.

Hardin, Smith & Namba (1990). The impact of different patterns of dispersal on population persistence and extinction is analysed by using integrodifference equations.

Hastings (1982). The stabilizing influence of diffusion in a spatially varying environment is analysed for single species models of density-dependent population growth.

Hastings (1991). Analysis establishes the possibility of diffusive instability for single-species age-structured population dynamics.

Kishimoto (1982). Lotka-Volterra models of three species systems (3 competitors, 2 predators and 1 prey, or 1 predator and 2 prey) are analysed for the conditions that lead to diffusive instabilities.

Ludwig, Aronson & Weinberger (1979). Reaction-diffusion models are used to ask what is the critical size of a forest patch required to support a spruce budworm outbreak, and how wide a barrier is needed to contain a budworm outbreak.

McMurtrie (1978). A wide variety of reaction–diffusion models are analysed for critical patch size phenomena, diffusive instabilities, and potential stabilizing effects of patchiness. In addition to simple diffusion various forms of density-dependent and biased movement are considered.

Mimura (1984). Diffusive instabilities in three and four-species competitive interactions are analysed.

Mimura & Kawasaki (1980). Spatial segregation for two competing species is established assuming the presence of a cross-diffusion term, which means that individuals of one species 'repel' individuals of the other species.

Mimura & Murray (1978). The conditions that lead to diffusive instabilities in pairwise predator–prey or plant–herbivore interactions are identified.

Murray, Stanley & Brown (1986). The rate at which 'waves' of rabies will spread through fox populations is analysed by using a simple epidemiological model plus diffusion.

Namba (1980). The implications of density-dependent dispersal for the stationary spatial distribution of populations is explored.

Namba (1989). Reaction–diffusion models of competitive interactions are analysed, with special attention to repulsive movement (cross-diffusion and density-dependent diffusion) in a spatially heterogeneous environment.

Namba & Mimura (1980). Reaction–diffusion models are used to examine competition between two species in a spatially heterogeneous environment.

Okubo, Murray & Williamson (1989). A reaction–diffusion model is used to describe the invasion of grey squirrels into England, and their 'wavelike' displacement of red squirrels.

Shigesada, Kawasaki & Teramoto (1987). The speed with which an invading species expands its range in a heterogeneous environment is calculated.

Skellam (1951). A pioneering analysis of waves of invasion, critical habitat size, and competition in heterogeneous environments by using reaction–diffusion equations.

Turchin (1989). The implications of aggregative movement for plant-herbivore interactions are explored by using reaction–diffusion models.

van den Bosch, Zadoks & Metz (1989). The waves of invasion for a wide variety of age-structured populations are analysed by using integral equations.

Yachi, Kawasaki, Shigesada & Teramoto (1989). A reaction–diffusion model of rabies epidemics in which all classes of hosts (susceptible, infected, infectious) are allowed to disperse, creating irregular oscillating travelling waves of infection.

dynamics. I will first review the theory, emphasizing the structure and assumptions of models, as well as major recurring predictions. This leads naturally to a discussion of data, and the extent to which 'spatial theory' has been vindicated by field observations or experiments. To show the difficulties in designing experiments to test spatial models, I will highlight two investigations from my own laboratory. Finally, I will end with an agenda for future research, both empirical and theoretical.

AN OVERVIEW OF MATHEMATICAL MODELS CONCERNING ECOLOGICAL INTERACTIONS IN PATCHY ENVIRONMENTS

The burgeoning literature concerning metapopulation models, patch dynamics, and spatially distributed species interactions has spawned several superb reviews (Levin 1976; Taylor 1988, 1990; Hastings 1990; Reeve 1990; Harrison 1991). To avoid duplicating those earlier reviews, I will gloss over mathematical details and emphasize instead what I feel are the major messages of importance for experimentalists.

(*a*) *How do models represent spatially complex environments?*

The most fundamental distinction between different models of heterogeneous environments is the manner in which the spatial dimension is represented (table 1). The most common approach is to imagine the world subdivided into a large collection of patches, each of which include internal dynamics and are collectively coupled together through one common pool of dispersers. Such models do not have an explicit spatial dimension because there is no specification of the relative distances between patches, instead, all patches are equally accessible to one another. Movement occurs when individuals leave patches at some rate, enter a 'bath' of dispersers, and are then redistributed among patches (usually randomly). The key feature of these '*island models*' is population subdivision. By contrasting population fluctuations is an ensemble of islands as opposed to a single island, it is possible to learn how spatial subdivision alters a system's behaviour. One can also use island models to investigate the effect of between-patch dispersal on ensemble dynamics. However, it is important to remember that because island models have no explicit spatial dimension, dispersal rates refer to the fraction of individuals that move, not to the distances that individuals move.

A second way of representing the spatial environment involves '*stepping-stone models*', in which the world is again divided into patches, but the patches now have fixed spatial coordinates. Stepping-stone models are thus effectively island models with an explicit spatial dimension. Because patches are assigned actual positions in stepping-stone models, these models can be used to contrast the consequences of long-range versus short-range dispersal. The presence of a spatial coordinate system also enables one to introduce environmental variability with spatial structure (i.e. spatial autocorrelation).

The final way of treating the spatial dimension is to represent space with a continuous coordinate system along which populations interact and disperse. These so-called 'continuum models' typically take the form of partial differential equations, or more specifically reaction-diffusion models. Here, 'reaction' refers to the mathematical description of local population growth, and 'diffusion' refers to the mathematical description of dispersal. In most reaction-diffusion models the environment is assumed to be homogeneous, and the questions of interest concern what types of spatio-temporal patterns in population density emerge as a result of combining dispersal with local dynamics. To address this issue, reaction-diffusion models have typically used standard continuous-time Lotka-Volterra representations of local dynamics, with diffusion occurring at a constant rate (which corresponds to an assumption of random motion). However, several recent models have relaxed the assumption of purely diffusive dispersal and examined the consequences of movement that varies with population densities (see, for example, Turchin (1988); Shigesada *et al.* (1979)) or habitat quality (see, for example, Shigesada *et al.* (1987)). Continuum spatial models have also been developed for organisms that reproduce at discrete intervals (Kot & Schaffer 1986; van den Bosch *et al.* 1989).

Each of the above three representations of space has its own advantages and disadvantages: island models lend themselves to analytic solutions, stepping-stone models are probably most easily applied to field data, reaction-diffusion models provide a particularly powerful and compact notation. In some cases different approaches can be formally related to one another; for example, reaction-diffusion models can arise as limiting cases of stepping-stone models (Okubo 1980, 1986). A major gap in the theoretical literature is the absence of comparisons between the different approaches when they are applied to identical ecological processes (exceptions are Chesson (1985); Comins & Noble (1985); Fujita (1983)).

Although the language surrounding spatial models emphasizes 'patchiness' and 'heterogeneity', most models assume that all patches in an ensemble are the same; similarly, most reaction-diffusion models assume the environment is uniform in space. Population densities may vary from position to position, or from patch to patch, but such variation typically results from external perturbations that are equally likely everywhere, but happen to occur by chance in only a few locations at any one time. In other words, the vast theoretical literature on ecological interactions in 'patchy' environments generally does not treat habitats with consistent differences. The heterogeneity in these models results from the interplay of spatial subdivision and transient perturbations, not from permanent spatial heterogeneities. It would be more accurate if we described these models as analyses of 'subdivided' or 'spatially extensive' environments, rather than of 'patchy' or 'heterogeneous' environments (exceptions include Comins & Blatt (1978); Namba (1989); Namba & Mimura (1980); Shigesada *et al.* (1987); Pulliam (1989)).

(*b*) *Incorporating biological detail and environmental variability into spatial models*

After the mathematical portrayal of space has been selected, there still remain several fundamental decisions regarding model structure. A particularly key decision concerns the detail with which species or populations are described, ranging from recording presence or absence (Levins 1970; Slatkin 1974; Hanski 1983; Horn & MacArthur 1972), through keeping track of population densities (Hastings & Wolin 1989; Reeve 1988; Vance 1984), to a complete age-structured representation of populations (De Blasio & Lamberti 1979; Hastings 1990). Another crucial distinction between spatial models is whether environmental variation is introduced. Although the earlier models tended to be deterministic (Levin 1974), recent variants include chance extinctions and disturbances (Hastings & Wolin 1989), or allow parameters such as reproductive rates to vary randomly from patch to patch (Reeve 1988). Moreover, if environmental variation is featured in the model, the subsequent results are often determined by the degree to which the variation among patches is independent (as opposed to correlated) from patch-to-patch (Crowley 1977; Harrison & Quinn 1989). Since dispersal is the process that connects different patches (or different positions in a continuum), mathematical representations of the dispersal process play a central role in the development of spatial models. The major distinctions are whether dispersal depends on density, and whether the direction of movement is influenced by the quality of habitats.

The different ways of representing space, keeping track of populations, dealing with environmental variability, describing dispersal, and portraying dynamics within patches, together combine to yield an assortment of models that is overwhelming in its variety.

(*c*) *Some general results from island and stepping-stone models*

Many patch models are concerned with the dynamics and persistence of single-species populations that are subdivided to varying degrees. These models, often referred to as 'metapopulation theory', have been applied to longstanding debates about density-dependence and population regulation (den Boer 1970; Reddinguis & den Boer 1970; Kuno 1981), and have been consulted for guidance on matters of conservation in fragmented habitats (Burkey 1989; Doak 1989; Gilpin 1988; Quinn & Hastings 1987). When populations are subdivided, the risk of extinction is the result of two opposing forces: (i) because fragmentation creates smaller populations within each patch, demographic stochasticity and loss of genetic variability enhance extinction risks; (ii) because fragmentation may create statistically separate subpopulations, ensembles of subpopulations could enjoy a lower collective risk of total disaster (since it is unlikely that all fragments would suffer catastrophes simultaneously). The net effect of subdivision will depend on the magnitude of dispersal relative to environmental variation, the life history traits that determine the force of demographic stochasticity, and the likelihood of inbreeding depression. Given these contingencies, it is obvious that one cannot make blanket statements about the effects of fragmentation on a population's persistence. However, metapopulation models do make the generally useful point that we can expect to find vacant patches even if habitat is in short supply. Unfortunately, this idea has not always been appreciated by resource managers (USDA Forest Service 1988), who too often neglect the implications of a dynamic turnover in patches because of extinction and recolonization (Doak 1989).

Community ecologists have used island and stepping-stone models to explore possible contributions of the spatial dimension to competitive coexistence. If competitive interactions are subdivided into different patches and disturbances (or extinctions) are imposed randomly, a species that is an inferior competitor when confined to a single patch can gain an advantage if it is an especially good colonist of newly vacant patches (Horn & MacArthur 1972; Slatkin 1974; Hanski 1983). In general, the appropriate mix of subdivision (or patchiness) plus disturbance and dispersal can promote coexistence in competitive systems.

Patch models also have a long tradition of being applied to predator–prey interactions. Here, the central question is whether 'patchiness' can stabilize predator–prey systems that would otherwise fluctuate wildly. The original impetus for this theoretical inquiry were laboratory experiments, especially Carl Huffaker's (1958) investigation of predator–prey dynamics in a patchy laboratory microcosm. Huffaker found that a predator mite and its prey mite persisted longer in spatially complex than in simple environments. Although his experiment has found its way into most ecology textbooks, Huffaker's results are actually much weaker than is often realized: they were unreplicated, the persistence observed was for only three oscillations, and only one female mite made it through each trough in the oscillations. More convincing evidence of spatial effects was reported by Pimental *et al.* (1963) using an interaction between flies and parasites in which the number of small 'fly boxes' and their degree of connection were manipulated. Several other laboratory studies of predator–prey systems have documented the importance of physical refugia for prey persistence, a situation that may be interpreted as a special kind of 'patchiness' (see, for example, Gause (1934); Flanders & Badgley (1963)).

Although the experimental data may be ambiguous, models that these experiments have inspired make it unequivocally clear that spatial subdivision can stabilize predator–prey interactions (Chewning 1975; Comins & Blatt 1974; Hilborn 1975; Zielger 1977; Nachman 1987*b*). However, models also have made it clear that there are circumstances under which spatial subdivision has no effect on stability, or even reduces stability (Allen 1975; Crowley 1981; Reeve 1988). Fortunately, it is possible to sort out the different theoretical results attributed to spatial subdivision by considering exactly how the different subpopulations of predators and prey are connected. In particular, the key requirements for spatial subdivision to stabilize predator–prey systems are: (i) population densities must fluctuate asynchronously in different patches; (ii) predator and prey dispersal rates must be above some minimal rate (otherwise patches suffering on extinction would not be recolonized), and (iii) predators must not disperse so effectively that they inevitably find prey as soon as the prey colonize vacant patches (Reeve 1990; Taylor 1988). When these criteria are met, spatial subdivision promotes stability because it provides prey with a refuge from attack.

(*d*) *Key results from reaction–diffusion theory*

Reaction-diffusion equations have a rich tradition in mathematical biology (Murray 1989), but are not widely appreciated by field ecologists. This is unfortunate because these models can be used to generate baseline predictions of what to expect when populations interact and disperse without any complications (such as patchiness or subdivision) in the spatial dimension. In other words, reaction-diffusion equations represent a sort of null model for spatially distributed population dynamics; they inform us of the spatial patterns that develop because of random motion and population growth alone. Moreover, if we relax the assumption of constant diffusion and spatial homogeneity, reaction–diffusion models become tools for investigating heterogeneity in any form, not just as 'stepping stones' or 'islands' (see Banks *et al.* 1985, 1987).

One of the most intriguing predictions from reaction–diffusion theory is that species interactions in homogeneous environments can generate permanent spatial patterning. This patterning, or 'diffusive instability', arises when there is some sort of local activation due to one component of the system, and a longer range inhibition due to another component (Meinhardt 1982). The key idea is that an interaction

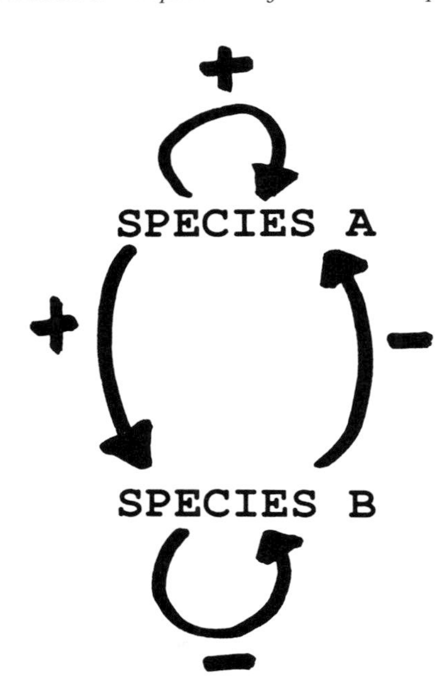

Figure 1. A graphical depiction of what is meant by an activator–inhibitor system. Each arrow shows how a change in the population at the tail of the arrow will alter the rate of population growth at the point of the arrow, all evaluated at equilibrium ((+) represents an increase in the rate of population growth and (−) represents reduction in the rate of population growth). In more technical jargon, the signs on the arrows correspond to the signs of terms in the jacobian matrix for the system.

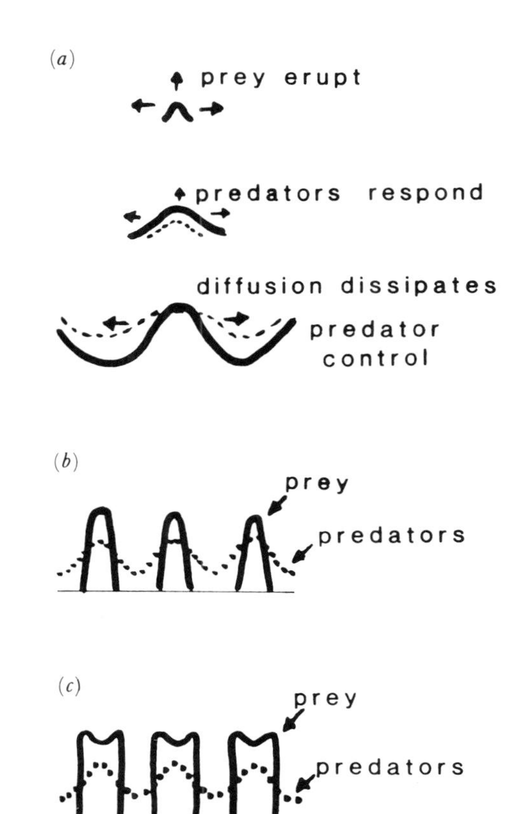

Figure 2. (*a*) How a diffusive instability gets initiated following a small perturbation for a predator–prey system. (*b*) One example of the sort of final stable spatial patterning that is possible for predator–prey diffusive instability. (*c*) Another example of spatial patterning due to diffusive instability in predator–prey systems. The only difference between figures (*b*) and (*c*) is that in (*c*) the prey are 0.01 times as mobile as in (*b*). For many geometries of habitat and ratios of dispersal rates, the patterning could involve much sharper peaks.

which is stable in the absence of dispersal can produce regular or bizarre spatial variations in densities when dispersal occurs. For pairwise species interactions, there is a simple rule of thumb for deciding whether diffusive instability is plausible, the interaction must be an activator–inhibitor system (see figure 1), and the inhibitor must diffuse substantially faster than the activator. To understand how diffusive instability generates spatial patterning, it is useful to consider a concrete example, such as predator–prey interactions in which the prey represents an activator, and the predator an inhibitor. First of all, note that to satisfy the 'activator–inhibitor criterion' (figure 1), increases in prey density above the equilibrium must promote further production of prey and predator; whereas any increases in predator density above the equilibrium must reduce further production of predators and of prey. (We should expect this criterion to be naturally met by many predator–prey systems.) In the absence of dispersal, the positive feedback from the prey (i.e. activation) and the negative feedback from the predators (i.e. inhibition) counteract one another to produce a stable equilibrium. However, when dispersal (diffusion) is added to the system, and the predators wander away from localized prey eruptions, the system can be destabilized. In particular, as prey densities increase and predators respond, some of the predator's inhibition is dissipated because the predators diffuse away from prey eruptions; the key to this 'dissipation' is sufficiently more rapid predator than prey diffusion (see figure 2*a*). It is important to emphasize that for such patterns to arise predator movement need not be purely random or diffusive (although this is usually the assumption pursued in the models); spatial patterns can erupt even if predators tend to move in a biased fashion toward peaks in prey density, as long as there is still some diffusive leakage of predators away from regions of high prey density (Kareiva & Odell 1987). The net result can be fixed spatial predator–prey patterns as shown in figures 2*b* and 2*c*.

The convenient notion of an activator–inhibitor system breaks down for communities of species that cannot be summarized by a simple diagram such as figure 1 (Evans 1980). None the less, when we try to understand the diffusive instabilities that arise in multispecies interactions, we can usually identify some sort of 'inhibition' that is dissipated away from perturbations by dispersal (Mimura 1984; Mimura & Kawasaki 1980; Kishimoto 1982). For example, diffusive instabilities arise in competitive systems when there are two species that cannot coexist unless a third species is also present; that third species is the 'inhibitor' that reduces the other two species to sufficiently low densities that they can coexist, and it is the inhibition of this third species that is dissipated by diffusion (Mimura 1984). In the absence of dispersal, these particular three-species competitive systems attain a stable equilibrium, but with dispersal and a perturbation complex spatial patterns can be generated. Similarly, the diffusive instability that arises in single population models with age-structure involves the dissipation of a stabilizing influence because a particular age class is especially mobile (Hastings 1991). Although this discussion is abstract, the possibility of diffusive instabilities can be shown for a wide variety of ecological systems, ranging from age-structured density-dependent population growth, to predator–prey interactions, to competitive systems that involve three or more species (two-species competitive interactions cannot produce diffusive instabilities unless they include age-structure). I suspect the possibility of diffusive instabilities can also be established for plant–herbivore and host–pathogen interactions. What makes diffusive instability so interesting is that it provides a mechanism for creating persistent spatial variation in population densities without any underlying environmental variation. Indeed, the opportunities for diffusive instability are widespread in nature; the interesting question is, to what extent does this mechanism for pattern formation contribute to the clumped patterns of population dispersion we observe in natural populations? Perhaps much of the patterning attributed to underlying environmental mosaics could be more parsimoniously explained by diffusive instabilities. Or more realistically, perhaps environmentally driven spatial variation in population density is amplified by the mechanism of diffusive instability.

A second phenomenon of interest in reaction–diffusion models is the existence of travelling waves of population densities, or trains of travelling waves. Travelling waves occur when organisms invade a new habitat and then reproduce and disperse (Andow *et al.* 1991; Murray 1988; Okubo *et al.* 1989), or when a disease epidemic is initiated from a single point in space and spreads outward (Murray 1986; Murray *et al.* 1988; Yachi *et al.* 1989). It is often possible to express the speed at which a travelling wave of an invading population should spread as a function of basic life history traits; this calculation may thus usefully summarize the invasiveness of a species in terms of measurable parameters (Andow *et al.* 1991). Predator–prey interactions with diffusion can also produce what look like waves of predators chasing prey, even though both predators and prey move randomly (Dunbar 1983; Murray 1989). Especially complex propagating waves appear when reaction–diffusion models include what is called 'cross-diffusion', or diffusion at a rate that depends on the product of species densities (Murray 1989); cross-diffusion can be used to model predators that actively pursue prey and prey that attempt to evade predators.

A final feature of reaction–diffusion systems is that they generally predict spatial gradients in population densities as a function of habitat geometry. For instance, if a population is confined to a small habitat with diffusive losses across the boundaries of the habitat, there will be a critical patch size below which the population cannot sustain itself (Kierstead & Slobodkin 1954; Skellam 1951; McMurtrie 1978). Moreover, for larger habitats, population densities will attain spatially heterogeneous profiles that are functions of the shape of habitats, the size of habitats, and movement behaviour at habitat boundaries (Turchin 1988; Okubo 1980). The initiation and final patterning due to diffusive instabilities will also be strongly influenced by the size and geometry of habitats. For instance, if we imagine species interacting in a habitat whose boundaries are relatively impermeable (i.e. no-flux boundary conditions), then it is possible for habitats to be too small to allow for the development of spatial patterning through diffusive instability (Murray 1989). In sufficiently large habitats with complex geometries, diffusive instability is likely to produce remarkably intricate chequerboards of peak and trough densities, ranging from 'spots' to 'bands' to 'paisley' mosaics (Murray 1989).

(*e*) *'Pseudospatial' models: aggregation and species interactions*

Several models deal with heterogeneity by assuming a particular pattern of spatial covariation in species densities, and then attempting to deduce consequences for temporal dynamics. Usually these models have focused on the consequences of aggregation for competitive (Atkinson & Shorrocks 1981; Ives 1988, 1991*a*; Shorrocks *et al.* 1979) or host–parasitoid interactions (Pacala *et al.* 1990; Hassell & Pacala, this symposium; Murdoch & Oaten 1989; Ives 1991*b*). I do not view these aggregation models in the same spirit as the theory discussed above because they assume one panmictic population. Instead of analysing spatially distributed population dynamics, aggregation models examine the temporal consequences of a behaviour that happens to be expressed in the spatial dimension. There is no parameter that reflects a dispersal rate in aggregation models, and there is no sense in which population dynamics are subdivided or spread out over space. Thus when considering this class of models, it is important to realize that 'heterogeneity' refers to the experience of individuals in the interacting populations, but not to well-defined subpopulations. None the less, the approaches used to uncover the effects of aggregative behaviour may suggest ways in which explicit spatial models can be collapsed into more tractable difference equation and ordinary differential equation models.

(*f*) *The recurring theme of critical rates, critical scales and critical geometries*

Mathematical models make it clear that simply by adding a spatial dimension and dispersal to population processes (while holding all other aspects of the ecology the same), population dynamics can be altered fundamentally. Unstable interactions can be made stable (Reeve 1990), competitive exclusion can be thwarted (Hanski 1983), stable interactions can be made unstable (Allen 1975; Crowley 1981; Reeve 1988), and spatially homogeneous systems can be turned into highly patterned or spatially heterogeneous systems (Hastings 1991; Murray 1989). These different possibilities depend on the details of dispersal rates and the spatial scale over which population dynamics are considered. For example, if patchiness is to have any effect on predator–prey dynamics there must be sufficiently many patches that are far enough apart that their dynamics are somehow asynchronized; on the other hand the patches cannot be so widely separated that dispersal is inadequate to recolonize empty patches. Similarly, for patchiness to alter the outcome of competitive interactions, inferior competitors must be sufficiently more mobile than superior competitors, and disturbances must interrupt local within-patch dynamics sufficiently frequently relative to the speed of the exclusion process (Hanski 1983). Finally, for the spatial dimension to produce spatial patterning, the relative diffusion rates of interacting species must exceed some critical ratio (Murray 1989). One clear message of this theory is that although experimentalists can expect important changes in population dynamics due to spatial subdivision or the spatial dimension, such effects will be attained only at certain spatial scales and only for certain dispersal rates.

AN OVERVIEW OF EMPIRICAL EXAMINATIONS OF SPATIAL THEORY

Theoreticians are not the only ones interested in habitat subdivision. Indeed, field biologists often attribute key features of natural populations to the influences of spatially heterogenous environments. For example, in a review of field studies that focused on arthropods and were published between 1975 and 1989, Doak *et al.* (1991) found 62 'data papers' concerning the impact of patchiness on species interactions. Unfortunately, the data in these papers do not offer much insight regarding the applicability of island and stepping-stone models. One problem is that over half of the reviewed studies failed to report the number of patches, the size of patches and the distance between patches; since theory predicts that the effects of patchiness depend on its spatial scale, this lack of information concerning the scale of the investigations hinders our ability to interpret the data that were collected (Addicott *et al.* 1987). A more forgivable shortcoming of patch studies is the absence of between-patch dispersal rates; since collecting these data is a major undertaking it is not surprising that researchers rarely assess dispersal. None the less, without quantitative measurements of dispersal we can never know whether so-called patches reflect genuinely subdivided populations.

Field evaluations of reaction–diffusion models are almost unheard of, probably because the theory itself remains esoteric. The best that has been done with reaction–diffusion theory is to estimate parameters in simple invasion models, and ask whether the models effectively predict observed rates of spread for invading populations of red squirrels (Okubo *et al.* 1989), or cabbage butterflies and muskrats (Andow *et al.* 1991). The problems with these analyses include large gaps in the data such that some critical parameter must be guessed (Okubo *et al.* 1989), matches between model and reality that are often ambiguous (Andow *et al.* 1991), and an assumption of a uniform environment that cannot possibly hold at the scale relevant to most ecological invasions (Andow *et al.* 1991). Ironically, reaction–diffusion models receive much more empirical attention in developmental biology (see, for example, Meinhardt (1982)) than in ecology, even though there is no consensus about what chemical 'species' should be measured when studying developmental processes. In contrast, applying reaction–diffusion models to predator–prey interactions should be comparatively straightforward, since it is obvious that the densities, 'reaction kinetics' (birth, death, consumption, etc.) and diffusion rates of the predator and prey are what need to be evaluated.

Ecologists are proud of the fact their science has become experimental. However, when it comes to even the best-known theory of species interactions in subdivided environments, the number of field experiments can be counted on one hand. This is unfortunate, because varying the spatial arrangement of habitats in some controlled manner is the only way to determine whether subdivision fundamentally alters the dynamics of populations. The few field studies that include such manipulations have generally found an 'effect of subdivision'. For example, when Hanski (1987) divided a fixed amount of liver into either 1, 2, 4, 8, or 16 patches, he found that coexistence occurred more frequently among competing carrion flies in the highly subdivided treatment than when the liver was undivided. Quinn & Robinson (1987) subdivided annual grassland into 2, 8 or 32 subunits and observed higher diversity of flowering plants in the subdivided grasslands (summed over all patches) compared to the single large patches (although total area remained the same). Kareiva (1987) found that aphids attained localized outbreaks far more frequently in subdivided than in continuous strips of habitat. The only experimental investigation that has failed to detect an effect of subdivision was a study of a subtidal snail whose population dynamics were unaltered by the degree to which artificial habitat plates were fragmented (Quinn *et al.* 1989).

In addition to controlled experiments, there are several observational studies that have taken advantage of naturally occurring variation in habitat mosaics to assess the significance of habitat division and geometry (Pokki 1981; Fahrig & Merriam 1985; Jennersten 1988; Franklin & Forman 1987; Quinn & Harrison 1988; Sousa 1979; Solbreck & Sillen-Tullberg 1990). In all but one of these studies (Solbreck & Sillen-Tullberg 1990), the authors conclude that the degree to which habitats are subdivided or isolated from one another, significantly alters the density, dynamics or diversity of residents. Further support for the importance of spatial structure comes from a handful of quantitative investigations that have documented a pattern of within-patch extinctions and between-patch dispersal events that could only be captured by an island or metapopulation model (Addicott 1978; Hanski & Ranta 1983; Bengtsson 1989).

I believe that the above collection of field experiperiments and observational analyses make a strong case for the importance of spatial subdivision in population and community dynamics. However, none of the evidence is compelling vindication of any particular model or theoretical prediction. Simply documenting an effect of patchiness is a long way from testing models that predict particular shifts in dynamical behaviour as a result of changes in dispersal rates, or in the geometry of a spatial mosaic. Before one could convincingly test any of the existing island or stepping-stone models, one would need to know a good deal about dispersal rates, within-patch dynamics, frequencies and dispersion of catastrophes, and the spatial structure of environmental variability. Moreover, when we turn our attention to real systems, rarely will one model cover the full range of processes that are likely to interact with habitat subdivision; instead we are likely to find ingredients at work that are the focus of many different spatial models (e.g. stabilization due to asynchronous fluctuations, bizarre inhomogeneities due to diffusive instability, altered vulnerability to extinction because of demographic stochasticity, and so forth). To make concrete the problems that arise when one attempts to apply 'spatial models' to field studies, we now examine in some detail two experiments involving manipulations of habitat subdivision.

TWO CASE STUDIES THAT HIGHLIGHT THE DIFFICULTIES TO BE EXPECTED WHEN 'TESTING' PATCH MODELS

(*a*) *Manipulating habitat subdivision in an insect predator–prey interaction*

From 1982 to 1985 I manipulated the degree of habitat subdivision in a predator–prey interaction by mowing monocultures of goldenrod into either continuous strips of vegetation or fragmented rows of patches (figure 3). The interaction examined was between an aphid that specializes on goldenrod (living its whole life in goldenrod fields) and a ladybird beetle that specializes on aphids, but not necessarily the goldenrod aphid. I initially analysed these experiments with respect to effects on mean aphid density and the likelihood of aphid outbreaks (Kareiva 1987); instead, here I use the data to ask whether habitat subdivision alters the magnitude of temporal fluctuations in aphid density as predicted by predator–prey models. Since the experiment spanned at least eight predator generations and thirty prey generations, the timespan is adequate for quantification of temporal variability. I focus only on aphid densities because ladybird densities are tightly correlated with aphid numbers and yield identical patterns with respect to degrees of fluctuation.

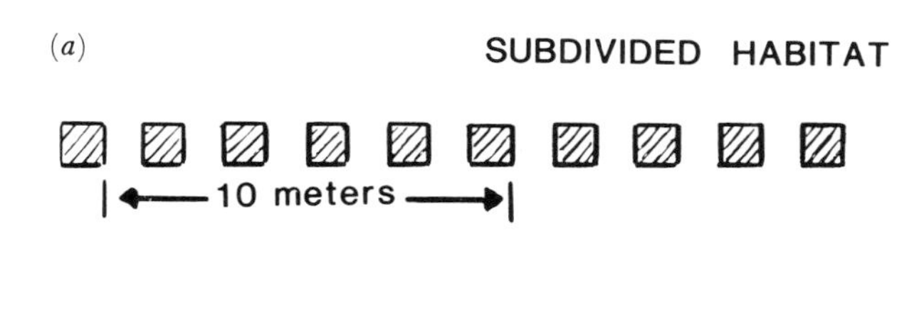

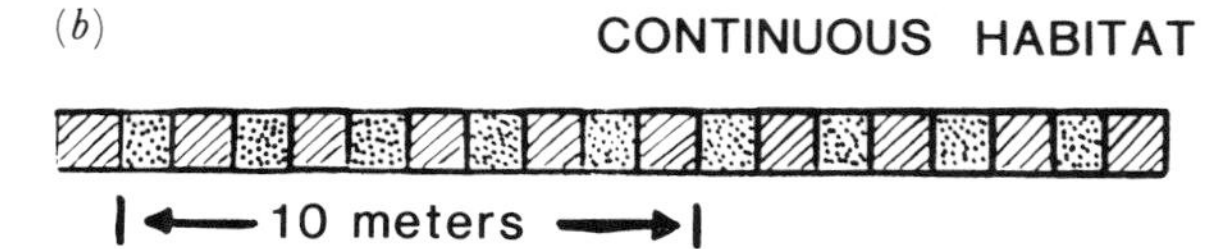

Figure 3. The two treatments used to manipulate habitat subdivision in a ladybird beetle–aphid interaction (from Kareiva 1987): (*a*) is subdivided treatment; (*b*) is undivided treatment. Crosshatched and stipled areas represent monocultures of goldenrod surrounded by mown grass. The distinction between crosshatched and stipled areas is that crosshatched areas were repeatedly censused over a four-year period, whereas the stipled areas are simply part of the treatment background (and were not censused).

My experiment included six ensembles: three ensembles each with ten 'goldenrod islands' (figure 3*a*), and three ensembles each with ten goldenrod quadrats embedded in a continuous goldenrod strip (figure 3*b*). The distance between islands or quadrats was one metre; the difference between the two treatments was the intervening vegetation, the absence of intervening goldenrod in the subdivided treatment hampered the dispersal of both aphids and ladybird beetles compared to their movement in the undivided treatment. By averaging together my censuses from all ten islands or quadrats, I can obtain an ensemble-wide density for aphids and thus a record of population fluctuations at the scale of entire ensembles. The magnitude of fluctuations was identical for subdivided versus undivided treatments (figure 4*a*, paired t-test on coefficients of variation, $p > 0.5$). At first this might seem like a contradiction of theory, since subdivision is generally supposed to be stabilizing. However, when we examine the pattern of fluctuations in individual metre-square islands or quadrats, the absence of any difference in population 'stability' at the ensemble level is no longer paradoxical. In particular, at the level of single patches or single quadrats, we see that in both types of habitat the aphid densities fluctuate asynchronously in space. Such asynchrony is exactly what is needed to promote stability in predator–prey interactions, and thus it appears that even when goldenrod occurs as continuous strips, there is enough asynchrony in dynamics along the strip of goldenrod that stability is promoted. This interpretation is bolstered by the observation that aphid populations do, in fact, appear to fluctuate only modestly at the level of entire ensembles; in particular, the coefficients of variation for ensembles are significantly lower than for patches or quadrats (contrast figure 4*a* with 4*b* and 4*c*; one-way ANOVA, $p < 0.05$). Interestingly, although patches or quadrats within the same field fluctuate out of synchrony, ensembles of patches in different fields fluctuate in remarkable harmony. Thus when aphids are abundant in one field, they are likely to be abundant in a neighbouring field that could be as far as 300 m away; on the other hand, when aphids are abundant in one square metre of goldenrod, that is no indication of likely aphid abundance as close as one metre away. I suspect that the synchronized fluctuations in different ensembles result from the common action of weather on aphid and ladybird demography, whereas the asynchrony among patches in the same fields is driven by the vagaries of aphid colonization,

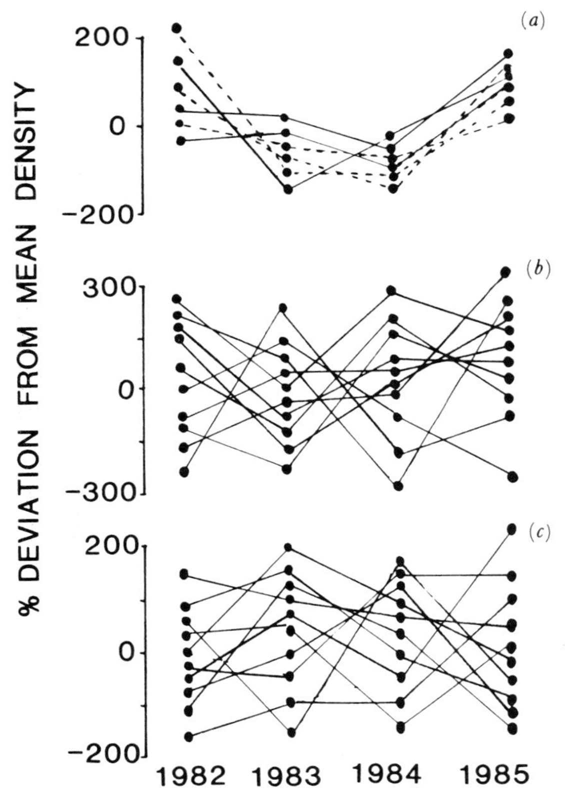

Figure 4. The pattern of temporal variation for aphid populations: (*a*) temporal variation at the level of entire ensembles of goldenrod patches or quadrats; each dashed line represents an ensemble in subdivided habitat, whereas solid lines are for ensembles in continuous habitat; (*b*) temporal variation at the level of single square metre patches of goldenrod from the subdivided habitat treatment; (*c*) temporal variation at the level of single square-meter quadrats of goldenrod from the undivided treatment. These single-patch (i.e. figure 4*b*) or single-quadrat (i.e. figure 4*c*) figures are from only one ensemble; the results from the remaining two ensembles are identical with respect to degree of asynchrony and overall variability as measured by coefficients of variation. To draw these plots only the aphid densities recorded during the week of maximum abundance each summer were used; the pattern is the same if each week's census is included, except a seasonal trend is then embedded in the year-to-year fluctuations.

ladybird searching and heterogeneous goldenrod phenology or chemistry. Clearly scale in terms of the area over which densities are assessed plays a major role in the amount of synchrony detected. Finally, even though subdivision did not appear to alter the severity of fluctuations for ensembles, it did enhance fluctuations at the level of single patches (the coefficient of variation was significantly higher in the thirty patches from subdivided goldenrod than in the thirty patches from undivided goldenrod). The explanation for this effect is that subdividing goldenrod interferes with ladybird searching behaviour, and gives aphids a temporary escape from predation, thereby facilitating localized aphid eruptions (Kareiva 1984, 1986, 1987).

There are several weaknesses in using data from the above experiment to address theories of habitat subdivision. First of all, mowing the goldenrod into patches did more than simply alter the degree of subdivision, it also reduced the total area of goldenrod and changed the competitive circumstances faced by goldenrod plants (which could have changed their quality as aphid food). One option would have been to mow the goldenrod into patches that summed to the same total area as the continuous strips of vegetation; however, pursuing this option would have nearly doubled the spatial dimension spanned by the subdivided goldenrod (i.e. the length of each row of patches). The diameter or span of an experimental unit could alone alter predator–prey dynamics (because span determines the scale over which an environment is sampled as well as opportunities for diffusive instabilities). Finally, the scale at which the experiment contrasts subdivided versus undivided habitats did not produce marked differences in what theoreticians often refer to as 'the connectedness' of patches. In other words, from the perspective of dispersal as a process that can synchronize population fluctuations (Taylor 1988), the subdivided and undivided treatments did not differ substantially. (Although, from the perspective of dispersal as a component of foraging behaviour, the subdivided treatment greatly inhibited foraging efficiency.)

One final intriguing feature of this ladybird beetle–aphid interaction is that aphid outbreaks seemed to possess consistently the same 'wavelength' or diameter, with a halo of predators aggregated at each aphid peak yet also spilling over to the surrounding vegetation (Kareiva 1984; Kareiva & Odell 1987). It is difficult to determine whether this pattern is stable because aphid populations collapse in late August as goldenrod senesces, making the concept of stability ambiguous. None the less, regular aphid peaks with ladybird halos is exactly what is expected from diffusive instabilities; moreover, the jacobian matrix for the interaction and the ratio of predator–prey diffusion rates are such that diffusive instability is expected (Kareiva 1984). However, because the interaction includes taxis as well as diffusion, a standard analysis does not apply. Numerical simulations verify that one could reproduce the patterns observed in the field with a simple reaction–diffusion–taxis model (Kareiva & Odell 1987), but the results are sensitive to the initial perturbations in aphid density. These data suggest, but do not prove, that diffusive instability occurs in this system.

(*b*) *Resource subdivision in a successional community*

Peter Turchin and I have examined the effects of habitat subdivision at a small scale in an extremely severe environment (Turchin & Kareiva 1989; Kareiva & Turchin, manuscript). In the regions that surround Mount St Helens, we have asked whether dividing fireweed (*Epilobium augustifolium*) into series of isolated stems as opposed to clumps of stems alters the densities of *Aphis varians*, an aphid for which fireweed is the only foodplant in the vicinity of the Mt St Helens blast zone. The question could be framed in terms of the effects of patch size (small versus large) or in terms of divided (many small) versus undivided (one

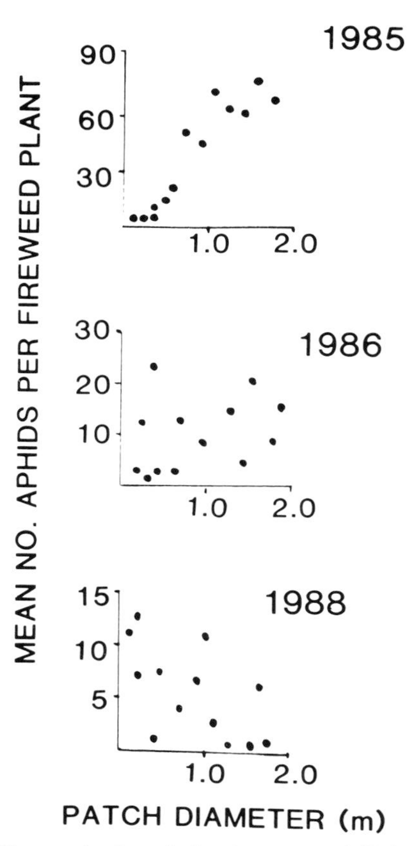

Figure 5. Changes in the relation between aphid density and patch size as succession proceeds at Mount St Helens.

large) habitats. We were initially motivated by reaction–diffusion models that predict strong relations between patch size and the densities of habitat specialists, assuming that the environment surrounding habitat patches is so harsh that individuals wandering out of patches are likely to die (Okubo 1980; McMurtrie 1979). This assumption holds for aphids at Mount St Helens, and preliminary data provided support for the predicted relation between patch size and aphid density (figure 5*a*). This led us to perform experiments in which we manipulated patch size ourselves, as well as continued sampling naturally occurring patches of differing sizes. Our hypothesis that aphids would be more abundant per fireweed stem in large fireweed patches than on isolated fireweed stems was dramatically rejected, but was rejected in a way that is worth examining with respect to tests of spatial theory. First, the pattern we detected by simply sampling fireweed reversed itself between 1985 and 1988 (figure 5*a* versus 5*c*). Secondly, the results of manipulative experiments also changed through time, such that no effect was attributed to patch size in 1986, and a significantly higher aphid density was found on single fireweed stems in 1988 (Kareiva & Turchin, manuscript). We suspect this reversal of patterns is due to the entry of a third species into the fireweed–aphid association. In 1985, colonization of the blast zone was just getting started and ladybird beetles were relatively scarce; however, by 1988 ladybird beetles (mainly *Hippodamia convergens*) were so abundant that they seemed to be the dominant factor regulating *Aphis varians* density (Morris 1990). Thus what was initially a plant–aphid interaction switched to a plant–aphid–predator system. We believe the addition of the third trophic level fundamentally altered the aphid's relation with fireweed patchiness because predation pressure is higher on patches of fireweed than on isolated stems of fireweed (Kareiva & Turchin, manuscript). Of course, to establish rigorously this hypothesis we need to repeat our manipulations in both the presence and absence of predation (an experiment that would require predator removals at a vast spatial scale). None the less, our results suggest that the effects of habitat fragmentation on particular pairwise interactions can disappear or be contradicted when additional species are taken into consideration, a discouraging scenario if we must rely exclusively on experiments to understand the interplay of habitat subdivision and species interactions.

AN AGENDA FOR FUTURE RESEARCH

(*a*) *Empirical challenges*

First, it is worth emphasizing that experimental tests of theory are not tests to see whether theoreticians have done their calculations correctly. They are also not tests of model assumptions, since model assumptions are always 'wrong' to some extent (simplifying assumptions are a key part of model-building). Rather when we test models, we often are asking whether the phenomenon predicted by models emerge as important in natural situations, where numerous other confounding and competing forces act upon the species under study. For instance, a model could be 'mathematically correct', yet offer minimal insight into the workings of real systems simply because the trends it predicted were usually swamped out by factors not included in the model. Thus certain spatial models may fail because they address pairwise species interactions in a world that cannot be described by theory of such a limited scope. Notice that even in failing, models can be instructive (as long as they and their associated experiments are not so silly as to be trivially doomed to failure) – if our existing spatial theory repeatedly fails because its emphasis on pairwise interactions is misplaced, then it is straightforward (albeit cumbersome) to build multispecies and multitrophic level spatial models (see, for example, Mimura 1984; Kishimoto 1982). Good models make it clear in which situations they are most likely to yield insights, or when their predictions will most likely be sharply exhibited. For example, no one conversant with spatial predator–prey theory would expect population subdivision at the level of leaves to have any effect on the stability of coccinellid–aphid interactions; but subdivision at the scale of leaves could have an effect on the microbial predator–prey interactions that take place on plant surfaces. The point is that one does not use experiments to 'keep score' of how many times some model 'wins' or 'loses', a model might repeatedly 'lose' because it was being applied in circumstances where the theory itself predicted ambiguous or weak effects. Rather, one tests a model to gain an appreciation of its range of applications, and to learn directions in which the model ought to be modified.

The key idea in testing a model is to manipulate experimentally some factor identified by the model as important, and to then record responses in terms of the appropriate variables.

The obvious first step is evaluating spatial theory is to manipulate factors that are supposed to alter spatiotemporal dynamics, and to then observe how the manipulation influences patterns of population persistence, of species coexistence, of density fluctuations, or of dispersion. Since all of the existing theory pinpoints dispersal as a process that controls the effects of the spatial environment, manipulations that alter the dispersal rates of species are especially desirable. However, simply performing these manipulations is not enough. The experiments need to be accompanied by measurements of dispersal rates so that it is possible to check whether resulting patterns are in accord with interpretations based on the interplay of local dynamics and dispersal. A particularly clearcut result might be a qualitative shift in the dynamics of a system (e.g. from stable to unstable, or from coexistence to exclusion, and so forth) in the direction predicted by theory upon some manipulation of subdivision. For instance, one's faith in the applicability of spatial models would be greatly enhanced if an unstable predator–prey system could be stabilized by reducing rates of between patch-movement (which in turn converted synchronized fluctuations in subpopulations into asynchronized fluctuations).

Ideally, when we manipulate experimentally habitat subdivision or adjust rates of dispersal, we should impose our manipulations at a hierarchy of scales. Multiple scales are necessary because it will be difficult to know in advance whether one has identified the appropriate scale for detecting a response (Heads & Lawton 1983). Conservationists are already debating the interpretation of habitat fragmentation experiments because of differing opinions about what is the 'right scale' (Murphy 1989). Theory makes it clear that it is not so much a matter of the right scale, as identifying over what scale different spatially mediated processes are likely to be exhibited. For instance, one scale of fragmentation might highlight effects due to changes in predator–prey dynamics, whereas a different scale might reveal changes in extinction due to demographic stochasticity, and yet a different scale would reveal differences in persistence in the face of random catastrophes.

So far, empirical investigations of habitat subdivision have dealt exclusively with single-species dynamics, with pairwise interactions, or with interactions among many species but on one trophic level. Our results concerning fireweed aphids dramatize the problem with such a narrow scope. It may be impossible to extend the understanding we gain from studies of simple interactions to the effects of subdivision on food webs and entire communities. The only way to resolve this problem is to observe the response of entire communities to replicated manipulations of habitat fragmentation. It may be that theory can predict which species should be affected (and which should not) on the basis of dispersal rates and strengths of interactions with other species.

Although the experiments will be difficult, investigations of diffusive instability would be extremely interesting. These would need to be performed in relatively uniform environments (e.g. agricultural fields) and would require substantial modelling before a useful experiment could be designed. Ideally, one would like to have an estimate of the critical size of habitat necessary for patterning before embarking on a manipulative field study. Given such knowledge, one might be able to show that no patterning occurs if the habitat is too small, whereas consistent patterning arises in habitats sufficiently large.

In general, the most fruitful avenue for empirical studies of spatially distributed dynamics is an approach that entails a tight connection between the experiments and specific models (inspired by more general theory). A superb example of what can be gained by this approach is the analyses of competition between annual plants by Pacala & Silander (1990). Pacala (1986, 1987) first analysed the dynamics of competition by using models of varying degrees of abstraction, and found situations that did require spatial models, as well as situations for which dynamics could be captured by a simple representation of mean population densities changing through time. Pacala & Silander (1990) then applied both spatial and non-spatial models to field studies of competition; they found that indeed, spatial models were not necessary for the species under study because spatial clumping was weak and plant performance was especially plastic. The net result is not only well-developed spatial theory for plant competition, but a good understanding of when that theory is necessary and when it is not needed.

Finally, because field experiments concerning habitat subdivision or diffusive instability present vast logistical obstacles and are difficult to replicate properly, there is room for using laboratory microcosms as a testing ground (Forney & Gilpin 1989). This is especially true because the so-called classic experiments in this area (Huffaker 1958) are much more ambiguous than is commonly realized, and did not have the benefit of theory as a guide to their design.

(*b*) *Theoretical challenges*

So far, theoreticians have been busy examining yet another model in which habitat subdivision could alter dynamics, or yet another reaction-diffusion system in which diffusive instabilities could arise. We now need to reconcile this vast diversity of theory into a more unified form and to better address applications of theory to particular systems. More attention to highlighting when spatial theory is not needed would be a great aid to empiricists. It would also be especially useful if a common interpretation could be developed to approximate different 'clusters' of theoretical results, as has been so elegantly accomplished for host–parasitoid aggregation theory (Hassell & Pacala, this symposium). For instance, perhaps some measure of spatial asynchrony might summarize whether habitat subdivision could contribute to predator–prey stability. Another area in which it would pay to synthesize seemingly disparate results is the meeting ground between the spatial models I have reviewed and so-called 'disturbance models' (Fahrig 1989). Although 'spatial theory' and 'disturbance theory' tend to ask different questions, dispersal rates are key in both types of models (as the factor determining how connected are spatially separated population dynamics, or as the factor determining the rate at which disturbances are colonized). A final synthesis that has been repeatedly called for yet never achieved, is linking models of individual foraging behaviour (which predict how animals should search and use heterogeneous environments) with models for population interactions in heterogeneous environments (Hassell & May 1985).

There is need for original theoretical inquiry as well as consolidating existing results. Three relatively unexplored phenomena of potential importance are: (i) the observation that ecological interactions include ingredients acting at multiple scales (Powell 1989); (ii) the implications of nonrandom movement and mortality while animals actively search for habitats in fragmented landscapes, and (iii) the complications that might emerge in multispecies (or food web) interactions as opposed to pairwise interactions. Processes at multiple scales will be especially important in predator–prey interactions, where we often find long-range dispersal in response to regional variation in prey densities layered on top of foraging behaviour that adjusts to prey patchiness at a fine scale. Non-random search becomes an issue when we attempt to apply metapopulation models to conservation questions; such applications need to consider the extent to which details of 'island' geography determine the overall metapopulation dynamics (since it is the detail of habitat arrangement rather than total amount of habitat that is the key scientific question in conservation). Finally, it is important to establish how food web dynamics change in subdivided habitats, since single-species and pairwise species models cannot possibly capture the range of possible effects attributed to spatial subdivision. The limitations of pairwise theory is dramatically shown by analyses of diffusive instability in competitive interactions; whereas diffusive instability can be shown to be theoretically impossible for pairwise competitive interactions, diffusive instabilities easily arise in competitive systems that involve three or more species (Mimura 1984).

CONCLUSIONS

Theoretical support for the importance of habitat subdivision and spatially distributed dynamics is overwhelming. In contrast, experimental evidence is at best suggestive, and not at all illuminating regarding the precise mechanisms by which the spatial environment alters population and community dynamics. In the face of all the other factors that shape population dynamics, we simply do not know the relative role of spatial factors. Only a vigorous and theoretically informed experimental programme will be able to address this question, a question that is central to ecology because of its implications for biodiversity (Pickett & Thompson 1978), resource management (MacCall 1989), and conservation (Wilcove 1987; Quinn & Hastings 1988; Gilpin 1988).

This manuscript was prepared while the author was a visitor at the Centre for Mathematical Biology at Oxford University and the Centre for Population Biology at Silwood Park. I thank Professors James Murray (Oxford) and John Lawton (Silwood) for hosting my visit and providing a stimulating environment. The author was supported by a Guggenheim Fellowship and a grant from the National Science Foundation. Fritzi Grevstad helped compile the bibliography, and Greg Dwyer and William Morris commented on the manuscript. I am especially grateful to my incorrigible colleague, R. T. Paine, for defending the record of empiricists and questioning the virtue of the theory I review, as usual, he lost the argument, but he did force me to express myself more clearly.

REFERENCES

Addicott, J. 1978 The population dynamics of aphids on fireweed: a comparison of local populations and metapopulations. *Can J. Zool.* **56**, 2554–2564.

Addicott, J. F., Aho, J. M., Antolin, M. F., Padilla, D. K., Richardson, J. S. & Soluk, D. A. 1987 Ecological neighborhoods: scaling environmental patterns. *Oikos* **49**, 340–346.

Allen, J. 1975 Mathematical models of species interactions in time and space. *Am. Nat.* **109**, 319–342.

Andow, A. A., Kareiva, P. M., Levin, S. A. & Okubo, A. 1990 Spread of invading organisms. In *Landscape Ecology*. (In the press.)

Andreasen, V. & Christiansen, F. B. 1989 Persistence of an infectious disease in subdivided population. *Math. Biosci.* **96**, 239–253.

Atkinson, W. D. & Shorrocks, B. 1981 Competition on a divided and ephemeral resource: a simulation model. *J. Anim. Ecol.* **50**, 461–471.

Banks, H. T., Kareiva, P. & Lamm, P. 1985 Estimation of temporally and spatially varying coefficients in models for insect dispersal. *J. Math. Biol.* **22**, 259–277.

Banks, H. T., Kareiva, P. & Murphy, K. 1987 Parameter estimation techniques for interaction and redistribution models: a predator–prey example *Oecologia* **74**, 356–362.

Bengtsson, J. 1989 Interspecific competition increases local extinction rate in a metapopulation system. *Nature, Lond.* **340**, 713–715.

Burkey, T. V. 1989 Extinction in nature reserves: the effect of fragmentation and the importance of migration between reserve fragments. *Oikos* **55**, 75–81.

Chesson, P. L. 1981 Models for spatially distributed populations: the effect of within-patch variability. *Theor. Popul. Biol.* **19**, 288–325.

Chesson, P. L. 1984 Persistence of a markovian population in a patchy environment. *Z. Wahrscheinlichkeitstheorie verw. Gebiete* **66**, 97–107.

Chesson, P. L. 1985 Coexistence of competitors in spatially and temporally varying environments: a look at the combined effects of different sorts of variability. *Theor. Popul. Biol.* **28**, 263–287.

Chewning, W. C. 1975 Migratory effects in predator–prey models. *Math. Biosci.* **23**, 253–262.

Comins, H. N. and Blatt, D. W. E. 1974 Prey–predator models in spatially heterogeneous environments. *J. theor. Biol.* **48**, 75–83.

Comins, H. N. & Noble, I. R. 1985 Dispersal, variability, and transient niches: species coexistence in a uniformly variable environment. *Am. Nat.* **126**, 706–723.

Crowley, P. H. 1977 Spatially distributed stochasticity and the constancy of ecosystems. *Bull. Math. Biol.* **39**, 157–165.

Crowley, P. H. 1978 Effective size and the persistence of ecosystems. *Oecologia* **35**, 185–195.

Crowley, P. H. 1981 Dispersal and the stability of predator-prey interactions. *Am. Nat.* **118**, 673–701.

den Boer, P. J. 1970 Stabilization of animal numbers and the heterogeneity of the environment: the problem of the persistence of sparse populations. *Proc. Adv. Study. Inst. Dynamics Numbers Popul.* pp. 77–79. Wageningen: PUDOC.

Diekmann, O. 1978 Thresholds and travelling waves for the geographical spread of infection. *J. math. Biol.* **6**, 109–130.

Diekmann, O., Metz, J. A. J. & Sabelis, M. W. 1988*a* Mathematical models of predator–prey–plant interactions in a patchy environment. Report AMR 8804 *Centrum voor Wiskunde en Informatica.* Amsterdam.

Diekmann, O., Metz, J. A. J. & Sabelis, M. W. 1988*b* Reflections and calculations on a prey-predator-patch problem. Report AMR 8805 *Centrum voor Wiskunde en Informatica.* Amsterdam.

Doak, D. 1989 Spotted owls and old growth logging in the Pacific Northwest. *Cons. Biol.* **3**, 389–395.

Doak, D., Marino, P. & Kareiva, P. 1991 The implications of spatial scale when examining the threat posed by habitat fragmentation for conservation. *Theor. Popul. Biol.* (In the press.)

Dunbar, S. 1983 Travelling wave solutions of diffusive Lotka-Volterra equations. *J. Math. Biol.* **17**, 11–32.

Evans, G. T. 1980 Diffusive structure: counterexamples to any explanation? *J. Theor. Biol.* **82**, 313–315.

Fahrig, L. 1989 Interacting effects of disturbance and dispersal on individual selection and population stability. *Comm. Theor. Biol.* **1**, 275–298.

Fahrig, L. & Merriam, G. 1985 Habitat patch connectivity and population survival. *Ecology* **66**, 1762–1768.

Flanders, S. E. & Badgley, M. E. 1963 Prey-predator interactions in self-balanced laboratory populations. *Hilgardia* **35**, 145–183.

Forney, K. A. & Gilpin, M. E. 1989 Spatial structure and population extinction: a study with *Drosophila* flies. *Cons. Biol.* **3**, 45–51.

Franklin, J. F. & Formann, R. T. T. 1987 Creating landscape patterns by forest cutting: ecological consequences and principles. *Landscape Ecol.* **1**, 5–18.

Fujita, K. 1983 Systems analysis of an acarine predator-prey system II: interactions in discontinuous environment. *Res. Popul. Ecol.* **25**, 387–399.

Gause, G. 1934 *The struggle for existence.* Baltimore: Williams & Wilkins.

Gilpin, M. E. 1988 A comment on Quinn and Hastings: extinction in subdivided habitats. *Cons. Biol.* **2**, 290–292.

Gurney, W. S. C. & Nisbet, R. M. 1978 Predator–prey fluctuations in patchy environments. *J. Anim. Ecol.* **47**, 85–102.

Gurtin, M. & McCamy, C. 1977 On the diffusion of biological populations. *Math. Biosci.* **33**, 35–49.

Hanski, I. 1983 Coexistence of competitors in patchy environment. *Ecology* **64**, 493–500.

Hanski, I. 1987 Carrion fly community dynamics: patchiness, seasonality and coexistence. *Ecol. Ent.* **12**, 257–266.

Hanski, I. & Ranta, E. 1983 Coexistence in a patchy environment: three species of *Daphnia* in rock pools. *J. Anim. Ecol.* **52**, 263–279.

Harrison, S. 1991 Local extinction in a metapopulation context: an empirical evaluation. *Ann. Zool. Fennica* (In the press.)

Harrison, S. & Quinn, J. F. 1989 Correlated environments and the persistence of metapopulations. *Oikos* **56**, 1–6.

Hassell, M. & May, R. 1985 From individual behaviour to population dynamics. In *Behavioural ecology*, (ed. R. Sibley & R. Smith), pp. 3–32. Oxford: Blackwells Publications.

Hassell, M. P. & May, R. M. 1988 Spatial heterogeneity and the dynamics of parasitoid-host systems. *Ann. Zool. Fennica* **25**, 55–61.

Hastings, A. 1977 Spatial heterogeneity and the stability of predator–prey systems. *Theor. Popul. Biol.* **12**, 37–48.

Hastings, A. 1978 Spatial heterogeneity and the stability of predator-prey systems: predator-mediated coexistence. *Theor. Popul. Biol.* **14**, 380–395.

Hastings, A. 1980 Disturbance, coexistence, history, and competition for space. *Theor. Popul. Biol.* **18**, 363–373.

Hastings, A. 1990 Spatial heterogeneity and ecological models. *Ecology* **71**, 426–428.

Hastings, A. 1991 The effect of density dependence, age and spatial structure on stability and chaos. *Theor. Popul. Biol.* (In the press.)

Hastings, A. & Wolin, C. 1989 Within-patch dynamics in a metapopulation. *Ecology* **70**, 1261–1266.

Heads, P. & Lawton, J. 1983 Studies of the natural enemy complex of the holy leaf miner: the effects of scale on the detection of aggregative responses and the implications for biological control. *Oikos* **40**, 267–276.

Hilborn, R. 1975 The effect of spatial heterogeneity on the persistence of predator–prey interactions. *Theor. Popul. Biol.* **8**, 346–355.

Horn, H. & MacArthur, R. 1972 Competition among fugitive species in a harlequin environment. *Ecology* **53**, 749–752.

Huffaker, C. B. 1958 Experimental studies on predation: dispersion factors and predator–prey oscillations. *Hilgardia* **27**, 343–383.

Ives, A. R. 1988 Aggregation and the coexistence of competitors. *Ann. Zool. Fennica* **25**, 75–88.

Ives, A. R. 1991 Aggregation and coexistence in a carrion-fly community. *Ecol. Monogr.* (In the press.)

Jennersten, O. 1988 Pollination in *Dianthus deltoides* (Caryophyllaceae): effects of habitat fragmentation on visitation and seed set. *Cons. Biol.* **2**, 359–366.

Kareiva, P. 1984 Predator–prey interactions in spatially structured populations: manipulating dispersal in a coccinellid–aphid interaction. *Lect. Notes Biomath.* **54**, 368–389.

Kareiva, P. 1986 Patchiness, dispersal and species interactions. In *Community ecology* (ed. J. Diamond & T. Case), pp. 192–206, New York: Harper & Row.

Kareiva, P. 1987 Habitat fragmentation and the stability of predator-prey interactions. *Nature, Lond.* **326**, 388–390.

Kareiva, P. 1990 The spatial dimension in pest-enemy interactions. In *Critical issues in biological control* (ed. Mackauer, Ehler & Roland), pp. 213–228, Andover, U.K: Intercept Press.

Kareiva, P. & Odell, G. 1987 Swarms of predators exhibit preytaxis if individual predators use area restricted search. *Am. Nat.* **130**, 233–270.

Kierstead, H. & Slobodkin, L. 1953 The size of water masses containing plankton bloom. *J. mar. Res.* **12**, 141–147.

Kishimoto, K. 1982 The diffusive Lotka-Volterra system with three species can have a stable non-constant equilibrium solution. *J. math. Biol.* **16**, 103–112.

Kot, M. & Schaffer, W. 1986 Discrete-time growth dispersal models. *Math. Biosci.* **80**, 109–136.

Kuno, E. 1981 Dispersal and the persistence of populations in unstable habitats: a theoretical note. *Oecologia* **49**, 123–126.

Levin, S. A. 1974 Dispersion and population interactions. *Am. Nat.* **108**, 207–228.

Levin, S. 1976 Population dynamics models in heterogeneous environments. *Ann. Rev. Ecol. Syst.* **7**, 287–310.

Levins, R. 1970 Extinction. *Lect. Math. Life Sci.* **2**, 75–107.

Levins, R. & Culver, D. 1971 Regional coexistence of species and competition between rare species. *Proc. Natn. Acad. Sci.* **68**, 1246–1248.

Ludwig, D., Aronson, D. G., & Weinberger, H. F. 1979 Spatial patterning of the spruce budworm. *J. math. Biol.* **8**, 217–258.

MacCall, A. 1989 *Dynamic geography of marine fish populations.* Seattle: University of Washington Press.

Meinhardt, H. 1982 *Models of biological pattern formation.* New York: Academic Press.

Milne, B. T., Johnston, K. M. & Forman, R. T. T. 1989 Scale-dependent proximity of wildlife habitat in a spatially-neutral Bayesian model. *Landscape Ecol.* **2**, 101–110.

Mimura, M. 1984 Spatial distribution of competing species. *Lect. Notes Biomath.* **54**, 492–501.

Mimura, M. & Kawasaki, K. 1980 Spatial segregation in competitive interaction-diffusion equations. *J. math. Biol.* **9**, 49–64.

Mimura, M., & Murray, J. D. 1978 On a diffusive prey–predator model which exhibits patchiness. *J. theor. Biol.* **75**, 249–262.

Murdoch, W. & Stewart-Oaten, A. 1989 Aggregation by parasites and predators: effects on equilibrium and stability. *Am. Nat.* **134**, 288–310.

Murphy, D. D. 1989 Conservation and confusion: wrong species, wrong scale, wrong conclusions. *Cons. Biol.* **3**, 82–84.

Murphy, D. D. & Wilcox, B. A. 1986 On island biogeography and conservation. *Oikos* **47**, 385–389.

Murray, J. D. 1988 Spatial dispersal of species. *TREE* **3**, 307–309.

Murray, J. 1989 *Mathematical biology.* New York: Springer–Verlag.

Murray, J. D., Stanley, E. A. & Brown, D. L. 1986 On the spatial spread of rabies among foxes. *Proc. R. Soc. Lond.* B **229**, 111–151.

Nachman, G. 1987*a* Systems analysis of acarine predator–prey interactions. I. A stochastic simulation model of spatial processes. *J. Anim. Ecol.* **56**, 247–265.

Nachman, G. 1987*b* Systems analysis of acarine predator-prey interactions. II. The role of spatial processes in system stability. *J. Anim. Ecol.* **56**, 267–281.

Namba, T. 1980 Density-dependent dispersal and spatial distribution of a population. *J. theor. Biol.* **86**, 351–363.

Okubo, A. 1980 *Diffusion and ecological problems: mathematical models.* New York: Springer–verlag.

Okubo, A. 1986 Dynamical aspects of animal grouping. *Adv. Biophysics* **22**, 1–94.

Okubo, A., Murray, J. & Williamson, M. 1989 On the spatial spread of grey squirrels in Britain. *Proc. R. Soc. Lond.* B **238**, 113–125.

Othmer, H., Dunbar, S. & Alt, W. 1988 Models of dispersal in biological systems. *J. Math. Biol.* **26**, 263–298.

Pacala, S. 1986 Neighborhood models of plant population dynamics. IV. single and multispecies models of annuals with dormant seeds. *Am. Nat.* **128**, 859–878.

Pacala, S. W., Hassell, M. P. & May, R. M. 1990 Host–parasitoid associations in patchy environments. *Nature, Lond.* **344**, 150–153.

Pacala, S. & Silander, S. 1990 Field tests of neighborhood population dynamics models of two annual weed species. *Ecol. Monogr.* **60**, 113–134.

Pacala, S. & Silander, J. 1985 Neighborhood models of plant population dynamics. I. Single-species models of annuals. *Am. Nat.* **125**, 385–411.

Pickett, S. & Thompson, J. 1978 Patch dynamics and the design of nature reserves. *Biol. Cons.* **13**, 27–37.

Pimentel, D., Nigel, W. & Madden, J. 1963 Space-time structure of the environment and the survival of host-parasite systems. *Am. Nat.* **97**, 141–166.

Pokki, J. 1981 Distribution, demography, and dispersal of the field vole, Microtus agrestis, in the Tvarminne Archipelago, Finland. *Acta Zool. Fenn.* **164**, 1–48.

Powell, T. M. 1989 Physical and biological scales of variability in lakes, estuaries, and the coastal ocean. In *Perspectives in ecological theory* (ed. Roughgraden, May, & Levin), pp. 157–176. Princeton: Princeton University Press.

Pulliam, R. 1988 Sources, sinks, and population regulation. *Am. Nat.* **132**, 652–661.

Quinn, J. F. & Harrison, S. P. 1988 Effects of habitat fragmentation and isolation on species richness: evidence from biogeographic patterns. *Oecologia* **75**, 132–140.

Quinn, J. F. & Hastings, A. 1987 Extinction in subdivided habitats. *Cons. Biol.* **1**, 198–208.

Quinn, J. F. & Hastings, A. 1988 Extinction in subdivided habitats: reply to Gilpin. *Cons. Biol.* **2**, 293–296.

Quinn, J. F. & Robinson, G. R. 1987 The effects of experimental subdivision on flowering plant diversity in a California annual grassland. *J. Ecol.* **75**, 837–856.

Quinn, J. F., Wolin, C. L. & Judge, M. L. 1989 An experimental analysis of patch size, habitat subdivision, and extinction in a marine intertidal snail. *Cons. Biol.* **3**, 242–251.

Readshaw, J. L. 1965 A theory of phasmatid outbreak release. *Aust. J. Zool.* **13**, 475–490.

Reddingius, J. & de Boer, P. J. 1970 Simulation experiments illustrating stabilization of animal numbers by spreading of risk. *Oecologia* **5**, 240–284.

Reeve, J. D. 1988 Environmental variability, migration, and persistence in host–parasitoid systems. *Am. Nat.* **132**, 810–836.

Reeve, J. 1990 Stability, valiability, and persistence in host-parasitoid systems. *Ecology* **71**, 422–426.

Sabelis, M. W. & Laane, W. E. M. 1986 Regional dynamics of spider-mite populations that become extinct locally because of food source depletion and predation by Phytoseiid mites (Acarina: Tetranychidae, Phytoseiidae). In *Dynamics of physiologically structured populations* (ed. J. A. J. Metz & O. Diekmann), pp. 345–375. New York: Springer–Verlag.

Seger, M. 1989 *Spatial components of plant disease epidemics.* Englewood Cliffs, New Jersey: Prentice Hall.

Shigesada, N. 1984 Spatial distribution of rapidly dispersing animals in heterogeneous environments. *Lect. Notes Biomath.* **54**, 478–491.

Shigesada, N., Kawasaki, K. & Teramoto, E. 1979 Spatial segregation of interacting species. *J. theor. Biol.* **79**, 83–99.

Shigesada, N., Kawasaki, K. & Teramoto, E. 1987 The speeds of travelling frontal waves in heterogeneous environments. In *Mathematical topics in population biology, morphogenesis and neurosciences* (ed. E. Teramoto & M. Yamaguti), pp. 88–97. Springer–Verlag.

Shorrocks, B., Atkinson, W. & Charlesworth, P. 1979 Competition on a divided and ephemeral resource. *J. Anim. Ecol.* **48**, 899–908.

Skellam, J. G. 1951 Random dispersal in theoretical populations. *Biometrika* **38**, 196–218.

Slatkin, M. 1974 Competition and regional coexistence. *Ecology* **55**, 128–134.

Solbreck, C. & Sillen-Tullberg, B. 1990 Population dynamics of a seed feeding bug, *Lygaeus equestris*. 1. Habitat patch structure and spatial dynamics. *Oikos* **58**. (In the press.)

Sousa, W. 1979 Disturbance in marine intertidal boulder fields: the nonequilibrium maintenance of species diversity. *Ecology* **60**, 1225–1239.

Takafuji, A., Tsuda, Y. & Miki, T. 1983 System behaviour in predator–prey interaction, with special reference to acarine predator–prey system. *Res. Popul. Ecol.* **3**, 75–92.

Taylor, A. 1988 Large-scale spatial structure and population dynamics in arthropod predator–prey systems. *Ann. Zool. Fenn.* **25**, 63–74.

Taylor, A. 1990 Metapopulations, dispersal, and predatory-prey dynamics: an overview. *Ecology* **71**, 429–436.

Turchin, P. 1988 Population consequences of aggregative movement. *J. Anim. Ecol.* **58**, 75–100.

Turchin, P. & Kareiva, P. 1989 Aggregation in *Aphis varians*: an effective strategy for reducing predation risks. *Ecology* **70**, 1008–1016.

USDA Forest Service. 1988 *Final supplement to the environmental impact statement for an amendment to the Pacific Northwest Regional Guide.* Portland: USDA Forest Service.

Vance, R. R. 1980 The effect of dispersal *Popul.* on population size in a temporally varying environment. *Theor. Popul. Biol.* **18**, 343–362.

Vance, R. 1984 The effect of dispersal on population stability in one-species, discrete-space population growth models. *Am. Nat.* **123**, 230–254.

Wilcove, D. S. 1987 From fragmentation to extinction. *Nat. Areas J.* **7**, 23–29.

Wilcox, B. A. & Murphy, D. D. 1985 Conservation strategy: the effects of fragmentation on extinction. *Am. Nat.* **125**, 879–887.

Yachi, S., Kawasaki, K., Shigesada, N., & Teramoto, E. 1989 Spatial patterns of propagating waves of fox rabies. *Forma* **4**, 3–12.

Zeigler, B. P. 1977 Persistence and patchiness of predator–prey systems induced by discrete event population exchange mechanisms. *J. theor. Biol.* **67**, 687–713.

Discussion

A. Mullen (*Renewable Resource Assessment Group, Imperial College, London, U.K.*). In Professor Kareiva's paper published in the American Naturalist in 1987 he reported a behavioural mechanism whereby ladybirds tend to aggregate in regions of high prey abundance, through net migration. Wouldn't that stabilizing mechanism create a similar 'halo effect' to that he had described today arising from diffusive instability?

P. Kareiva. You are correct in pointing out that our continuum model of predator aggregation due to area restricted search (Kareiva & Odell 1987) does indeed produce a halo of predators surrounding aphid outbreaks. However, if you look carefully at that model you will see that predator movement has two components: (i) a diffusive component, and (ii) a taxis towards regions of increasing prey density. It is the diffusive component that produces the halo (if there were taxis alone predators would all end up piled on top of local peaks in aphid abundance). In fact, I think it is useful to think of something like diffusive instabilities arising in models with aggregation; in such models it will take more than a small perturbation to escape the inhibition due to predators, in part because the 'flux' or movement component of the model contributes to the inhibition. However, just as is the case with conventional diffusive instabilities, diffusion dissipates some of the inhibition and may allow the emergence of spatial patterning. Finally, it is worth emphasizing that most behavioural-based derivations of predator aggregation will yield partial differential equation models that include both a diffusion term and a taxis term.

R. Southwood (*Department of Zoology, University of Oxford, U.K.*). The model described by Professor Kareiva was also proposed by Readshaw (1964) as the mechanism underlying outbreaks of stick insects in Australian forests.

Reference

Readshaw, J. L. 1964 A theory of phasmatid outbreak release *Aust. J. Zool.* **13**, 475–490.

The regulation of gastrointestinal helminth populations

R. J. QUINNELL[1], G. F. MEDLEY[2] AND A. E. KEYMER[1]

[1] *Department of Zoology, University of Oxford, South Parks Road, Oxford OX1 3PS, U.K.*
[2] *Department of Biology, Imperial College of Science, Technology and Medicine, Prince Consort Road, London SW7 2BB, U.K.*

SUMMARY

One quarter of the world's human population suffers infection with helminth parasites. The population dynamics of the ten or so species, which cause disease of clinical significance have been well characterized by epidemiological field survey. The parasites are in general highly aggregated between hosts, and their populations seem to be temporally stable and to recover rapidly from perturbation, including interventions designed to alleviate disease.

This paper reviews current understanding of the population regulation of helminth species of medical significance. Both empirical (field and laboratory) and theoretical results are included, and we attempt to interpret the findings in the broader context of the population ecology of free-living species. We begin by considering the evidence for regulation from field data concerning the temporal stability of helminth populations within communities and from the results of perturbation experiments. The detection of regulatory processes is then discussed (with regard to statistical and logistical considerations), and the evidence from both the field and laboratory studies reviewed.

Deterministic models are described to investigate the possible consequences of regulation imposed at different points in the parasite life-cycle. The causes and consequences of parasite aggregation are considered, and a stochastic model used to investigate the impact of different combination of regulatory processes and heterogeneity generating mechanisms.

1. INTRODUCTION

Helminths are very common parasites of humans, infecting a sizeable proportion of the population; *Ascaris*, for instance, is estimated to infect some 1000 million people. The most abundant species are *Ascaris lumbricoides*, the hookworms *Necator americanus* and *Ancylostoma duodenale*, *Trichuris trichiura*, *Enterobius vermicularis* and *Schistosoma* spp., and it is on these that the paper concentrates.

There is a large amount of literature associated with the epidemiology of these species (see, for example, Anderson & May (1985*a*); Bundy & Cooper (1989); Crompton *et al.* (1985); Rollinson & Simpson (1987)). There are, however, noticeable gaps in the available data. These arise mainly from the ethical requirement to treat people where possible, thus making long-term observational studies unacceptable, and the necessity to spend considerable effort persuading subjects to collaborate, thus reducing sample sizes. Experimental studies on humans are, to a large extent, impossible. Consequently we have taken examples from helminths of non-human animals (both laboratory and agricultural), in an attempt to derive conclusions concerning the population dynamics of human infections.

We begin by examining the available evidence for the stability of helminth populations from long-term population studies and perturbation experiments. We discuss the regulatory mechanisms that may be operating in the light of experimental systems that manipulate host nutrition and host immunity and studies of complete parasite transmission systems. The difficulties associated with the detection of density dependence in the field are considered with particular respect to helminth fecundity, size and the host immune response. Finally, we present some preliminary results from a simulation model that examines the interaction between the generation of heterogeneity in helminth burdens between hosts and different density-dependent mechanisms.

2. EVIDENCE FOR POPULATION REGULATION

Evidence for the regulation of helminth populations has come largely from long-term studies of helminth infection without intervention, and from perturbation experiments, in which the population of a helminth parasite is reduced by antihelminthic treatment and the population size followed after the cessation of control.

(*a*) *Long-term population studies*

The persistence of populations for long periods of time is clearly suggestive of regulation. Ethical considerations mean that there have been few long-term studies of human parasite populations without intervention. Such studies often show a remarkably constant population size over time, both within host populations and within individual hosts (Anderson & May 1985*a*; Anderson 1986). One of the longest runs of data is of the prevalence of *Taenia saginata* in its cattle intermediate host in Kenya. This shows a more variable pattern, the prevalence of heavy infections varying between 0.5% and 19.1% during a 55-year period (Froyd 1965).

Evidence for stability from long-term studies of helminth populations in non-human animals is less conclusive. Although temporal constancy is often observed (Anderson 1979), there are also studies which have recorded large temporal changes in parasite population size. For instance, epizootics of acanthocephalans of many species have been reported (reviewed by Nickol (1985)) and Montgomery & Montgomery (1990) recorded large changes in the intensity of nine helminth species in wood mice *Apodemus sylvaticus* over a five-year period. In each of six study sites at least one helminth species was lost or gained.

(*b*) *Perturbation experiments*

The strongest evidence for the regulation of animal populations comes from perturbation experiments (Murdoch 1970). Parasitologists are thus fortunate that, at least for human helminth infections, a considerable body of data on the effects of perturbation is available (Anderson & May 1985*a*). These data come from studies of the effect of antihelminthic treatment on all the five major gastrointestinal helminth infections of man.

Mathematical modelling of reinfection has revealed several factors that will influence the rate of reinfection (Anderson & Medley 1985).

(1) Reinfection is related to the lifespan of the parasite, with short-lived parasites showing the greatest percentage reinfection.

(2) Percentage reinfection will be greater in children than in adults, since the time over which reinfection occurs will be a proportionately larger fraction of the time children had been exposed to infection before treatment.

(3) Reinfection rate will depend on both the basic reproductive rate of the parasite and on the degree of reduction in work burden achieved by drug treatment (i.e. on the efficacy of the drug and the proportion of the population treated).

(4) Variation in parasite transmission with time will clearly be important. This may result from the provision of increased sanitation at the time of treatment (see, for example, Hill (1926)) or from climatic variation affecting transmission (Wilkins 1989).

Figure 1 shows the results from studies of reinfection with *Ascaris*, hookworm and schistosomes. The choice of study was limited to those that encompassed the whole age range of the population. In all studies treatment was followed by relatively rapid reinfection, and the intensity in some cases approached the pretreatment level after several months or years. Unfortunately, none of the studies was continued for a long enough period to show conclusively that intensity returns to, and is maintained at, the pretreatment level.

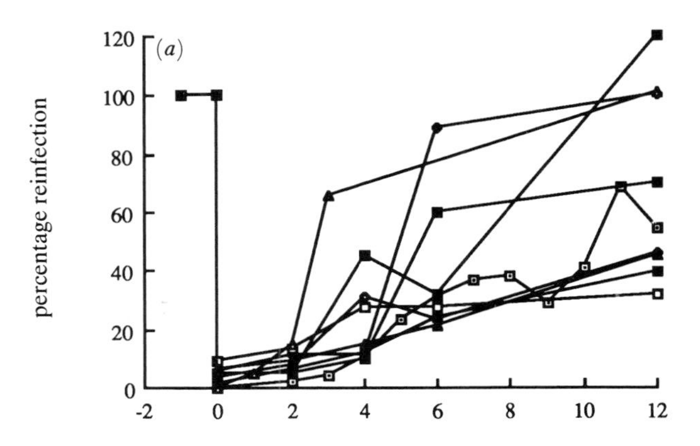

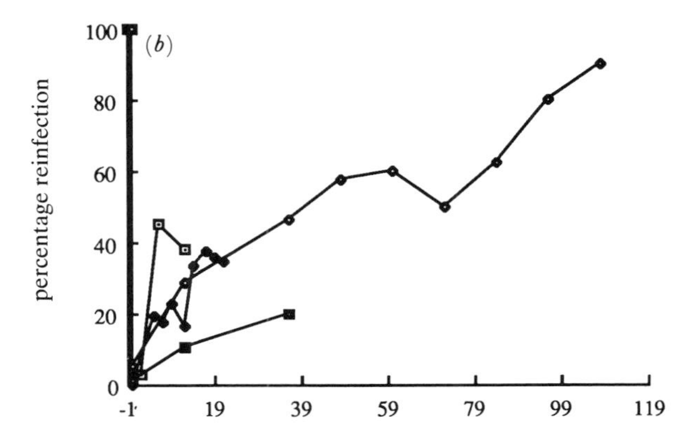

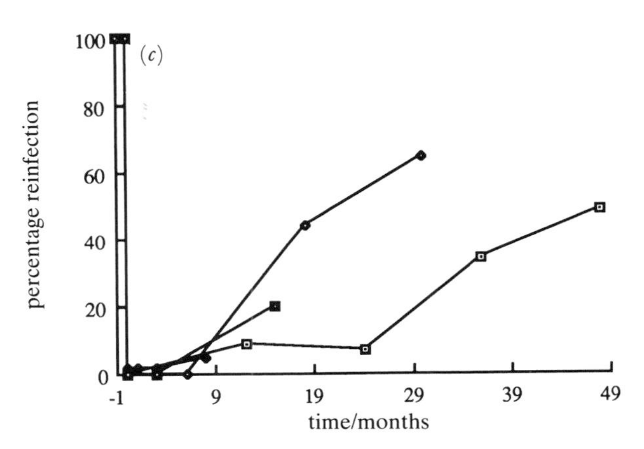

Figure 1. Reinfection after treatment of human helminth infection. Proportion reinfection was assessed as the proportion of the pretreatment faecal egg count (or worm burden, Seo *et al.* (1980)), or as the proportion of the faecal egg count of an untreated control group (Roux *et al.* 1975; Duke & Moore 1976). (*a*) *Ascaris* infection. Data from Arfaa & Ghadirian (1977); Croll *et al.* (1982); Seo *et al.* (1980); Thein-Hlaing *et al.* (1987). (*b*) Hookworm infection. Data from Hill (1926); Sweet (1925); Schad & Anderson (1985); Haswell-Elkins *et al.* (1988). (*c*) Schistosome infection. Data from Roux *et al.* (1975); Duke & Moore (1976); Wilkins *et al.* (1987); Sturrock *et al.* (1987). The first three studies were of *Schistosoma haematobium*, the last of *S. mansoni*.

Repeated measurements of parasite intensity, such as those shown in figure 1, can only be done indirectly using faecal egg counts. These may not always accurately reflect intensity, as parasite fecundity has

Table 1. *Reinfection with gastrointestinal helminths*

(The initial intensity of infection, the percentage reinfection after a certain number of months and the age range of the host population studied. The estimated lifespan of the parasite (from Anderson & May (1985*a*)) is given in years.)

references	species	initial intensity	percentage reinfection	time (months)	age (years)	lifespan (years)
Haswell-Elkins *et al.* (1988)	Hookworm	2.2	35	11	—	2–4
Bundy *et al.* (1985)	*Trichuris*	114.9	173	7	2–8	1–3
Bundy *et al.* (1987)	*Trichuris*	54.2	44	17	—	—
Holland *et al.* (1989)	*Ascaris*	11	32	6	5–16	1–2
Thein-Hlaing *et al.* (1987)	*Ascaris*	10.7	37	6	—	—
	Ascaris	8.7	76	12	—	—
Elkins *et al.* (1988)	*Ascaris*	9.9	55	11	—	—
Forrester *et al.* (1990)	*Ascaris*	10	120	6	2–10	—
Seo *et al.* (1979)	*Ascaris*	1.2	33	6	—	—
Seo *et al.* (1980)	*Ascaris*	0.6	5	2	—	—
		1.4	45	4	—	—
		1.3	32	6	—	—
		0.8	120	12	—	—
Haswell-Elkins *et al.* (1987)	*Enterobius*	25.7	156	11	—	0.1–0.2

been observed to increase after treatment (Haswell-Elkins *et al.* 1988; Elkins *et al.* 1988). Studies using only faecal egg counts may thus overestimate the amount of reinfection. Accurate measurement of parasite intensity can only be achieved by destructive sampling of the parasite. Details of those studies which incorporate direct worm counts both before and after treatment are given in table 1. Again, rapid reinfection occurred in all studies, but the timescale was generally too short to demonstrate a return to pretreatment levels. The degree of reinfection was clearly related to the lifespan of the parasite, and in two cases reinfection burdens exceeded those pretreatment.

The available evidence is thus consistent with the hypothesis that helminth populations are regulated, and the universal rise in parasite burdens after treatment is strongly suggestive of the relaxation of density-dependent constraints on population growth. The degree of population stability is less obvious, and there is a clear need for more long-term studies of reinfection after chemotherapy.

3. REGULATORY MECHANISMS

Density-dependent constraints are usually considered to act on the parasitic stages of the life cycle, where they may influence parasite establishment, maturation, survival or fecundity. Free-living stages and transmission may also be subject to density-dependent constraints, although these have received little attention.

Several possible causes of density dependence have been identified. The most important of these are likely to be intraspecific competition for resources, such as nutrients or space, and the effects of the host immune response. Additionally, parasite-induced host mortality and host resistance due to parasite-induced pathology may also be density dependent; the former, however, is probably of little importance in human hosts, and the latter is probably only significant in trematode (fluke) infections. Most helminth infections of humans appear to be regulated by either competition or immunity.

The mechanism underlying density dependence may have important epidemiological consequences. The degree of intraspecific competition will depend on current parasite density, which can be approximated by parasite burden. The magnitude of the host immune response, however, will depend not on parasite burden, but on the host's cumulative exposure to infection (strictly speaking on the cumulative exposure weighted according to how long ago exposure occurred). The population dynamics produced by regulation by the host immune response can thus be very different to those due to competition. All forms of density dependence acting within the primary host and related to current worm burden produce only stable dynamics, with one non-zero equilibrium point. When dependence on some measure of cumulative past exposure is introduced (essentially a distributed time delay), oscillatory behaviour may result (Anderson & May 1985*b*).

Many laboratory studies have shown density dependence, typically by examining the effects of parasite density in primary infection. However, the results from primary infections do not allow inferences to be made about the causative mechanisms of density dependence. In a primary infection parasite density will be highly correlated with exposure to the parasite; whatever the generative mechanism, similar patterns will be expected. Disentangling the relative importance of competition or immunity requires experimental protocols that manipulate either the resources available to the parasite or the host's immune response.

(*a*) *Manipulation of nutrient supply*

As endoparasites obtain all their nutrients from the host, their nutrient supply can be readily manipulated by altering the host's diet. This is likely to have a marked and immediate effect on gastrointestinal parasites which obtain nutrients from the intestinal contents, such as cestodes, acanthocephalans and some nematodes. It may have less effect on parasites that feed on host tissues.

The effects of host diet on density dependence have been best studied on infections of the tapeworm *Hymenolepis diminuta* and the acanthocephalan *Moniliformis moniliformis* in rats. These studies have concentrated on manipulation of the carbohydrate content of the diet, and have shown that the effects of increasing parasite density can be mimicked by the feeding of a carbohydrate-deficient diet (Read 1959; Keymer *et al.* 1983).

Evidence for a role of competition for food in the regulation of nematode and trematode infections is less clear. The feeding of nutrient-deficient diets (such as protein, protein-energy or vitamin A deficient diets) to schistosome infected mice is known to result in decreased worm burden, worm size and worm fecundity, but how this relates to density dependence is not clear. In contrast, many nematode species generally do better in hosts fed protein-deficient diets, presumably as a result of impaired host immunity (Dobson & Bawden 1974; Slater & Keymer 1986).

(*b*) *Manipulation of host immunity*

A large number of studies have shown that there is an immune response to helminths, and that the immune response can act to decrease worm establishment, development, survival and fecundity (reviewed by Wakelin (1986)). The magnitude of the immune response is dependent on the immunising dose, and thus immunity clearly has regulatory potential.

If density dependence is due at least partly to immunity, then its severity should be reduced in immunocompromised hosts, and increased in immunized hosts. *Heligmosomoides polygyrus* in the laboratory mouse is an ideal system for the study of the effects of host immunity, as adult worms immunosuppress the host and the immune response is ineffective in primary infection. In many strains of mice there appear to be no density-dependent constraints during primary infection, and there may in fact be inverse density dependence: the survival of worms is prolonged in heavy infections, as the degree of immunosuppression increases with parasite burden (Robinson *et al.* 1989). Similarly, Slater & Keymer (1986) found no density-dependent constraints on parasite population growth during repeated infection of protein-malnourished hosts which lack a functional immune response. In contrast, populations in well-nourished hosts were strongly regulated by the host immune response.

(*c*) *Studies of complete systems*

Density dependence during a primary infection in the laboratory may have little relevance to the population dynamics of the parasite in the field. Yet it is important to identify the underlying mechanisms in the field, and the stages in the parasite life cycle they act on. One approach to this problem is to study population dynamics during repeated infection. If the dynamics of different parasitic life-cycle stages are known, then mathematical modelling may allow an assessment of the importance of regulatory processes, and the mechanism behind them. This approach has been successfully used to study two cattle nematodes, *Ostertagia ostertagi* (Grenfell *et al.* 1987) and *Haemonchus contortus* (Smith 1988).

Table 2. *Empirical descriptions of density-dependent constraints on the parasitic stages of* Ostertagia ostertagi *and* Haemonchus contortus — *the factors important in influencing parasite establishment, arrestment, mortality and fecundity during repeated infections*

(E, cumulative exposure to infective stages; t, time since start of infection; N, parasite burden (density). From Grenfell *et al.* (1987); Smith *et al.* (1987) and Smith (1988))

	O. ostertagi	*H. contortus*
establishment	e^{-at}	$\frac{e^{(a-bt)}}{1+e^{a-bt)}}$
arrestment	none	$a+bt$
mortality	$a+bE$	$a+bE$
fecundity	e^{-aNt}	—

The results of these studies are shown in table 2. The mechanisms behind the regulatory processes can be inferred from their dependence on parasite density (implying competition) or exposure (implying immunity). Mortality and fecundity appear to be exposure-dependent, suggesting that the host immune response is of overriding importance. Surprisingly, establishment and arrestment appear to be independent of both current density and exposure, although they are time dependent.

Models based on laboratory data have also been produced for *Taenia hydatigena* and *Echinococcus granulosus* infection. The only regulatory process identified in *E. granulosus* infections is the immune response in the ovine intermediate host, and there appears to be no density-dependence in the canine final host (Gemmell *et al.* 1986). *T. hydatigena* can be regulated in the final host by density-dependent constraints on growth and fecundity, but in the field immunity in the intermediate host apparently keeps worm burdens below the level at which these constraints start to operate (Gemmell *et al.* 1987). Neither species affects the mortality of either the final or intermediate host. Models incorporating intermediate host immunity as the only regulatory process give an accurate representation of population dynamics in the field, and the consequences of intervention (Roberts *et al.* 1986, 1987).

4. THE DETECTION OF REGULATORY PROCESSES IN THE FIELD

The mechanism responsible for regulation will have important consequences for the detection of density dependence in the field. Since cumulative exposure to helminths has not been measured in the field, the detection of regulatory processes has relied on the demonstration of dependence on current parasite density. This will allow the detection of regulation caused by intraspecific competition. However, the results of laboratory experiments discussed above suggest that the host immune response may be the most important regulatory mechanism, particularly for nematode populations. Detection of regulation by host immunity will only be possible if there is a correlation between current parasite density and exposure.

The relation between density and exposure is likely to be complex, and to depend on such factors as overall transmission rates and host age. A host with a low worm burden may either have been exposed to few worms (and have no immunity) or have developed an effective immune response to past exposure. Only in very young hosts, which can be considered as experiencing a primary infection, are density and exposure likely to be closely related. An additional complicating factor is the likelihood that there will be genetic heterogeneity among hosts in their ability to mount an effective immune response (reviewed by Wakelin (1988)). In this case, the host's immune response may not be related to exposure in a consistent manner. High exposure will lead to high worm burdens and yet little immunity in low responder hosts, but low worm burdens and high immunity in high responders. There may thus be a negative correlation between current density and host immunity. If parasite fecundity, survival or size is related to immunity, a positive relation between these factors and worm burden would then be expected (Keymer & Slater 1987). Such a positive relation between fecundity and worm burden has been demonstrated in a captive population of wood mice exposed to natural infection with *H. polygyrus* (Quinnell 1990).

(*a*) *Density-dependent fecundity*

The detection of density-dependent processes operating on free-living organisms comes mainly from the analysis of life-table data (reviewed by Sinclair (1989)). The collection of life-table data requires assessments of the numbers of various different life-cycle stages at various time points. For helminth species the only stages of the life cycle that can be quantified are adults (and then only at a single time point) and the total production of eggs or larvae from a single host. Life-table analysis is thus restricted to examining the effect of adult parasite density on parasite fecundity.

A negative correlation between average fecundity and worm burden has been reported for all the major helminth parasites of humans (Anderson & May 1985*a*). However, the variables in such an analysis are not independent, and errors in the measurement of worm burden may bias the data towards a negative correlation. In the field such errors are likely to be large (Anderson & Schad 1985), and Monte Carlo simulations show that artefactual density dependence can be produced (R. J. Quinnell & G. F. Medley, unpublished data).

Problems such as these in the detection of density-dependence in field data can be minimised by the use of regression analysis (Varley & Gradwell 1960; Eberhardt 1970). In the present case, the regression should be that of total egg production on worm burden, which are measured independently. If estimates of the errors in the two variables can be made,

Table 3. *The analysis of density-dependence data*

(Data on human hookworm fecundity (eggs/g faeces) from Karkar I., Papua New Guinea, were analysed (*a*) by correlation between mean fecundity and hookworm burden and (*b*) by regression of total egg production on hookworm burden. All data were log(x) transformed before analysis, and only egg- and worm-positive data were included.)

(*a*) Analysis of mean fecundity data. Pearson's product-moment correlation coefficient (r) for the relationship between mean fecundity and hookworm burden (either females only or total).

	n	r	p
females	93	−0.309	0.0026
total	93	−0.299	0.0036

(*b*) Regression analysis. The slope (b) and 95 % confidence limits of the regression line between total egg production and hookworm burden, estimated from the structural relations model with an error variance ratio of 3.05.

	n	b	95 % c.l.
females	93	0.953	0.628–1.30
total	93	0.828	0.526–1.14

the structural relations model will give a maximum likelihood estimate of the slope of the regression line (Sprent 1966; Rayner 1985). Error variances have recently been estimated from a study of human hookworm infection (Pritchard *et al.* 1990). The structural relations model provided no evidence for density-dependent fecundity, despite a significant negative correlation between average fecundity and worm burden (table 3) Analysis of a further five data sets did not produce firm evidence for density dependence, although this analysis was hampered by the lack of estimates of error variances from these studies.

An alternative approach to the detection of density-dependent fecundity is to use an independent measure of worm fecundity, such as the uterine egg content of female worms. This approach has been used in experimental studies of helminth infection (see, for example, Coadwell & Ward (1982); Shaw & Moss (1989)). Its use in the field has been limited to a single study of hookworm infection, when there was no relation between uterine egg count and parasite density (R. J. Quinnell *et al.* unpublished data).

The available data thus do not provide any firm evidence that fecundity is dependent on current parasite density. It is possible that further studies involving estimates of error variances will allow the detection of a significant relation. However, variation between hosts and between parasites, coupled with the measurement errors discussed above, is likely to make detection difficult (Medley 1989). Either way, the possibility that the fecundity of human helminths in the field is dependent not on current parasite density, but on previous exposure to the parasite, should not be ignored. It is also possible that regulatory processes may not act on fecundity, but rather on other stages of the parasite life cycle.

(*b*) *Parasite size*

The relation between worm burden and worm size or mass in the field is simple to examine. Although density-dependent constraints on parasite size need not affect population dynamics, they are likely to be associated with effects on parasite development rate, mortality or fecundity. Several studies have investigated the relation between worm size and worm burden in humans (*Ascaris* (Mello 1974; Cho 1977; Martin *et al.* 1983; Elkins & Haswell-Elkins 1989); *Necator* (R. J. Quinnell *et al.* unpublished data); *Trichuris* (Burrows 1950)). In none of these studies was there a significant relation.

(*c*) *The detection of host immunity*

In view of the importance of host immunity in laboratory infections, it may not be surprising that direct effects of parasite density are not detectable in the field. Attempts to detect immunity in the field have taken three forms. Epidemiological evidence for immunity can be obtained from studies of reinfection rates, providing some estimate of current exposure is made. This has provided strong epidemiological evidence for the role of immunity in schistosome infections (reviewed by Hagan 1987; Butterworth *et al.* 1988*a*), as has a comparison of age-intensity curves from different areas (Anderson 1987). Evidence for immunity to other parasite species is weak and equivocal (reviewed by Anderson 1986; Behnke 1987; Bundy 1988).

Direct immunological correlates of infection or resistance to reinfection have been sought in studies of schistosomes (see above reviews) and hookworm (Pritchard *et al.* 1990). Such studies have mostly failed to show a clearly protective immune response; immunological parameters generally reflect, rather than determine, worm burdens. The complexity of the antibody response to helminth infection, which may include protective, non-protective and blocking antibodies to a wide range of antigens, makes the detection of protective responses difficult. Increasing knowledge of anti-schistosome responses has recently led to the identification of possibly protective anti-schistomulum IgG antibodies (Butterworth *et al.* 1988*b*).

Finally, it is possible, at least for non-human animals, to obtain an empirical measure of host resistance by challenging animals taken from the field. This technique has been used to quantify resistance to reinfection with *Heligmosomoides polygyrus* after either natural infection (Slater 1988) or trickle infection (Keymer *et al.* 1990). Further experiments on the same parasite in wood mice have shown a negative correlation between resistance to reinfection and the worm burden of naturally infected mice, indicating a role for the immune response in determining parasite burdens in a semi-natural situation.

5. AGGREGATION OF PARASITES WITHIN HOSTS

Heterogeneity in worm burdens within human hosts is ubiquitous. As noted previously, any process that depends on the degree of aggregation (as all density-dependent processes will) implies that consideration must be given to the distribution pattern of worms among hosts. This distribution must be viewed as a dynamic as opposed to a static entity, created by a plethora of factors derived from hosts, environment and parasites. Chemotherapy will inevitably alter the distribution. We examine the interaction between heterogeneity and different density-dependent processes by using a simulation model in which the distribution of worms is created rather than assumed to be of some particular form.

A recent study (Guyatt *et al.* 1990) has shown that the aggregation pattern of *Ascaris* in different communities can be described by using the negative binomial distribution with the degree of aggregation decreasing with increasing mean worm burden. The fact that there is a consistent pattern of heterogeneity in different communities leads to the tentative hypothesis that the generative factors are to be found in the biological association between parasite and host and that they are independent of specific environmental factors.

(*a*) *Consequences of heterogeneity*

One immediate effect of heterogeneity is that clinical disease, which is thought to be positively associated with worm burden, is limited to a small proportion of the human population. By using an estimate of the distribution of the worm population throughout the host population, H. Guyatt & D. A. P. Bundy (unpublished data) suggest that this empirical relation can be used to rank communities according to the level of disease caused by *Ascaris* on the basis of prevalence data alone.

More difficult to study is the effect that heterogeneity has on the worm population itself. Most worms live in environments populated with other worms of the same species. The population genetic consequences of aggregation have not been considered in any detail, although Anderson *et al.* (1989) show that it may have a significant effect on the evolution of resistance against chemotherapy. Depending on the methods of infection, it is possible that there is a significant degree of genetic relatedness between worms within an individual host. This aspect has not been considered to our knowledge.

(*b*) *The effect of heterogeneity generation on dynamics*

The details and some results of the simulation model are available elsewhere (Anderson & Medley 1985; Medley 1988, 1989). One of the important results is that if differences in exposure to infective stages is assumed to generate the aggregation observed in adult parasites, then the exact mechanism can significantly alter the dynamics of the parasite population in the

Table 4. *A brief summary of the assumptions associated with the simulation model discussed in the text*

(Four parasite developmental stages are assumed: immature and mature (within host), free-living uninfective stage and free-living infective stage. The free-living stages are modelled deterministically, and the remaining rates are modelled stochastically using the Monte Carlo technique. The processes are summarized below, where $M_{(i,t)}$ is the number of mature worms in individual host i at time t.)

process	assumption
mature worm death	constant mean rate with 1 year life-expectancy
immature worm death	constant mean rate with 1 year life-expectancy
maturation	density-independent:
	constant mean rate with 6 week average maturation time.
	density-dependent:
	mean rate given by $\sigma \exp\{-a.M(i,t)\}$, where σ is the pristine maturation rate (= 1–6 weeks) and a is a parameter chosen to give the required mean worm burden
fecundity	density-independent:
	constant mean rate with 5 799 500 eggs per week
	density-dependent:
	mean rate given by $\lambda \exp\{-\gamma M(i,t)\}$ where λ is the pristine fecundity rate and γ is a parameter
transmission	
transmission is divided into two processes: contact with and pick-up of infective stages. The contact rate is calculated as $\beta.I(t).s_i$, where $I(t)$ is the number of infective stages at time t and β is a transmission coefficient	
susceptibility:	the s_i are random numbers drawn from a gamma distribution with mean unity, and a single worm is picked up on each contact
environmental:	the s_i are all unity, and the number of worms establishing is drawn from a logarithmic distribution with mean unity
host population	constant at 250 hosts
free-living uninfective death	constant mean rate with mean life expectancy 8 weeks
free-living uninfective maturation	constant mean rate with mean time to maturation 3 weeks
free-living infective death	constant mean rate with mean life expectancy 8 weeks

face of perturbation (= chemotherapy). We do not repeat a description of the structure of the simulations here, but include a compendium of relevant biological assumptions shown in table 4. The results presented here are by no means exhaustive, but are intended to make a few salient points regarding patterns that may arise, and data that may provide clues to population dynamic mechanisms.

The generation of heterogeneity can have two causes: differences in exposure to infective stages, and differences in parasite population parameters within different hosts. The former mechanism encapsulates a spectrum, from which we take two illustrative examples. The first heterogeneity generating hypothesis, termed susceptibility, assumes that the immigration rate varies systematically between hosts. Each host has a factor designated at birth and constant thereafter which determines that host's relative resistance/susceptibility to infection compared with the remainder of the population. The second heterogeneity-generating hypothesis, termed environmental, assumes that the immigration rate does not differ systematically between hosts. Each host will experience the same average immigration rate over time, but at any point the values between hosts will vary randomly. Essentially, noise is added to immigration.

The alternative possibility (termed host variation) is that heterogeneity is generated by differences between hosts in their ability to mount effective, density-dependent, immunological/pathological responses to immature parasites. The average rate of maturation, σ, is modified by the factor exp $\{-s_i M(i,t)\}$, where $M(i,t)$ is the number of mature worms in individual i and time t, and s_i is a factor peculiar to individual host i drawn from a log-normal distribution. In this case we assume that there is no heterogeneity (other than that arising from chance) in host exposure to infectives. Note also that we assume an effect of the current worm burden, not some measure of accumulated past exposure.

We combine these heterogeneity-generating mechanisms with two choices for the site of density dependence: fecundity and maturation. Maturation is defined here as the transition from one stage of development to the next, and not necessarily confined to the development of sexual maturity. For example, the major density-dependent effects may occur during some phase of migration through the host tissues. Only two parasite stages are considered: immature and mature. Thus, we have five different models: susceptibility plus fecundity or maturation density dependence, environmental plus fecundity or maturation density dependence and host variation in maturation density dependence (the combination of host variation on density-dependent fecundity will result in population regulation, but not heterogeneity.

(*c*) *Results*

The results are summarized with respect to two aspects: the equilibrium situation which may be detected through a single, horizontal survey of an infected community, and the response of the parasite population to a single round of chemotherapy (perturbation).

At equilibrium, the two features of interest are the relative numbers of immature and mature worms within individual hosts, and their frequency distribution between hosts. When density-dependent maturation is operating equally for all hosts regardless of which hypothesis is generating heterogeneity (susceptibility or environment), the number of immature worms is greater in relation to the number of mature worms than when maturation is constant (figure 2). When maturation occurs at a constant rate, the ratio of mature to immature worms within individual hosts is approximated by the ratio of the maturation rate to the adult worm death rate, and will be typically less than unity. If the maturation rate decreases with increasing mature worm burden then not only will the ratio be much higher, but will increase with increasing mature worm burdens. If there is significant variation in the abilities of hosts to delay maturation, then a reverse pattern is seen: those hosts with few mature worms have few because they prevent maturation, and consequently have higher immature burdens and vice versa.

The introduction of density-dependent maturation tends to decrease the frequency of hosts with very high burdens, and decrease the numbers of hosts with zero and low worm burdens. When exposure to infective stages is high, a large number of immature worms accumulates in response to the high number of mature worms. As these mature worms die, they are replaced from the pool of immatures, consequently generating a mode of observations close to the mean. This has not been the typical observation from the field, and shows that either some modification of survival of immature or mature worms is likely to complement or replace density dependence within the life cycle. During the exposure to infective stages is more extreme than assumed here.

Figure 3 shows differences in the recovery of the parasite population with different assumptions following a single round of chemotherapy. After an initial period of five years, 80% of the human community was chosen at random, and all the worms (mature and immature) removed from them. Density-dependent maturation induces faster recovery of the parasite population than when density-dependent fecundity is the limiting process. This is because of the site of density dependence within the life cycle. During the first 1–2 years following chemotherapy, those individual hosts that will develop large worm burdens can do so without density-dependence acting. Also, when large mature burdens are developed in some hosts the rate of recovery in the remaining hosts is unaltered, unlike the effect of fecundity. Thus the slower rate of recovery for the case of susceptibility generated heterogeneity with density-dependence fecundity.

The reacquisition of heterogeneity causes the two slowest population recovery rates, host variation and environmentally generated heterogeneity with density-dependent fecundity. Environmental heterogeneity

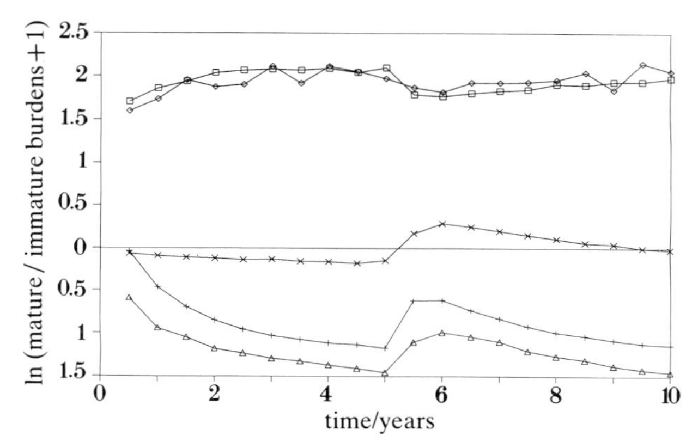

Figure 2. The ratio of mature:immature worms in response to a single round of random, mass chemotherapy. Note the logarithmic scale. The chemotherapy was applied at year five, after five years of equilibration, and killing all worms (immature and mature) in 80% of hosts chosen at random. Each line is the ratio of the mean mature to mean immature burdens of 10 simulations of 250 hosts each. The lines are: (*a*) (□), susceptibility generated heterogeneity, density-dependent fecundity; (*b*) (+), susceptibility generated heterogeneity, density-dependent maturation; (*c*) (◇), environmental generated heterogeneity, density-dependent fecundity: (*d*) (△), environmental generated heterogeneity, density-dependent maturation, and (*e*) (×), heterogeneity generated by variation in maturation rates.

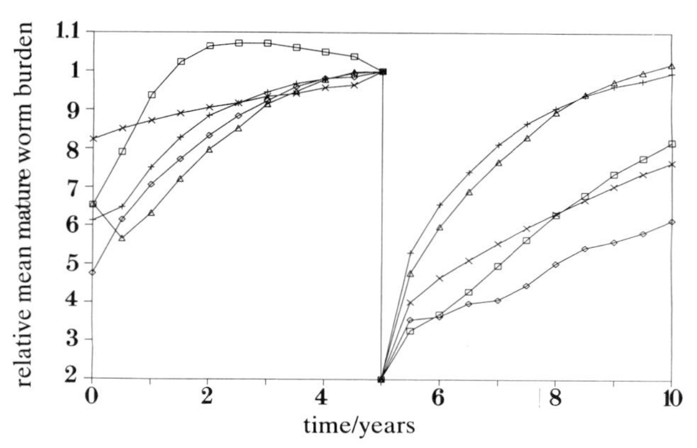

Figure 3. The rate of recovery of the mature parasite population following a single round of random mass chemotherapy. The chemotherapy regime applied was as shown in figure 2. Each line is the mean of 10 simulations of 250 hosts each, and have been adjusted by a factor such that the mature burden at chemotherapy is unity. The lines are: (*a*) (□), susceptibility generated heterogeneity, density-dependent fecundity; (*b*) (+), susceptibility generated heterogeneity, density-dependent maturation; (*c*) (◇), environmental generated heterogeneity, density-dependent fecundity; (*d*) (△), environmental generated heterogeneity, density-dependent maturation, and (*e*) (×), heterogeneity generated by variation in maturation rates.

immediately reintroduces the same degree of overdispersion (compared with susceptibility generated heterogeneity in which those hosts with large burdens must reacquire them over time). Likewise, in the host variation case, the heterogeneity is generated immediately by different maturation rates between hosts. Consequently, density-dependent restraint is reintroduced more quickly.

Figure 2 shows the response of the ratio of mature to immature worm burdens through the same chemotherapy régime as previously applied. Those cases where density-dependent maturation is operating show an increase in the ratio following chemotherapy, as opposed to a decrease when parasite development is constant. This is again because of the large number of immature worms available for maturation when relatively small mature burdens are removed.

Increasing our knowledge of the 'demography' of the parasite within the definitive host will greatly increase our knowledge of the population dynamics of the parasite. Some progress has been made in this direction by looking at the size distribution of parasites and the changes induced by chemotherapy (Elkins & Haswell-Elkins 1989). These data show a dramatic shift in weight distribution of *Ascaris* in children after chemotherapy. At treatment, smaller worms outnumber larger worms, although the reverse is true following nine months reinfection. Interestingly, this change does not occur in adults. This may suggest that some form of density-dependent maturation or development is occurring in children but not adults.

By considering the developmental stages within the definitive host, both in cross-sectional and perturbation surveys, some conclusions may be drawn regarding the population dynamics of the parasites. One remaining problem is the definition of immature worms in this context. It is likely that they constitute the tissue migratory stages rather than sexually immature stages within the gut lumen. Consequently, indirect methods of investigation will be required in human hosts, although direct investigations on animal models may be useful. It is also possible that developmental stages may be distinguished by the molecular markers that they exhibit. What is required is a molecular or immunological score of intensity of infection that is specific to a stage in the parasite's development and which is quantitative enough to distinguish relatively small changes in worm burden.

6. CONCLUSIONS

This brief review of the evidence suggests two main conclusions. First, there is strong evidence that many helminth populations are subject to regulation, especially in human hosts (Anderson & May 1985*a*). Even so, there is a need for more field studies: for instance, there is evidence for the stability of only one acanthocephalan species (Kennedy 1985). Secondly, evidence for the importance of particular regulatory mechanisms or processes is much less conclusive.

An important difference between the ecology of parasites and that of free-living organisms is the influence of host defences, particularly the immune response. If the host immune response is important, there may be very different ecological and evolutionary constraints on parasitic helminths than on their free-living counterparts. Laboratory studies suggest that immunity can be of overriding importance, yet our knowledge of the importance of immunity in the field is so sketchy that we cannot rule out the possibility that it may be largely irrelevant (Wakelin 1984). This lack of knowledge stems partly from the difficulties in assessing immunity, particularly in human hosts, but this is because of the relatively little attention that has been paid to the identification of the regulatory processes affecting helminth populations. Although there have been a large number of laboratory studies, very few studies have attempted to combine field and laboratory work.

Knowledge of the mechanisms regulating helminth populations is also of considerable applied importance, both for the design of chemotherapy control programmes (Anderson & May 1985*b*) and for the development of vaccines. Combined with simple economic arguments, our current, limited understanding of helminth population dynamics is enough to aid in the design of control programmes. However, it is important to continue to collect appropriate data, both to satisfy academic curiosity and to improve health-care policies.

Discussion

P. J. Grubb (*Botany School, Cambridge University, U.K.*). Do I understand that the failure to detect density dependence in the fecundity of the intestinal worms in wild populations results from the great variation in background conditions, e.g. densities of competitors, or is it overwhelmingly a result of variable control by the immune response system?

A. E. Keymer. The detection of density-dependent fecundity will always be hampered by unavoidable errors in the assessment of both fecundity and density, and by variability in both host (genetics, nutrition, presence of other parasite species, etc.) and parasite populations. Unfortunately, it is not possible to decide between the two possibilities outlined by Dr Grubb: that the failure to detect density-dependence is only because of these effects, or that the effect of the host immune response means that there really is no relationship between fecundity and current parasite density.

REFERENCES

Anderson, R. M. 1979 The influence of parasitic infection on the dynamics of host population growth. In *Population dynamics* (ed. R. M. Anderson, B. D. Turner & L. R. Taylor), pp. 245–281. Oxford: Blackwell.

Anderson, R. M. 1986 The population dynamics and epidemiology of intestinal nematode infections. *Trans. R. Soc. trop. Med. Hyg.* **80**, 686–696.

Anderson, R. M. 1987 Determinants of infection in human schistosomiasis. In *Schistosomiasis: clinics in tropical medicine and diseases* (ed. A. A. F. Mahmoud), pp. 279–300. London: Bailliere Tindall.

Anderson, R. M. & May, R. M. 1985*a* Helminth infections of humans: mathematical models, population dynamics and control. *Adv. Parasitol.* **24**, 1–101.

Anderson, R. M. & May, R. M. 1985*b* Herd immunity to helminth infection: implications for parasite control. *Nature, Lond.* **315**, 493–496.

Anderson, R. M., May, R. M. & Gupta, S. 1989 Non-linear phenomena in host-parasite interactions. *Parasitology* **99**, S59–S79.

Anderson, R. M. & Medley, G. F. 1985 Community control of helminth infections of man by mass and selective chemotherapy. *Parasitology* **90**, 629–660.

Anderson, R. M. & Schad, G. A. 1985 Hookworm burdens

and faecal egg counts: an analysis of the biological basis of variation. *Trans. R. Soc. trop. Med. Hyg.* **79**, 812–825.

Arfaa, F. & Ghadirian, E. 1977 Epidemiology and mass-treatment of ascariasis in six rural communities in central Iran. *Am. J. trop. Med. Hyg.* **26**, 866–871.

Behnke, J. M. 1987 Do hookworms elicit protective immunity in man? *Parasitol. Today* **3**, 200–206.

Bundy, D. A. P. 1988 Population ecology of intestinal helminth infections in human communities. *Phil. Trans. R. Soc. Lond.* B **321**, 405–420.

Bundy, D. A. P. & Cooper, E. S. 1989 *Trichuris* and trichuriasis in humans. *Adv. Parasitol.* **28**, 108–173.

Bundy, D. A. P., Cooper, E. S., Thompson, D. E., Didier, J. M., Anderson, R. M. & Simmons, I. 1987 Predisposition to *Trichuris trichiura* infection in humans. *Epidem. Inf.* **98**, 65–71.

Bundy, D. A. P., Thompson, D. E., Golden, M. H. N., Cooper, E. S., Anderson, R. M. & Harland, P. S. E. 1985 Population distribution of *Trichuris trichiura* in a community of Jamaican children. *Trans. R. Soc. trop. Med. Hyg.* **79**, 232–237.

Burrows, R. B. 1950 On the estimation of *Trichuris* worm burdens in patients. *J. Parasitol.* **36**, 227–231.

Butterworth, A. E., Dunne, D. W., Fulford, A. J. C., Capron, M., Khalife, J., Capron, A., Koech, D., Ouma, J. H. & Sturrock, R. F. 1988*b* Immunity in human schistosomiasis mansoni: cross-reactive IgM and IgG2 anti-carbohydrate antibodies block the expression of immunity. *Biochimie* **70**, 1053–1063.

Butterworth, A. E., Fulford, A. J. C., Dunne, D. W., Ouma, J. H. & Sturrock, R. F. 1988*a* Longitudinal studies on human schistosomiasis. *Phil. Trans. R. Soc. Lond.* B **321**, 495–511.

Cho, S. Y. 1977 Study on the quantitative evaluation of reinfection of *Ascaris lumbricoides*. *Korean J. Parasitol.* **15**, 17–29.

Coadwell, W. J. & Ward, P. F. V. 1982 The use of faecal egg counts for estimating worm burdens in sheep infected with *Haemonchus contortus*. *Parasitology* **85**, 251–256.

Croll, N. A., Anderson, R. M., Gyorkos, T. W. & Ghadirian, E. 1982 The population biology and control of *Ascaris lumbricoides* in a rural community in Iran. *Trans. R. Soc. trop. Med. Hyg.* **76**, 187–197.

Crompton, D. W. T., Nesheim, M. C. & Pawlowski, Z. S. (eds) 1985 *Ascariasis and its public health significance.* London: Taylor & Francis.

Dobson, C. & Bawden, R. J. 1974 Studies on the immunity of sheep to *Oesophagostomum columbianum*: effects of low-protein diet on resistance to infection and cellular reactions in the gut. *Parasitology* **69**, 239–255.

Duke, B. O. L. & Moore, P. J. 1976 The use of a molluscicide in conjunction with chemotherapy to control *Schistosoma haematobium* at the Barombi Lake foci in Cameroon. II. Urinary examination methods, the use of niridazole to attack the parasite in man, and the effect of transmission from man to snail. *Tropenmed. Parasit.* **27**, 489–504.

Eberhardt, L. L. 1970 Correlation, regression and density-dependence. *Ecology* **51**, 306–310.

Elkins, D. B. & Haswell-Elkins, M. 1989 The weight/length profiles of *Ascaris lumbricoides* within a human community before mass treatment and following reinfection. *Parasitology* **99**, 293–299.

Elkins, D. B., Haswell-Elkins, M. & Anderson, R. M. 1988 The importance of host age and sex to patterns of reinfection with *Ascaris lumbricoides* following mass anthelmintic treatment in a South Indian fishing community. *Parasitology* **96**, 171–184.

Forrester, J. E., Scott, M. E., Bundy, D. A. P. & Golden, M. H. N. 1990 Predisposition of individuals and families in Mexico to heavy infection with *Ascaris lumbricoides* and *Trichuris trichiura*. *Trans. R. Soc. trop. Med. Hyg.* **84**, 272–276.

Froyd, G. 1965 Bovine cysticercosis and human taeniasis in Kenya. *Ann. trop. Med. Parasitol.* **59**, 169–180.

Gemmell, M. A., Lawson, J. R., Roberts, M. G., Kerin, B. R. & Mason, C. J. 1986 Population dynamics in echinococcosis and cysticercosis: comparison of the response of *Echinococcus granulosus*, *Taenia hydatigena* and *T. ovis* to control. *Parasitology* **93**, 357–369.

Gemmell, M. A., Lawson, J. R. & Roberts, M. G. 1987 Population dynamics in echinococcosis and cysticercosis: evaluation of the biological parameters of *Taenia hydatigena* and *T. ovis* and comparison with those of *Echinococcus granulosus*. *Parasitology* **94**, 161–180.

Grenfell, B. T., Smith, G. & Anderson, R. M. 1987 The regulation of *Ostertagia ostertagi* populations in calves: the effect of past and current experience of infection on proportional establishment and parasite survival. *Parasitology* **95**, 363–372.

Guyatt, H., Bundy, D. A. P., Medley, G. F. & Grenfell, B. T. 1990 The relationship between the frequency distribution of *Ascaris lumbricoides* and the prevalence and intensity of infection in human communities. *Parasitology* **101**, 139–143.

Hagan, P. 1987 Human immune response. In *The biology of schistosomes* (ed. D. Rollinson & A. J. G. Simpson), pp. 295–320. London: Academic Press.

Haswell-Elkins, M. R., Elkins, D. B., Manjula, K., Michael, E. & Anderson, R. M. 1987 The distribution and abundance of *Enterobius vermicularis* in a South Indian fishing community. *Parasitology* **95**, 323–338.

Haswell-Elkins, M. R., Elkins, D. B., Manjula, K., Michael, E. & Anderson, R. M. 1988 An investigation of hookworm infection and reinfection following mass anthelmintic treatment in the South Indian fishing community of Vairavankuppam. *Parasitology* **96**, 565–577.

Hill, R. B. 1926 Hookworm reinfestation for three years after treatment in a sanitated area in Porto Rico, and its bearing on permanent hookworm control in the group studied. *Am. J. Hyg.* **6**, 103–117.

Holland, C. V., Asaolu, S. O., Crompton, D. W. T., Stoddart, R. C., Macdonald, R. & Torimiro, S. E. A. 1989 The epidemiology of *Ascaris lumbricoides* and other soil-transmitted helminths in primary school children from Ile-Ife, Nigeria. *Parasitology* **99**, 275–285.

Kennedy, C. R. 1985 Regulation and dynamics of acanthocephalan populations. In *Biology of the Acanthocephala* (ed. D. W. T. Crompton & B. B. Nickol), pp. 385–416. Cambridge University Press.

Keymer, A. E., Crompton, D. W. T. & Walters, D. E. 1983 Parasite population biology and host nutrition: dietary fructose and *Moniliformis*. *Parasitology* **87**, 265–278.

Keymer, A. E., Roberts, M. G. & Tarlton, A. B. 1990 The population dynamics of acquired immunity to *Heligmosomoides polygyrus* in the laboratory mouse: strain, diet and exposure. *Parasitology*. (In the press.)

Keymer, A. E. & Slater, A. F. G. 1987 Helminth fecundity: density dependence or statistical illusion? *Parasitol. Today* **3**, 56–58.

Martin, J., Keymer, A. E., Isherwood, R. J. & Wainwright, S. M. 1983 The prevalence and intensity of *Ascaris lumbricoides* infections in Moslem children from northern Bangladesh. *Trans. R. Soc. trop. Med. Hyg.* **77**, 702–706.

Medley, G. F. 1988 The role of theoretical research in the design of programmes for the control of infectious disease agents. In *Ecodynamics* (ed. W. Wolff, C. J. Soeder F. R. Drepper). New York: Springer–Verlag.

Medley, G. F. 1989 Theoretical studies of the epidemiology

and control of human helminths. Ph.D. Thesis, University of London.
Mello, D. A. 1974 A note on the egg production of *Ascaris lumbricoides*. *J. Parasitol.* **60**, 380–381.
Montgomery, S. S. J. & Montgomery, W. I. 1990 Structure, stability and species interactions in helminth communities of wood mice, *Apodemus sylvaticus*. *Int. J. Parasitol.* **20**, 225–242.
Murdoch, W. W. 1970 Population regulation and population inertia. *Ecology* **51**, 497–502.
Nickol, B. B. 1985 Epizootiology. In *Biology of the Acanthocephala* (ed. D. W. T. Crompton & B. B. Nickol), pp. 307–346. Cambridge University Press.
Pritchard, D. I., Quinnell, R. J., Slater, A. F. G., McKean, P. G., Dale, D. D. S., Raiko, A. & Keymer, A. E. 1990 The epidemiological significance of acquired immunity to *Necator americanus*: humoral responses to parasite collagen and excretory-secretory antigens. *Parasitology* **100**, 317–326.
Quinnell, R. J. 1990 The epidemiology of gastrointestinal nematode infection in mammals. Ph.D. thesis, University of Oxford.
Rayner, J. M. V. 1985 Linear relations in biomechanics: the statistics of scaling functions. *J. Zool., Lond.* A **206**, 415–439.
Read, C. P. 1959 The role of carbohydrates in the biology of cestodes. VIII. Some conclusions and hypotheses. *Expl. Parasitol.* **8**, 365–382.
Roberts, M. G., Lawson, J. R. & Gemmell, M. A. 1986 Population dynamics in echinococcosis and cysticercosis: mathematical model of the life-cycle of *Echinococcus granulosus*. *Parasitology* **92**, 621–641.
Roberts, M. G., Lawson, J. R. & Gemmell, M. A. 1987 Population dynamics in echinococcosis and cysticercosis: mathematical models of the life-cycles of *Taenia hydatigena* and *T. ovis*. *Parasitology* **94**, 181–197.
Robinson, M., Wahid, F. N., Behnke, J. M. & Gilbert, F. S. 1989 Immunological relationships during primary infection with *Heligmosomoides polygyrus* (*Nematospiroides dubius*): dose-dependent expulsion of adult worms. *Parasitology* **98**, 115–124.
Rollinson, D. & Simpson, A. J. G. (ed.) 1987 *The biology of schistosomes*. London: Academic Press.
Roux, J., Picq, J.-J., Lafaye, A. & Sellin, B. 1975 Protection des bilharziens en zone d'endemie a *S. haematobium* par une cure reduite immunisante de niridazole. *Med. Trop.* **35**, 377–387.
Schad, G. A. & Anderson, R. M. 1985 Predisposition to hookworm infection in humans. *Science, Wash.* **228**, 1537–1540.
Seo, B. S., Cho, S. Y. & Chai, J. Y. 1979 Seasonal fluctuation of *Ascaris* reinfection incidences in a rural Korean population. *Korean J. Parasitol.* **17**, 11–18.
Seo, B. S., Cho, S. Y., Chai, J. Y. & Hong, S. T. 1980 Comparative efficacy of various interval mass treatment on *Ascaris lumbricoides* infection in Korea. *Korean J. Parasitol.* **18**, 145–151.
Shaw, J. L. & Moss, R. 1989 The role of parasite fecundity and longevity in the success of *Trichostrongylus tenuis* in low density red grouse populations. *Parasitology* **99**, 253–258.
Sinclair, A. R. E. 1989 Population regulation. In *Ecological concepts: the contribution of ecology to an understanding of the natural world* (ed. J. M. Cherret), pp. 197–242. Oxford: Blackwell.
Slater, A. F. G. 1988 The influence of dietary protein on the experimental epidemiology of *Heligmosomoides polygyrus* (Nematoda) in the laboratory mouse. *Proc. R. Soc. Lond.* B **234**, 239–254.
Slater, A. F. G. & Keymer, A. E. 1986 *Heligmosomoides polygyrus* (Nematoda): the influence of dietary protein on the dynamics of repeated infection. *Proc. R. Soc. Lond.* B **229**, 69–83.
Smith, G. 1988 The population biology of the parasitic stages of *Haemonchus contortus*. *Parasitology* **96**, 185–186.
Smith, G., Grenfell, B. T. & Anderson, R. M. 1987 The regulation of *Ostertagia ostertagi* populations in calves: density-dependent control of fecundity. *Parasitology* **95**, 373–388.
Sprent, P. 1966 A generalized least-squares approach to linear functional relationships. *J. R. Stat. Soc.* B **28**, 278–297.
Sturrock, R. F., Bensted-Smith, R., Butterworth, A. E., Dalton, P. R., Kariuki, H. C., Koech, D., Mugambi, M., Ouma, J. H. & Arap Siongok, T. K. 1987 Immunity after treatment of human schistosomiasis mansoni. III. Long-term effects of treatment and retreatment. *Trans. R. Soc. trop. Med. Hyg.* **81**, 303–314.
Sweet, W. C. 1925 Hookworm reinfection; an analysis of 8,239 Ceylon egg counts. *Ceylon J. Science, Section D* **1**, 129–140.
Thein Hlaing, Than Saw & Myint Lwin. 1987 Reinfection of people with *Ascaris lumbricoides* following single, 6-month and 12-month interval mass chemotherapy in Okpo village, rural Burma. *Trans. R. Soc. trop. Med. Hyg.* **81**, 140–146.
Varley, G. C. & Gradwell, G. R. 1960 Key factors in population studies. *J. Anim. Ecol.* **29**, 399–401.
Wakelin, D. 1984 *Immunity to parasites*. London: Edward Arnold.
Wakelin, D. 1986 The role of the immune response in helminth population regulation. In *Parasitology – quo vadit?* (ed. M. J. Howell), pp. 549–558. Canberra: Australian Academy of Science.
Wakelin, D. 1988 Helminth infections. In *Genetics of resistance to bacterial and parasitic infection* (ed. D. Wakelin & J. M. Blackwell), pp. 153–225. London: Taylor & Francis.
Wilkins, H. A. 1989 Reinfection after treatment of schistosome infections. *Parasitol. Today* **5**, 83–87.
Wilkins, H. A., Blumenthal, U. J., Hayes, R. J. & Tulloch, S. 1987 Resistance to reinfection after treatment of urinary schistosomiasis. *Trans. R. Soc. trop. Med. Hyg.* **81**, 29–35.

Heterogeneity and the dynamics of host–parasitoid interactions

M. P. HASSELL[1] AND S. W. PACALA[2]

[1] *Department of Biology and Centre for Population Biology, Imperial College at Silwood Park, Ascot, Berks SL5 7PY, U.K.*
[2] *Department of Ecology and Evolutionary Biology, University of Connecticut, Storrs, Connecticut 06268, U.S.A.*

SUMMARY

This paper is concerned with the dynamical effects of spatial heterogeneity in host–parasitoid interactions with discrete generations. We show that the dynamical effects of any pattern of distribution of searching parasitoids in such systems can be assessed within a common, simple framework. In particular, we describe an approximate general rule that the populations of hosts and parasitoids will be regulated if the coefficient of variation squared (CV^2) of the distribution of searching parasitoids is greater than one. This criterion is shown to apply both generally and in several specific cases. We further show that CV^2 may be partitioned into a density-dependent component (direct or inverse) caused by the response of parasitoids to host density per patch, and a density independent component. Population regulation can be enhanced as much by density independent as by density-dependent heterogeneity. Thus the dynamical effects of any pattern of distribution of searching parasitoids can be assessed within the same common framework. The paradoxical impact of density-independent heterogeneity on dynamics is especially interesting: the greater the density independence, and thus the more scattered the data of percent parasitism against local host density, the more stable the populations are likely to be. Although a detailed analysis of host–parasitoid interactions in continuous time has yet to be done, evidence does not support the suggestion of Murdoch & Oaten (1989) that non-random parasitism may have quite different effects on the dynamics of continuous-time interactions. There appears to be no fundamental difference in the role of heterogeneity in comparable discrete- or continuous-time interactions.

A total of 65 data sets from field studies have been analysed, in which percentage parasitism in relation to local host density have been recorded. In each case, estimated values of CV^2 have been obtained by using a maximum likelihood procedure. The method also allows us to partition the CV^2 into the density dependent and density-independent components mentioned above. In 18 out of the 65 cases, total heterogeneity was at levels sufficient (if typical of the interactions) to stabilize the interacting populations (i.e. $CV^2 > 1$). Interestingly, in 14 of these it is the host-density-independent heterogeneity that contributes most to the total heterogeneity.

Although heterogeneity has often been regarded as a complicating factor in population dynamics that rapidly leads to analytical intractability, this clearly need not necessarily be so. The $CV^2 > 1$ rule explains the consequences of heterogeneity for population dynamics in terms of a simple description of the heterogeneity itself, and provides a rough rule for predicting the effects of different kinds of heterogeneity on population regulation.

INTRODUCTION

A relatively recent and major concern in population ecology has been to determine the effects of heterogeneity on population dynamics. Earlier, theoretical work tended, for convenience of analysis, to assume completely homogeneous populations in which each individual has the same chance of reproducing, and the same risk of dying. The wealth of evidence to the contrary has prompted the development of theoretical frameworks for most types of species interactions that take explicit account of heterogeneity between individuals. The common conclusion, whether one considers competing species (see, for example, Atkinson & Shorrocks (1981); Hanski (1981); Ives & May (1985)), plant–herbivore interactions (see, for example, Crawley (1983)), hosts and pathogens (see, for example, Anderson & May (1984) or predators and prey (Comins & Hassell (1987)), is that heterogeneity promotes the persistence of the interacting populations. This paper considers the effects of heterogeneity on the dynamics of one particular kind of predator–prey system, that of insect parasitoids and their hosts.

Within the broad class of predatory metazoans, the parasitoid lifestyle predominates; one recent estimate of the number of parasitoid species is as high as 15×10^6 (Hochberg & Lawton 1990). Parasitoids occur in a number of different insect groups (but mostly in the Hymenoptera). They are recognized by their characteristic life cycle which has aspects in common with that of both predators and parasites. The adult female lays her eggs on, in or close to the body of another

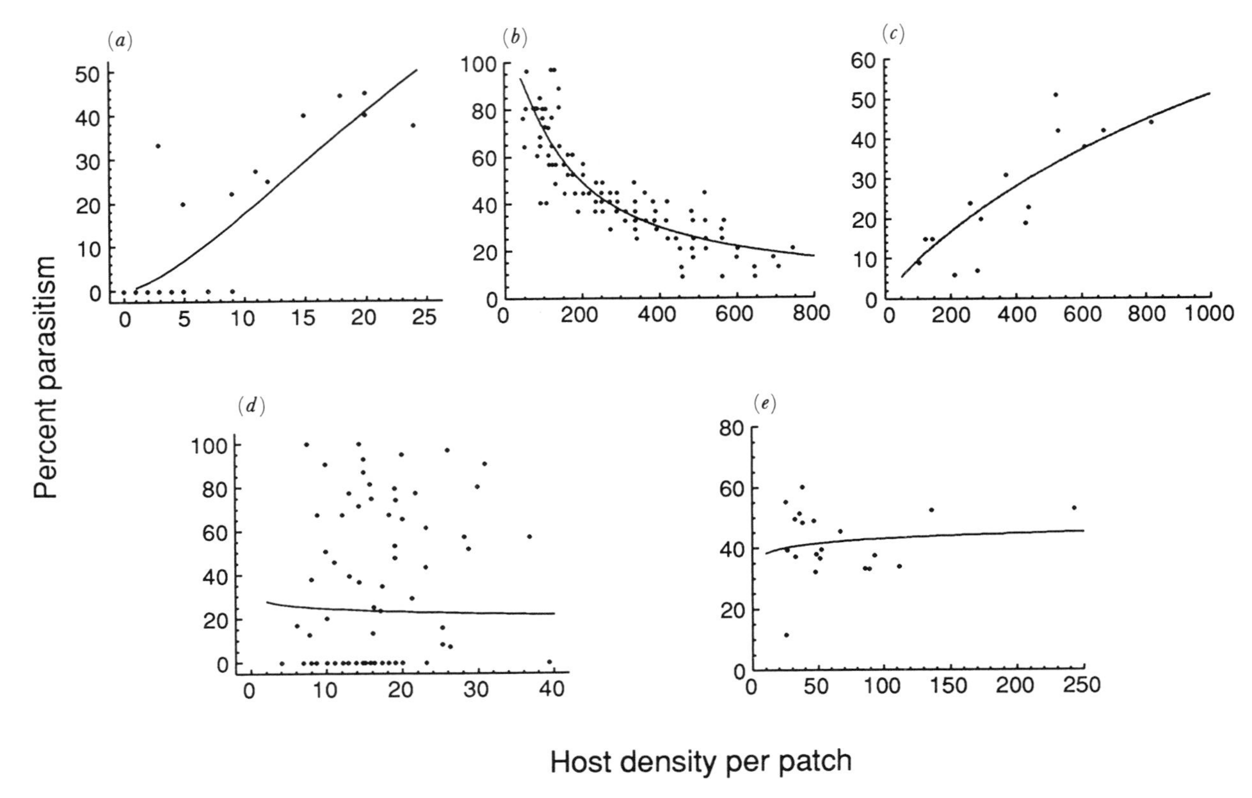

Figure 1. Examples of field studies showing percentage parasitism as a function of host density. Curves are means predicted from equation (3) evaluated at maximum likelihood estimates. (*a*) Direct density dependence from data set 25 in tables 2 and 3; (*b*) inverse density dependence from data set 9; (*c*) direct density dependence from data set 33; (*d*) density independent heterogeneity from data set 14; (*e*) density independent heterogeneity from data set 39. (From Pacala & Hassell 1990.)

arthropod (usually an insect), which is eventually killed by the feeding parasitoid larva. In their effect on a prey population, they are, in effect, predators where the act of 'predation' involves oviposition rather than direct consumption of the prey.

The two attributes of: (i) only the females searching for hosts, and (ii) parasitism defining reproduction, combine to make parasitoids particularly appropriate subjects for the development of generalized predator–prey models such as those of Lotka (1926) and Volterra (1926) and Nicholson & Bailey (1935). In recent years, the emphasis in this work has increasingly been to determine how various forms of heterogeneity can affect host–parasitoid population dynamics. Such heterogeneity can arise in many ways, but most often has been viewed in terms of a patchily distributed host population with different probabilities of parasitism from patch to patch (where the patch may be an arbitrary sampling unit or some clearly recognizable discrete unit of the habitat, such as a leaf or plant).

The widespread interest in the effects of heterogeneity on host–parasitoid dynamics has led many workers to record the distribution of parasitism in the field in relation to the local density of hosts per patch. Of 194 different examples listed in the recent reviews of Lessells (1985), Stiling (1987) and Walde & Murdoch (1988), 58 show variation in attack rates among patches depending directly on host density (figure 1*a*, *c*), 50 show inversely density-dependent relations (figure 1*b*) and 86 show variation uncorrelated with host density (density independent (figure 1*d*, *e*).

A popular interpretation of these data, guided by the earlier theoretical literature (see, for example, Hassell & May (1973, 1974); Murdoch & Oaten (1975)), has been that only the direct density-dependent patterns promote the stability of the interacting populations. This, however, is not the true picture. Both inverse density-dependent patterns (Hassell 1984; Walde & Murdoch 1988), and variation in parasitism that is independent of host density (Chesson & Murdoch 1986; Hassell & May 1988; Pacala *et al.* 1990; Hassell *et al.* 1991), can in principle also be just as important to population regulation. The reasons for this are described below, but arise essentially because any variation in levels of parasitism from patch to patch has the net effect of reducing the mean parasitoid searching efficiency (measured over all hosts) as average parasitoid density increases (the so-called 'pseudointerference' effect of Free *et al.* (1977)). This has obvious implications for the design of field studies on host–parasitoid systems, because no longer can the effects of such heterogeneity be inferred simply from the shape of the relations between percentage parasitism and local host density.

Our aims in this paper are threefold: (i) to show that one can describe the effects of heterogeneity in levels of parasitism using relatively simple criteria; (ii) that this heterogeneity can be broken down into constituent parts that are density dependent and density independent, and (iii) that these measures can be readily applied to the kinds field data that are normally available.

HOST–PARASITOID MODELS IN PATCHY ENVIRONMENTS

Let us consider a habitat which is divided into discrete patches (e.g. food plants for an herbivorous insect) among which adult insects distribute their eggs. The immature stages of these insects are hosts for a specialist parasitoid species whose adult females forage across the patches according to some as-yet-unspecified foraging rule. We also assume that parasitism dominates host mortality such that the hosts are on average kept well below their carrying capacity.

Quite different modelling approaches have developed around this scenario, depending on whether the generations of host and parasitoid are discrete, or continuous with all stages overlapping. Difference equations, stemming back to the Nicholson–Bailey model, have been traditionally used to represent discrete-generation interactions, while differential equations in the Lotka–Volterra tradition have been used for interactions with overlapping generations. The principal difference between the two is that the continuous-time models should exhibit more stable dynamics than the corresponding models in discrete time (May 1974). This view has recently been challenged by Murdoch & Oaten (1989), and is further discussed below.

Yet a third approach to modelling host–parasitoid interactions has recently been developed by R. M. Nisbet and W. S. C. Gurney and colleagues by using systems of time-lagged differential equations incorporating age-structure and developmental rates (Nisbet & Gurney 1983; Gurney, Nisbet & Lawton 1983; Gurney & Nisbet 1985; Murdoch *et al.* 1987). An interesting property of such continuous models with time lags, first shown by Auslander *et al.* (1974), is the appearance of population cycles with a period of roughly one host generation interval. This has recently been more fully discussed by Godfray & Hassell (1987, 1989) who showed that, despite the continuous interaction, such models can show either stable populations with all stages overlapping, or cycles more-or-less recovering the discrete generation interactions described above. Which of these outcomes occur depends largely on the ratio of the lengths of the host and parasitoid life cycles.

(*a*) *Interactions in discrete time*

In this paper, we focus principally on discrete-time interactions and commence with a familiar framework for such interactions (Hassell 1978):

$$N_{t+1} = \lambda N_t f(N_t, P_t), \tag{1a}$$

$$P_{t+1} = w N_t (1 - f(N_t, P_t)). \tag{1b}$$

Here N and P are the host and parasitoid populations in successive generations t and $t+1$, λ is the host's finite rate of increase in the absence of the parasitoid (the fecundity per adult discounted by the average of all mortalities other than parasitism), and w is the average number of female parasitoids emerging from each host parasitized (henceforth assumed to be one). Finally, $f(N_t, P_t)$ is a function giving the average fraction of hosts that escape parasitism, and whose form depends upon all the factors that affect the rate of parasitism of hosts by the P_t searching adult parasitoids. An internal equilibrium of (1 *a, b*) is defined by: $\lambda f(N^*, P^*) = 1$ and $P^* = N^*(1-1/\lambda)$, and is locally stable if

$$-\lambda^2/(\lambda-1)\, P^* \partial f(N^*, P^*)/\partial P_t < 1, \tag{2a}$$

$$(\lambda-1)/\lambda(-\partial f(N^*, P^*)/\partial P_t) > \partial f(N^*, P^*)/\partial N_t. \tag{2b}$$

In a habitat composed of discrete patches, the term $f(N_t, P_t)$ in (1 *a*) and (1 *b*) represents the average, across all patches, of the fraction of hosts escaping parasitism. The distribution of hosts in such a patchy setting can either be random or vary in some other prescribed way. Similarly, the density of searching parasitoids in each patch can either be a random variable independent of local host density or a deterministic function of local host density. We call these patterns of heterogeneity in parasitoid distribution host density-independent heterogeneity (HDI) and host density-dependent heterogeneity (HDD), respectively (Pacala *et al.* 1990; Hassell *et al.* 1991). Comparable terms have been coined by Chesson & Murdoch (1986) who labelled models with randomly distributed parasitoids as pure error models and those with parasitoids responding to host density in a deterministic way as pure regression models.

A more biological interpretation of the stability criteria (2 *a*) and (2 *b*) is as follows. The process of parasitism generates negative covariance between the local (within-patch) abundance of parasitoids and the local abundance of surviving hosts. This is simply because high densities of parasitoids result in high levels of parasitism and correspondingly low densities of surviving hosts. Let C_e be the covariance, at equilibrium and at the end of a growing season, between the local densities of parasitoids and unparasitized hosts divided by the product of the equilibrium mean densities of parasitoids and unparasitized hosts. The division by the product of the means produces a scaled covariance in the same way that dividing a variance by the square of the mean produces a scaled variance (the square of the coefficient of variation). Similarly, let C_b be the scaled covariance of the local parasitoid and total (parasitized and unparasitized) local host densities at equilibrium. C_b may be thought of as the scaled covariance at the beginning of the growing season (before any parasitism takes place).

In the Appendix, we show, under quite general assumptions, that condition (2 *a*) is equivalent to $C_e + 1 < (C_b + 1)\, Z(\lambda)$, where $Z(\lambda)$ decreases from one to zero as λ increases from one to infinity (λ must be greater than one for the host to be able to persist even in the absence of parasitism). In the vicinity of $\lambda = 1$, $Z(\lambda)$ is approximately equal to $1/\lambda \approx 1$. Thus, as λ increases from one to infinity, condition (2 *a*) changes

Table 1. *A range of discrete-time models in which the $CV^2 > 1$ rule is either the exact or an approximate stability criterion*

model no.	HDD or HDI	brief description	$f(N_t, P_t)$	$CV^2 > 1$ rule
I	HDI	any host distribution and parasitoids uncorrelated with hosts (gamma distributed). Reduces to model of May (1978)	$f(P_t) = \int_0^\infty g(\epsilon)\, e^{-aP_t\epsilon}\, d\epsilon$	exact
II	HDD	gamma distributed hosts per patch and local parasitoid density a deterministic function of local host density	$f(P_t) = \int_0^\infty g(x)\, x\, e^{(-aP_t(x)^\mu)} dx$	approx.
III	HDI or HDD	a general HDI- and HDD-model of which I, II, and most other published discrete-generation models are special cases. Arbitrary host distribution and parasitoid distribution (either HDI or HDD) given by general functional forms	$f(N_t, P_t) = \int_0^\infty \phi(n) \int_0^\infty \gamma(\epsilon) \sum_{p=0}^{\infty} \frac{e^{-P_t g(n)\epsilon}[P_t g(n)\, \epsilon]^p}{p!} e^{-ap}\, d\epsilon\, dn$ where $\phi(n)$ is the host distribution, $\gamma(\epsilon)$ is the unit mean distribution of the residual ϵ, and $g(n)$ is the function governing density	approx.
IV	HDI	no spatial structure in the habitat but variation between hosts in their ability to encapsulate parasitoid progeny	$f(P_t) = \int_0^1 q(r)\, e^{(-aP_t(1-r))} dr$	approx.
V	HDI	evenly distributed hosts. Only a fraction of the parasitoid progeny emerging within a patch enter a 'pool' for subsequent redispersal according to a negative binomial distribution. The remainder stay on in the patches and reproduce there	(see text)	approx.

from $C_e < C_b$ to $C_e < -1$. For dynamics to be stable, the process of parasitism must reduce C_e to a level sufficiently below C_b. In other words, the process of parasitism must sufficiently spatially segregate parasitoids and unparasitized hosts. Obviously, small C_e implies that, at the end of a growing season, parasitoids are found primarily in patches that contain small numbers of uninfected hosts. This spatial segregation reduces the efficiency of the parasitoids and so acts to stabilize the host–parasitoid interaction. The level by which C_e must be reduced relative to C_b is an increasing function of the host's intrinsic rate of increase (λ). Host populations with large λ's rebound quickly and so are more difficult to stabilize than host populations with small λ's.

Stability condition (2*b*) is discussed fully in the Appendix. Suffice it to say here that the condition is always likely to be satisfied and (2*a*) is thus usually necessary and sufficient for stability.

Hassell & May (1988) suggested a very simple approximation of condition (2*a*) which states that interactions of the form of (1*a*, *b*) are stable if the distribution of parasitoids from patch to patch (measured as the square of the coefficient of variation, CV^2) is sufficiently heterogeneous. In particular, if the density of searching parasitoids in the *j*th patch is q_j ($j = 1, 2, \ldots, z$), the CV^2 of the q_j across the z patches should exceed one. More recently, Pacala *et al.* (1990) and Hassell *et al.* (1990) have extended this work and showed that a very similar criterion applies across a range of models in discrete time. Their criterion differs in that the density of searching parasitoids per patch is now weighted by the number of hosts in that patch. Thus, if p_i is the density of searching parasitoids in the vicinity of the *i*th host ($i = 1, 2, \ldots, y$), the stability criterion now becomes that the CV^2 of the p_i measured across all y hosts should exceed one. In what follows, 'CV^2' refers exclusively to the coefficient of variation squared of the p_i.

To show the generality of this criterion for discrete-generation host–parasitoid systems, we now survey a range of models in all of which the '$CV^2 > 1$ rule' applies either exactly or approximately. A brief summary is given in table 1 and full details in Hassell *et al.* (1990).

Model I

Consider a specific situation in which the host distribution across patches is arbitrary, and the distribution of parasitoids is unrelated to that of their hosts. It is, therefore, an HDI-model as the heterogeneity is host density independent, local densities of searching parasitoids are determined by chance and by responses to environmental cues that are uncorrelated with host densities per patch. More specifically, let us assume that parasitoid density varies as a gamma distributed random variable from patch to patch, such that the fraction of hosts that escape parasitism in (1*a*, *b*) is given by:

$$f(P_t) = \int_0^\infty g(\epsilon)\, e^{-aP_t \epsilon}\, d\epsilon. \tag{3}$$

Here $g(\epsilon)$ is the gamma probability density function for parasitoids per patch with unit mean and variance $1/\alpha$, where α is a positive constant governing the shape of the density function and a is the usual term for the per capita searching efficiency of the parasitoid. The term $\exp(-aP_t\epsilon)$ is thus the zero-term of a Poisson distribution with mean $aP_t\epsilon$. It gives the probability of a host being attacked zero times by parasitoids that search randomly within a patch containing searching parasitoids at density $P_t\epsilon$.

Hassell *et al.* (1990) show that equation (3) reduces to

$$f(P_t) = [\alpha/(\alpha + aP_t)]^\alpha.$$

A host–parasitoid equilibrium always exists if $\lambda > 1$, and will be locally stable if

$$\alpha[1 - 1/\lambda^{1/\alpha}] < 1 - 1/\lambda, \tag{4}$$

which will always be true if $\alpha < 1$. The CV^2 for gamma distributed parasitoids in this model is simply $1/\alpha$, so that the condition $\alpha < 1$ is identical to $CV^2 > 1$. Notice, incidentally, that this model is formally identical with the phenomenological model proposed by May (1978), in which parasitoid attacks are effectively distributed in a negative binomial manner with clumping parameter, $k (k = \alpha)$.

Model II

From the extreme of no correlation between the spatial distributions of parasitoids and hosts, the second example goes to the opposite extreme of a perfect correlation between the two. Local parasitoid density now deterministically tracks patch-to-patch variation in host density, presumably due to some deterministic foraging rule dominating parasitoid distribution. Specifically, let us suppose it is now the local host density (n) that varies from patch to patch as a gamma distributed random variable with mean N_t. Local parasitoid density is given by a regression function, $P_t(n/N_t)^\mu$, where μ is a constant governing the degree to which parasitoids aggregate in patches of high host density. The aggregation is directly density dependent if $\mu > 0$ and inversely density dependent if $\mu < 0$ (Hassell 1984). This expression has been widely used in host–parasitoid models (see, for example, Hassell & May (1973); Hassell (1984); Kidd & Mayer (1983), and provides better fits than an analogous linear function to the kind of data in figure 1, particularly in accounting for the curvature of inverse density dependence (figure 1*b*). The average fraction of hosts surviving parasitism can now be written as

$$f(P_t) = \int_0^\infty g(x)\, x\, e^{(-aP_t(x)^\mu)}\, dx, \tag{5}$$

where $g(x)$ is the unit mean gamma density and $x = n/N_t$.

Stability in model II is only affected by three parameters λ, μ and α. The relation of these to the CV^2 is shown by the numerical examples in figure 2. The $CV^2 > 1$ rule is now only the approximate condition for stability, but the approximation is good for values of λ near one and for highly aggregated host distributions (small values of α) (figure 2).

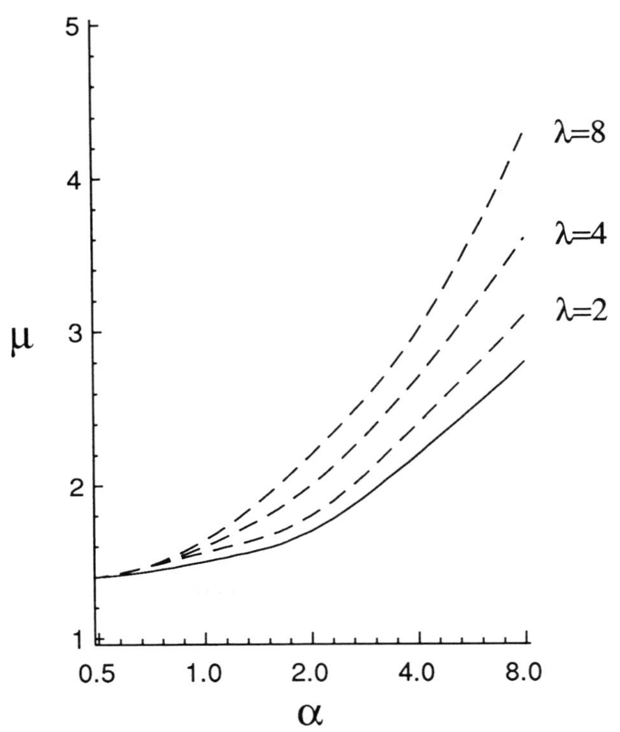

Figure 2. Stability criterion for model II. The internal equilibrium is stable for values of μ and α above the appropriate dashed curve, and is unstable for values below. The solid curve shows values of μ and α at which $CV^2 = 1$. $CV^2 > 1$ above the curve and $CV^2 < 1$ below. (From Hassell *et al.* 1990.)

Model III

Models I and II represent end-points of the continuum between HDI and HDD. They also include restrictive, though reasonable, assumptions about the different functional forms (e.g. gamma distributed populations or power-law dependence of local parasitoid on local host densities). We now consider a much more general model that relaxes these assumptions: spatial distributions are left unspecified and any degree of spatial covariance between parasitoid and host is allowed.

Specifically, we assume that the distribution of the relative numbers ($n = N/N_t$) of hosts in patches does not change with host density, but otherwise can be of any distribution whatsoever. We also assume that the mean density of parasitoids in a patch is given by the function $P_t g(n) U$. Here, g is an arbitrary function of host density that determines host density dependent aggregation by parasitoids. In contrast, U is a random variable with a mean of one that determines the level of HDI. This random variable creates heterogeneity among patches in their relative attractiveness to parasitoids that is independent of the host abundances in the patches. Finally, we assume that the actual number of parasitoids visiting a patch is Poisson distributed about the patch-specific mean, $P_t g(n) U$.

With these assumptions, the fraction of parasitoids that escape parasitism, $f(N_t, P_t)$, is given by $E(e^{-ap})$, where E is the expectation across all hosts and p is Poisson distributed about a mean, $P_t g(n) U$, which is itself a random variable (because of U) and a function of host density. It is important to note that between-patch variation in $P_t g(n) U$ reflects non-random aggregation by parasitoids because it reflects differences among patches to which parasitoids respond. In contrast, the Poisson variation in p reflects a purely random distribution of parasitoids.

The analysis of this general model is possible if λ is close to one. The derivation in Hassell *et al.* (1990) shows that the host–parasitoid interaction is stable if approximately $CV^2 - 1/p^* > 1$, where p^* is the average number (not density) of parasitoids that visit a patch at equilibrium. This average is again calculated with respect to a randomly chosen host. If, on average, there are several parasitoid visits per patch, then we approximately recover the $CV^2 > 1$ rule. If, on the other hand, the average visits per patch is very low (e.g. $\leqslant 1$) the value of CV^2 required for stability will increase. For example, with an average of one parasitoid visit per patch the stability criterion becomes $CV^2 > 2$. This arises because the term $1/p^*$ is the component of CV^2 that is caused solely by purely random (Poisson) variation in parasitoid abundance among patches. Because this quantity is subtracted from CV^2, Poisson variation does not contribute in any way to stability. Thus, host density independent heterogeneity only facilitates stability if it reflects non-random aggregation of parasitoids.

The next 2 models are used to demonstrate that the $CV^2 > 1$ rule is robust to at least some major changes in the biological assumptions underpining models I–III.

Model IV

The previous models have all been set in an explicitly patchy environment. We now turn to a model that breaks away from this mould and shows that the $CV^2 > 1$ rule can also apply when the heterogeneity arises in quite different ways. Insects in general possess a powerful haemocytic defence mechanism that enables them to encapsulate foreign objects, such as parasitoid eggs and larvae, recognized as 'non-self'. Heterogeneity now arises if there is variability between individual hosts in their ability to encapsulate parasitoids within them. Godfray & Hassell (1990) consider two cases: (i) all-or-none encapsulation where the probability of a host individual escaping parasitism by encapsulation is constant, irrespective of the number of parasitoid larvae it contains, and (ii) dosage-dependent encapsulation where the probability of a host surviving parasitism decreases with parasitoid load. Other forms of encapsulation (e.g. a threshold number of parasitoid larvae required to trigger the host's response) are possible, but were not considered.

In all-or-none encapsulation there are essentially two classes of host: those not encountered by parasitoids and those encountered one or more times. Within the framework of equation (1), this reduces to a straightforward refuge model whose properties are well-known (Hassell 1978). Furthermore, if heterogeneity is now introduced by individual hosts varying in their ability to encapsulate parasitoids Godfray & Hassell (1990) show this to have no affect on population dynamics.

With dosage-dependent encapsulation, however, explicit account must now be taken of hosts that are attacked once, twice, three times, etc. Assuming random search and a fixed probability, r, of a parasitoid being encapsulated leads directly back to the Nicholson–Bailey model with the searching parameter a reduced to $a' = a(1-r)$. The stability properties, which are independent of a', are thus unchanged. This is no longer the case, however, if heterogeneity in the ability of hosts to encapsulate is introduced. Let us assume that r now varies randomly among hosts with probability density function $q(r)$. If the number of attacks per host is a Poisson random variable, if one egg is laid per attack and if the probability that any one host successfully encapsulates n eggs is r^n, then we may write $f(\cdot)$ as

$$f(P_t) = \int_0^1 q(r)\, e^{(-aP_t\{1-r\})}\, dr. \tag{6}$$

In effect, each host is being viewed as a patch and, because of inter-host variation in r, there is inter-patch variation in the mean level of successful parasitism. Godfray & Hassell (1990) then derive a local stability criterion, whereby as long as λ is not too large, then the necessary and sufficient condition for local stability is approximately

$$\sigma_r^2/(1-\bar{r})^2 > 1, \tag{7}$$

where σ_r^2 is the variance of r, and $(1-r)$ is the mean. Thus, once again the $CV^2 > 1$ rule emerges as an approximate stability condition.

Model V

The preceding models assume that all hosts and parasitoids redistributed themselves each generation among the available patches. While there are many natural examples of this, particularly from univoltine species, there are also many cases of less complete mixing, where some of the hosts and parasitoids tend to remain within the patch from which they originated. This final model caters for this by allowing some hosts and parasitoids to stay in the patches from which they emerged, while the remainder enter a 'pool' to be redistributed anew in the next generation (Hassell & May 1988; Reeve 1988; Hassell *et al.* (1990)). There is thus a continuum from complete host or parasitoid mixing to no host mixing at all. Assuming, for simplicity, that the probability of leaving a patch is density independent and that there is no mortality associated with the movement, the hosts and adult parasitoids in the ith patch, N_i and P_i respectively, are now given by

$$N_i(t+1) = \lambda\left[S_i(1-x_i) + \alpha_i\left\{\sum_{j=1}^{n} S_j x_j\right\}\right], \tag{8a}$$

$$P_i(t+1) = N_{ai}(1-y_i) + \beta_i\left\{\sum_{j=1}^{n} N_{aj} y_j\right\}. \tag{8b}$$

Here α_i and β_i are the fractions of dispersing hosts and parasitoids, respectively, that enter the ith patch, S_i is the number of these hosts surviving from parasitism and N_{ai} the number of hosts parasitized $(N_i[1-\exp(-aP_i)])$. Finally, x_i and y_i are the fraction

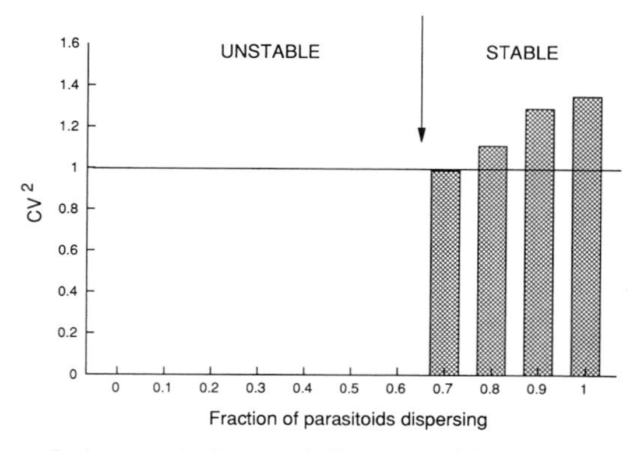

Figure 3. A numerical example from model V. The histogram bars show the values of CV^2 at equilibrium as a function of the fraction of parasitoids that disperse each generation (y in equation (8)). The model is stable, with $CV^2 \geqslant 1$, provided that at least 70% of the parasitoids disperse each generation. Below this value, the model shows limit cycles and CV^2 fluctuates with mean values less than one. Parameter values are: $\lambda = 2$, $a = 0.01$, k (negative binomial) $= 1$ and $n = 20$. (From Hassell *et al.* 1990.)

of host and parasitoid progeny, respectively, leaving the patch for subsequent redistribution in the next generation, and n is the total number of patches.

Numerical studies indicate that the CV^2 rule is a good indicator of stability for the model. For example, suppose that there is complete host mixing in each generation ($x_i = 1$), and that the fraction of parasitoids mixing varies from zero to one. We assume an even host distribution, and that those parasitoids that do disperse redistribute themselves according to a negative binomial distribution. The model is thus an HDI-model, not dissimilar to model I. Figure 3 shows a numerical example in terms of the CV^2 of the parasitoid distribution at equilibrium as the fraction of parasitoid mixing is changed. Once again the $CV^2 > 1$ rule is a good indicator of stability.

(*b*) *Parasitoid aggregation and the heterogeneity of risk*

Chesson & Murdoch (1986) define a quantity, ρ, the relative risk of parasitism for a host individual, as a means of unifying the analysis of both HDI-models (their 'pure-error' models) and HDD-models (their 'pure-regression' models). Heterogeneity in this risk of parasitism is the commodity that affects the stability of the populations. The '$CV^2 > 1$ rule', in contrast, is defined in terms of heterogeneity in the distribution of searching parasitoids. In models I–V (with assumed type I functional responses) this is generally the same as heterogeneity in the risk of parasitism. As Ives (1990) has recently pointed out, however, the distribution of parasitoids may differ from the distribution of risk if functional responses are sufficiently nonlinear. Thus in cases involving nonlinear functional responses, CV^2 refers specifically to heterogeneity in risk of parasitism. Further numerical work is needed to assess the generality of the $CV^2 > 1$ rule when functional responses are markedly nonlinear over realistic ranges of host densities per patch.

(*c*) *Continuous redistribution in discrete time*

Finite difference equation models such as (1) are probably most appropriate for species with non-overlapping generations. Such species are more typical of seasonal habitats, although Godfray & Hassell (1987, 1989) showed that developmental time-lags can, under some conditions, lead to cycles of approximately one generation period in species that would otherwise reproduce continuously. Even so, many continuously reproducing species do have overlapping generations and are probably best modelled by systems of differential equations.

Although restricted to species with non-overlapping generations, the discrete time model (1) can account for death and movement that occur continuously within a discrete generation. Models I–III, however, are most applicable to host species that are distributed amongst patches at the beginning of a growing season and do not disperse until the following season. The parasitoids in these models, however, may be viewed as either sedentary or highly dispersive (or anything in between). For example, model I is appropriate if parasitoids are sedentary within a season and are gamma distributed amongst patches, or if the time spent in each patch by each continuously dispersing parasitoid is gamma distributed.

In contrast, rapid within-season movement by hosts tends to expose each host to the same suite of parasitoid densities and hence to the same risk of parasitism. Rapid host movement can thus decrease heterogeneity (CV^2) caused by parasitoids. Godfray & Pacala (1990) show that with infinitely rapid within-season movement, and thus no heterogeneity of risk, the system collapses to the unstable Nicholson–Bailey model if parasitoid aggregation is independent of host density. If aggregation is density dependent, however, stability is possible if the host distribution is sufficiently clumped.

(*d*) *Overlapping generations in continuous time*

In this section we turn to continuous-time analogues of (1) and argue that, despite recent claims to the contrary (Murdoch & Oaten 1989), heterogeneity is also generally stabilizing in continuous-time host–parasitoid models.

Perhaps the most simple system of differential equations in a patchy environment analogous to (1) is:

$$\mathrm{d}N_i/\mathrm{d}t = bN_i - aP_i N_i - e_n N_i + z_i e_n \sum_{j=1}^{Q} N_j, \qquad (9a)$$

$$\mathrm{d}P_i/\mathrm{d}t = vaP_i N_i - \mathrm{d}P_i - e_p P_i + q_i e_p \sum_{j=1}^{Q} P_j, \qquad (1 = 1, 2, \ldots Q), \qquad (9b)$$

where P_i and N_i are the local densities of parasitoid and host in patch i, b is the density independent growth rate of the host, a is the mean searching efficiency of the parasitoid, v is the reproductive efficiency of the parasitoid, d is the death rate of the parasitoid, and e_n and e_p are the rates at which hosts and parasitoids disperse. Finally, z_i and q_i are the fractions of dispersing hosts and parasitoids, respectively, that enter patch i, where q_i may be either constant or a function of local host density.

The model (9) assumes that parasitized hosts that have not yet died have no effect on dynamics. The inclusion of living, but parasitized 'zombies' would make the system more directly analogous to the discrete-time models above (Godfray & Pacala 1991). Zombies affect dynamics because most parasitoids cannot perfectly distinguish infected and uninfected hosts. If parasitoids disperse to regions of high total host density, then parasitoids will occur primarily where uninfected hosts do not. This is one mechanism that reduces C_e (spatial covariance between uninfected hosts and parasitoids) and thus stabilizes the dynamics of discrete-time models (see above).

Host–parasitoid models such as (9) have received much less attention than corresponding models in discrete time. In a recent study, Godfray & Pacala (1991) showed that (9) has neutrally stable dynamics in several cases where there is no heterogeneity of risk at equilibrium – for example, when there is (1) no dispersal $e_n = e_p = 0$, (2) purely homogeneous dispersal ($z_i = z_j$ and $q_i = q_j$ for all i and j), and (3) infinite dispersal ($e_n \to \infty$, $e_p \to \infty$). If there is finite and heterogeneous dispersal, however, they showed that heterogeneity of risk caused by parasitoid aggregation is typically stabilizing, just as it is in discrete models.

In a recent study of continuous-time models, Murdoch & Oaten (1989) arrive at the opposite conclusion. Specifically, they contend that density independent aggregation has no effect on stability in continuous time and that the effect of density dependent aggregation depends on the rate at which the variance in local host density changes with the mean. If this variance is equal to AN^x, where N is the mean host density and A and x are constants then density dependent aggregation is stabilizing for $x > 2$ and destabilizing for $x < 2$. They argue that the latter case is more common in nature (Taylor *et al.* 1980). Interestingly, these results are identical to those obtained by Godfray & Pacala (1991) from their discrete-time model referred to above with infinitely rapid host and parasitoid movement within each season.

Murdoch & Oaten (1989) do not explicitly model the dynamics of each of several spatial patches. Rather, their model consists of a pair of differential equations governing the mean parasitoid and host abundances across all patches. They also assume that the covariance of local parasitoid and host abundances is given by one of several simple functions of the mean abundances across all patches. Godfray & Pacala (1990) show that Murdoch & Oaten's model is obtained from explicitly spatial models, such as (9), in the limit of infinitely rapid movement by both species (i.e. $e_n \to \infty$, $e_p \to \infty$), or if dispersal is governed by complex and biologically implausible density dependent rules.

Since the host stages attacked by parasitoids are generally relatively sedentary, and some degree of heterogeneity of risk is likely to be the rule in nature systems, models with infinite interpatch movement and no heterogeneity of risk would generally seem to be inappropriate. We conclude that aggregation by parasitoids generally acts to stabilize host–parasitoid interactions whether in continuous or discrete time.

ESTIMATING PARAMETERS FROM FIELD DATA

(*a*) *The method*

The $CV^2 > 1$ rule is in terms of the distribution of searching parasitoids. Such data, however, are rarely available from natural populations (see also Waage 1983; Casas 1988, 1989); most of the information is in the form of relationships between percent parasitism and host density per patch (figure 1). In this section we show how the parameters necessary to calculate the CV^2 can be estimated from such data. Full details of the technique are given in Pacala & Hassell (1990).

We begin by assuming that the local parasitoid density can be described by the regression function (see model II above) with multiplicative residual ϵ:

$$p = cP_t\left(\frac{n}{N_t}\right)^{\mu}\epsilon, \tag{10}$$

where n is the local host density, c and μ are constants, and ϵ is a unit mean gamma distributed random variable (as for model 1). Thus, host density dependent heterogeneity is described by the expression $(n/N_t)^{\mu}$, with the magnitude of the association between local parasitoid and local host density governed by the value of μ. The magnitude of any host density independent heterogeneity is determined by the gamma distributed random variable ϵ, with the magnitude of HDI increasing with the variance of ϵ. (Note that the methods described below also apply to functional forms of HDD other than (10). To extend the methods outlined below to another function $g(n)$, the steps are merely repeated after substituting the new $g(n)$ for $(n/N_t)^{\mu}$.)

With these assumptions

$$f(N_t, P_t) = \int_0^{\infty}\int_0^{\infty} \phi(n/N_t)\,\gamma(\epsilon)\,\frac{n}{N_t} \times \exp\left(-acP_t(n/N_t)^{\mu}\epsilon\right)\mathrm{d}\epsilon\,\mathrm{d}n, \tag{11}$$

where $\phi(n/N_t)$ is the distribution of n/N_t and $\gamma(\epsilon)$ is the unit-mean gamma density.

Pacala & Hassell (1991) show that $CV^2 > 1$ from this general model may be approximated as:

$$CV^2 \approx C_I C_D - 1. \tag{12}$$

Here $C_I = 1+\sigma^2$ and represents the component of heterogeneity that is independent of host density, and $C_D = 1 + V^2\mu^2$ and represents the component that depends on host density. The term σ^2 is the variance of ϵ and V is the coefficient of variation of the local host density calculated with respect to a randomly chosen host.

Notice from (12) that (1) $CV^2 > 1$ if either $C_I > 2$ or $C_D > 2$, (2) that C_I and C_D both affect $CV^2 > 1$ in the same way and (3) that because the slope-determining expression μ in C_D is squared, the effects of direct (positive slope) and inverse (negative slope) dependence on local host density are identical. Thus all of the patterns shown in figure 1 contribute to population regulation in the same way irrespective of whether the density dependence is positive or negative, or if there is no density dependence at all. The degree to which each affects stability just depends on the magnitude of the CV^2.

The method of obtaining CV^2 from field data involves estimating μ, σ^2 and acP_t from equation (11) by using a maximum likelihood procedure, and estimating V^2 for the host distribution directly from the data on local host densities. (Values of σ^2 were constrained to be $\geqslant 0.05$ to prevent numerical overflow during computation. As a result, estimates of σ^2 reported as 0.05 are actually between 0 and 0.05. In these cases, the calculated value of CV^2 is an overestimate, but the bias is less than 5% (Pacala & Hassell 1991).

(*b*) *The data*

The data set that we have analysed involves 65 examples from field studies reporting percent parasitism versus local host density per patch, for each of which we have obtained estimates of σ^2, μ and V^2, and thence calculated C_I, C_D and CV^2 (table 3). Of these, 32 are listed in two recent reviews by Stiling (1987); Walde & Murdoch (1988). We have also added a further 33 examples, mainly from unpublished studies. The full assemblage is listed in table 2. Other studies, listed but not used here, were found to be unsuitable for our analysis, usually because the per patch sample sizes were unreported. Several of these data sets are temporal or spatial replicates of others. We have, therefore, also produced a reduced assemblage of 26 data sets by choosing a single replicate at random and omitting all others (designated by an asterisk in the first column of the table 3). Each of this reduced set thus describes a different pair of species.

(*c*) *Results*

For each example in tables 2 and 3, the estimated value of the CV^2 can be evaluated in relation to the $CV^2 > 1$ rule and the relative importance of density dependent and density independent heterogeneity (HDD versus HDI) to the total heterogeneity determined. It is also possible in each case to predict the mean percent parasitism in relation to host density per patch from the expression:

$$100\left[1-\alpha\left(\alpha+acP_t\left(\frac{n}{N_t}\right)^{\mu}\right)\right]^{\alpha}, \tag{13}$$

where α, acP_t and μ are the maximum likelihood estimates referred to above. This is shown for the range of examples in figure 1 where each fitted curve represents the percent parasitism predicted by the mean of (11) evaluated at the maximum likelihood estimates. In all cases the correspondence is quite close between the predicted and actual mean levels of parasitism.

Figure 1*a* shows an example of a direct density dependent pattern of parasitism (data set 25 in tables

Table 2. *List of data sets analysed*

no.	host species (family)	parasitoid species (family)	author(s)
1	*Bupalus piniaria* (Geometridae)	*Dusona oxyacanthae* (Ichneumonidae)	N. Broekhuizen (unpublished data)
2	*B. piniaria*	*D. oxyacanthae*	N. Broekhuizen (unpublished data)
3	*B. piniaria*	*D. oxyacanthae*	N. Broekhuizen (unpublished data)
4	*B. piniaria*	*D. oxyacanthae*	N. Broekhuizen (unpublished data)
5	*B. piniaria*	*D. oxyacanthae*	N. Broekhuizen (unpublished data)
6*	*B. piniaria*	*D. oxyacanthae*	N. Broekhuizen (unpublished data)
7	*B. piniaria*	*D. oxyacanthae*	N. Broekhuizen (unpublished data)
8	*B. piniaria*	*D. oxyacanthae*	N. Broekhuizen (unpublished data)
9*	*Lymantria dispar* (Lymantriidae)	*Ooencyrtus kuwanai* (Encyrtidae)	Brown & Cameron (1979)
10*	*Trirhabda virgata* (Chrysomelidae)	Mirmithid nematodes	N. Cappuccino (unpublished data)
11*	*Microrhopala virtata* (Chrysomelidae)	Erythreid mites	N. Cappuccino (unpublished data)
12*	*Eurosta solidasinus* (Tephritidae)	*Eurytoma gigantea* *E. obtusiventris* (Eurytomidae)	N. Cappuccino (unpublished data)
13*	*Rhopalomyia californica* (Cecidomyiidae)	*Torymus baccaridis* (Torymidae)	Ehler (1986)
14*	*R. californica*	*Tetrasticus* sp.	Ehler (1986)
15*	*Silo pallipes* (Goeridae)	*Agriotypus armatus* (Agriotypidae)	Elliott (1983)
16*	*Sceliphron assimile* (Sphecidae)	*Mellittobia chalybii* (Eulophidae)	Freeman & Parnell (1983)
17	*Andricus quercuscalicis* (Cynipidae)	*Mesopolobus fuscipes* (Pteromalidae)	Hails (1988)
18*	*A. quercuscalicis*	*M. fuscipes*	Hails (1988)
19*	*A. quercuscalicis*	*Mesopolobus xanthocerus*	Hails (1988)
20	*A. quercuscalicis*	*M. xanthocerus*	Hails (1988)
21*	*Phytomyza ilicis* (Agromyzidae)	*Chrysocharis gemma* (Eulophidae)	Heads & Lawton (1983)
22*	*Papilio xuthus* (Papilionidae)	*Trichogramma papilionis* (Trichogrammatidae)	Hirose *et al.* (1976)
23	*Delia radicum* (Anthomyiidae)	*Trybliographa rapae* (Eucoilidae)	Jones & Hassell (1988)
24	*D. radicum*	*T. rapae* (Eucoilidae)	Jones & Hassell (1988)
25	*D. radicum*	*T. rapae*	Jones & Hassell (1988)
26	*D. radicum*	*T. rapae*	Jones & Hassell (1988)
27*	*D. radicum*	*T. rapae*	Jones & Hassell (1988)
28	*D. radicum*	*T. rapae*	Jones & Hassell (1988)
29	*D. radicum*	*T. rapae*	T. H. Jones (unpublished data)
30	*D. radicum*	*T. rapae*	T. H. Jones (unpublished data)
31	*D. radicum*	*T. rapae*	T. H. Jones (unpublished data)
32*	*Chirosia histricina* (Anthomyiidae)	Total parasitism	J. H. Lawton (unpublished data)
33*	*Fiorinia externa* (Diaspidae)	*Aspidiotiphagus citrinus* (Eulophidae)	McClure (1977)
34	*F. externa*	*A. citrinus*	McClure (1977)
35	*F. externa*	*A. citrinus*	McClure (1977)
36*	*Epinotia tedella* (Tortricidae)	*Apanteles tedellae* (Braconidae)	Munster-Swendsen (1980)
37*	*E. tedellae*	*Pimplopterus dubius* (Ichneumonidae)	Munster-Swendsen (1980)
38	*Parlatoria oleae*	*Aphytis paramaculicornis* (Aphelinidae)	Murdoch *et al.* (1984)
39*	*P. oleae*	*Coccophagoides utilis* (Aphelinidae)	Murdoch *et al.* (1984)
40*	*P. oleae*	*A. paramaculicornis*	Murdoch *et al.* (1984)
41	*Coleophora laricella* (Coleophoridae)	*Agathis pumila* (Braconidae)	R. B. Ryan (unpublished data)
42*	*C. laricella*	*A. pumila*	R. B. Ryan (unpublished data)
43	*C. laricella*	*A. pumila*	R. B. Ryan (unpublished data)
44	*C. laricella*	*A. pumila*	R. B. Ryan (unpublished data)

Table 2. (*cont.*)

45	*C. laricella*	*A. pumila*	R. B. Ryan (unpublished data)
46	*C. laricella*	*A. pumila*	R. B. Ryan (unpublished data)
47	*C. laricella*	*A. pumila*	R. B. Ryan (unpublished data)
48	*C. laricella*	*A. pumila*	R. B. Ryan (unpublished data)
49	*C. laricella*	*A. pumila*	R. B. Ryan (unpublished data)
50	*C. laricella*	*A. pumila*	R. B. Ryan (unpublished data)
51	*C. laricella*	*A. pumila*	R. B. Ryan (unpublished data)
52	*C. laricella*	*A. pumila*	R. B. Ryan (unpublished data)
53	*Aonidiella aurantii* (Diaspidae)	*Aphtyis melinus* (Aphelinidae)	Smith & Maelzer (1986)
54*	*A. aurantii*	*A. melinus*	Smith & Maelzer (1986)
55	*A. aurantii*	*A. melinus*	Smith & Maelzer (1986)
56	*A. aurantii*	*A. melinus*	Smith & Maelzer (1986)
57*	*Eupteryx cyclops* *E. urticae* (Cicadellidae)	*Anagrus* sp. (Mymaridae)	Stiling (1980)
58*	*Polistes exclamans* (Vespidae)	*Elasmus polistis* (Elasmidae)	Strassmann (1981)
59	*P. exclamans*	*E. polistis*	Strassmann (1981)
60*	*Icerya purchasi* (Margarodidae)	*Cryptochaetum iceryae* (Cryptochaetidae)	Thorarinsson (1990)
61	*I. purchasi*	*C. iceryae*	Thorarinsson (1990)
62	*I. purchasi*	*C. iceryae*	Thorarinsson (1990)
63	*I. purchasi*	*C. iceryae*	Thorarinsson (1990)
64*	*Trypargilum politum* (Sphecidae)	*Melittobia* sp. (Eulophidae)	Trexler (1985)
65	*Lymantria dispar* (Lymantriidae)	*Ooencyrtus kuwanai* (Encyrtidae)	Weseloh (1972)

* Reduced list with no spatial or temporal replicates.

2 and 3). The calculated CV^2 for this data set is 1.16, from which we can predict (assuming this to be a typical result in successive generations) that such levels of heterogeneity in parasitism would be just sufficient to stabilize the interaction. The relatively large value for C_D (2.06) and small value for C_I (1.05) indicate that virtually all of the stabilizing heterogeneity in this example comes from the host density dependent heterogeneity (HDD). This arises from two biological properties of the interacting populations: (i) the large spatial covariance of the parasitoids with local host density, indicated by the relatively large value of μ (= 1.37), and (ii) the highly aggregated spatial distribution of the host (indicated by the relatively large coefficient of variation of the host population (V^2 = 0.66).

Figure 1 *b* shows a further example (data set 9) where parasitism is strongly correlated with host density per patch, but in this case the pattern is inversely density dependent ($\mu = -0.95$). However, in contrast to the previous example, HDD is only slightly larger than HDI ($C_D = 1.29$; $C_I = 1.11$) and they thus combine to produce only a low level of total heterogeneity in parasitism ($CV^2 = 0.37$). The small effect of HDD, despite the relatively large value of μ, stems from the low level of spatial aggregation of the host population ($V^2 = 0.25$). In short, such levels of heterogeneity would be too small to contribute significantly to population regulation. This prediction is reassuring in the light of the spectacular quasi-periodic outbreaks sometimes shown by the gypsy moth (Myers 1988).

Figure 1 *c* (data set 33) provides a further example of pronounced spatial density dependence ($\mu = 0.87$) but a low value of CV^2 (= 0.34). Once again, there is too little aggregation in the host distribution ($V^2 = 0.21$) for the HDD to contribute appreciably to heterogeneity.

Figure 1 *d* (data set 14) shows a case that is quite different from the previous three. Density independent (HDI) variation in parasitism is now much more important than density dependent variation (HDD) ($C_I = 8.25$; $C_D = 1.01$). Furthermore, because C_I is significantly greater than 2, heterogeneity independent of host density is large enough by itself to stabilize the interacting populations. The prediction of stability thus does not depend at all on the host's spatial distribution, or on the existence of spatial covariance (beginning-of-season) between the parasitoid and host. Hence, although the data in 1 *d* appear erratic, they actually contain more evidence of factors that could stabilize dynamics than do the previous examples.

Finally, figure 1 *e* (data set 39) shows another example with virtually no stabilizing heterogeneity. There is no appreciable effect from HDI (the estimate of σ^2 is at the constrained lower bound), and there is no sign of the parasitoids responding to local host density ($\mu = 0.07$).

Rather than survey each of the 65 examples in table 2 in this one-by-one way, we now turn to a broader comparison of the range of CV^2 values and the relative contribution to these of HDD and HDI heterogeneity. In the first place, figure 4 shows all the examples plotted in relation to the values of C_I and C_D, with the $CV^2 = 1$ contour also overlaid. In 18 of the 65 cases, $CV^2 > 1$ indicating that heterogeneity at this level ought to be

Table 3. *Maximum likelihood estimates used in calculating C_D, C_I and CV^2*

no.	host species (family)	C_D (= HDD)	C_I (= HDI)	CV^2	μ (95 % limits)	σ^2 (95 % limits)	acP (95 % limits)	V^2 (95 % limits)
1	*Bupalus piniaria* (Geometridae)	1.04	1.26	0.30	0.34 (−0.46:1.18)	0.26 (0.05:20.0)	0.04 (0.02:0.06)	0.32
2	*B. piniaria*	1.04	1.11	0.10	0.53 (−0.13:1.19)	0.05 (0.05:0.26)	0.75 (0.60:1.01)	0.16
3	*B. piniaria*	1.05	1.06	0.12	0.60 (−0.52:1.61)	0.06 (0.05:20.0)	0.36 (0.20:0.60)	0.17
4	*B. piniaria*	1.02	1.60	0.63	0.22 (−0.49:1.02)	0.60 (0.05:2.81)	0.07 (0.05:1.12)	0.45
5	*B. piniaria*	1.08	1.05	0.14	−0.86 (−2.12:0.68)	0.05 (0.05:20.0)	0.47 (0.23:0.86)	0.11
6*	*B. piniaria*	1.99	2.30	3.48	1.87 (0.67:3.34)	1.25 (0.17:5.44)	0.06 (0.03:0.15)	0.29
7	*B. piniaria*	1.04	3.87	3.03	0.25 (−0.79:1.37)	2.87 (0.65:15.1)	0.02 (0.00:0.04)	0.68
8	*B. piniaria*	1.44	3.30	3.73	0.86 (−0.54:2.31)	2.30 (0.05:20.0)	0.02 (0.01:0.04)	0.61
9*	*Lymantria dispar* (Lymantriidae)	1.29	1.11	0.37	−0.95 (−1.0:0.87)	0.11 (0.05:0.15)	0.49 (0.47:0.52)	0.25
10*	*Trirhabda virgata* (Chrysomelidae)	1.00	2.63	1.64	0.77 (0.13:1.36)	1.63 (0.93:2.97)	0.52 (0.37:0.74)	0.01
11*	*Microrhopala virtata* (Chrysomelidae)	1.01	2.60	1.62	0.20 (−0.35:0.83)	1.60 (0.68:4.43)	0.04 (0.03:0.06)	0.26
12*	*Eurosta solidasinus* (Tephritidae)	1.18	1.05	0.24	−0.61 (−1.21:−0.10)	0.05 (0.05:0.50)	0.60 (0.39:0.88)	0.48
13*	*Rhopalomyia californica* (Cecidomyiidae)	1.01	4.00	3.02	0.25 (−0.93:1.00)	2.99 (1.91:4.74)	0.41 (0.28:0.61)	0.10
14*	*R. californica*	1.01	8.25	7.33	−0.24 (−1.23:0.71)	7.25 (4.70:11.2)	0.39 (0.26:0.61)	0.20
15*	*Silo pallipes* (Goeridae)	1.01	1.05	0.06	−0.20 (−0.84:0.47)	0.05 (0.05:0.19)	0.22 (0.15:0.30)	0.18
16*	*Sceliphron assimile* (Sphecidae)	1.04	1.23	0.28	0.20 (0.01:0.48)	0.23 (0.12:0.38)	0.70 (0.48:0.94)	1.03
17	*Andricus quercuscalicis* (Cynipidae)	1.00	1.92	0.92	−0.02 (−0.28:0.24)	0.92 (0.37:1.73)	0.09 (0.07:0.11)	0.57
18*	*A. quercuscalicis*	1.06	1.75	0.86	−0.27 (−0.48:−0.06)	0.75 (0.14:1.50)	0.07 (0.06:0.09)	0.86
19*	*A. quercuscalicis*	1.10	3.60	2.96	−0.42 (−0.72:−0.10)	2.60 (1.49:4.13)	0.10 (0.08:0.13)	0.57
20	*A. quercuscalicis*	1.08	1.66	0.79	−0.30 (−0.51:−0.10)	0.66 (0.07:1.30)	0.07 (0.06:0.09)	0.86
21*	*Phytomyza ilics* (Agromyzidae)	1.09	1.23	0.34	0.57 (0.12:1.03)	0.23 (0.20:0.53)	0.19 (0.15:0.24)	0.28

22*	*Papilio xuthus* (Papilionidae)	1.16	1.05	0.21	−0.69 (−1.56:0.09)	0.05 (0.05:2.00)	1.11 (0.64:1.83)	0.33
23	*Delia radicum* (Anthomyiidae)	1.62	1.05	0.70	1.59 (0.59:2.00)	0.05 (0.05:0.11)	0.11 (0.07:0.18)	0.24
24	*D. radicum*	1.47	1.11	0.63	0.91 (0.60:1.23)	0.11 (0.05:0.39)	0.33 (0.26:0.42)	0.57
25	*D. radicum*	2.06	1.05	1.16	1.37 (1.14:1.65)	0.05 (0.05:0.22)	0.09 (0.07:0.12)	0.56
26	*D. radicum*	1.36	1.05	0.43	0.88 (0.65:1.10)	0.05 (0.05:0.08)	0.28 (0.23:0.35)	0.47
27*	*D. radicum*	1.00	1.79	0.79	−0.10 (−0.55:0.39)	0.79 (0.31:1.79)	0.30 (0.21:0.44)	0.36
28	*D. radicum*	1.00	1.05	0.05	−0.09 (−0.54:0.37)	0.05 (0.05:0.13)	0.30 (0.23:0.38)	0.25
29	*D. radicum*	1.01	1.05	0.06	0.17 (−0.26:0.60)	0.05 (0.05:0.13)	0.27 (0.21:0.35)	0.28
30	*D. radicum*	1.27	1.05	0.34	0.51 (0.20:0.78)	0.05 (0.05:0.13)	0.13 (0.08:0.18)	1.05
31	*D. radicum*	1.01	1.05	0.06	0.13 (−0.65:0.84)	0.05 (0.05:0.11)	0.11 (0.07:0.17)	0.30
32*	*Chirosia histricina* (Anthomyiidae)	1.00	1.12	0.12	−0.05 (−0.55:0.48)	0.12 (0.05:0.96)	0.33 (0.22:0.49)	0.36
33*	*Fiorinia externa* (Diaspidae)	1.16	1.16	0.34	0.87 (0.52:1.12)	0.16 (0.08:0.37)	0.33 (0.27:0.40)	0.21
34	*F. externa*	1.02	1.08	0.10	0.40 (0.32:0.49)	0.08 (0.05:0.14)	0.38 (0.35:0.39)	0.15
35	*F. externa*	1.08	1.05	0.13	0.63 (0.45:0.80)	0.05 (0.05:0.06)	0.58 (0.54:0.64)	0.20
36*	*Epinotia tedella* (Tortricidae)	1.08	1.50	0.61	1.03 (0.10:2.00)	0.50 (0.05:0.68)	0.04 (0.03:0.05)	0.07
37*	*E. tedellae*	1.02	1.40	0.42	−0.43 (−1.39:0.39)	0.40 (0.40:0.49)	0.04 (0.03:0.05)	0.07
38	*Parlatoria oleae*	1.07	1.05	0.12	0.36 (0.09:0.64)	0.05 (0.05:0.19)	0.23 (0.19:0.28)	0.51
39*	*P. oleae*	1.00	1.05	0.05	0.07 (−0.14:0.28)	0.05 (0.05:0.15)	0.56 (0.49:0.63)	0.61
40*	*P. oleae*	1.05	1.05	0.10	0.51 (0.19:0.86)	0.05 (0.05:0.16)	0.65 (0.56:0.76)	0.18
41	*Coleophora laricella* (Coleophoridae)	1.06	2.40	1.54	−0.57 (−2.15:0.89)	1.40 (0.26:11.7)	0.10 (0.05:0.22)	0.17
42*	*C. laricella*	1.00	1.13	0.13	−0.02 (−0.73:0.74)	0.13 (0.05:0.55)	0.15 (0.11:0.21)	0.12
43	*C. laricella*	5.00	1.41	6.06	0.50 (−0.10:1.17)	0.41 (0.18:1.37)	0.24 (0.16:0.36)	16.19
44	*C. laricella*	1.00	2.05	0.05	−0.06 (−0.32:0.20)	0.05 (0.05:0.20)	1.78 (1.46:2.17)	0.39

Table 3. (*cont.*)

45	*C. laricella*	2.06	2.51	4.18	−1.97 (−4.24: −0.08)	1.51 (0.34:8.38)	0.16 (0.07:0.33)	0.27
46	*C. laricella*	1.02	1.05	0.07	−0.43 (−0.93:0.08)	0.05 (0.05:0.14)	1.79 (1.51:2.13)	0.11
47	*C. laricella*	1.01	1.05	0.06	0.12 (−0.67:0.90)	0.05 (0.05:1.00)	2.50 (1.40:4.53)	0.40
48	*C. laricella*	1.04	1.05	0.10	0.26 (−0.59:1.05)	0.05 (0.05:0.40)	0.04 (0.02:0.072)	0.66
49	*C. laricella*	1.85	1.33	1.47	1.30 (0.55:2.19)	0.33 (0.05:1.00)	0.13 (0.07:0.22)	0.50
50	*C. laricella*	2.09	1.38	1.88	0.44 (−0.57:1.57)	0.38 (0.15:1.29)	0.37 (0.25:0.56)	5.55
51	*C. laricella*	1.08	1.05	0.13	0.34 (−0.04:0.71)	0.05 (0.05:0.14)	0.79 (0.68:0.92)	0.69
52	*C. laricella*	1.00	1.05	0.05	−0.04 (−0.41:0.33)	0.05 (0.05:0.09)	2.22 (1.87:2.65)	0.15
53	*Aonidiella aurantii* (Diaspidae)	1.02	1.16	0.19	0.16 (−0.02:0.35)	0.16 (0.07:0.32)	0.53 (0.44:0.64)	0.87
54*	*A. aurantii*	1.00	3.00	2.00	−0.03 (−0.70:0.45)	2.00 (1.46:2.70)	0.30 (0.17:0.58)	0.48
55	*A. aurantii*	1.01	1.05	0.06	0.24 (−0.42:0.90)	0.05 (0.05:0.18)	0.57 (0.46:0.71)	0.10
56	*A. aurantii*	1.00	1.06	0.06	−0.04 (−0.77:0.73)	0.06 (0.05:0.37)	0.24 (0.18:0.32)	0.13
57*	*Eupteryx cyclops* *E. urticae* (Cicadellidae)	1.21	1.05	0.27	0.93 (0.39:1.50)	0.05 (0.05:0.19)	0.48 (0.35:0.64)	0.24
58*	*Polistes exclamans* (Vespidae)	1.00	2.72	1.73	0.08 (−0.40: −0.25)	1.72 (0.95:3.58)	0.24 (0.17:0.32)	0.46
59	*P. exclamans*	1.08	8.14	7.77	−0.46 (−0.87: −0.33)	7.14 (2.34:10.0)	0.03 (0.03:0.07)	0.36
60*	*Icerya purchasi* (Margarodidae)	1.00	1.05	0.05	0.01 (−0.22:0.25)	0.05 (0.05:0.18)	1.11 (0.92:1.33)	0.31
61	*I. purchasi*	1.01	1.05	0.06	0.05 (−0.23:0.34)	0.05 (0.05:0.20)	0.74 (0.57:0.94)	0.57
62	*I. purchasi*	1.03	1.30	0.34	−0.25 (−0.75:0.26)	0.30 (0.11:0.81)	1.42 (1.01:2.03)	0.48
63	*I. purchasi*	1.00	1.05	0.05	−0.01 (−0.16:0.14)	0.05 (0.05:0.19)	1.11 (0.92:1.34)	0.79
64*	*Trypargilum politum* (Sphecidae)	1.02	2.05	1.09	0.18 (−0.43:0.89)	1.05 (0.13:6.58)	0.08 (0.04:0.12)	0.56
65	*Lymantria dispar* (Lymantriidae)	1.07	1.09	0.17	−0.53 (−0.63: −0.38)	0.09 (0.07:0.13)	0.42 (0.39:0.43)	0.24

* Reduced list with no spatial or temporal replicates.

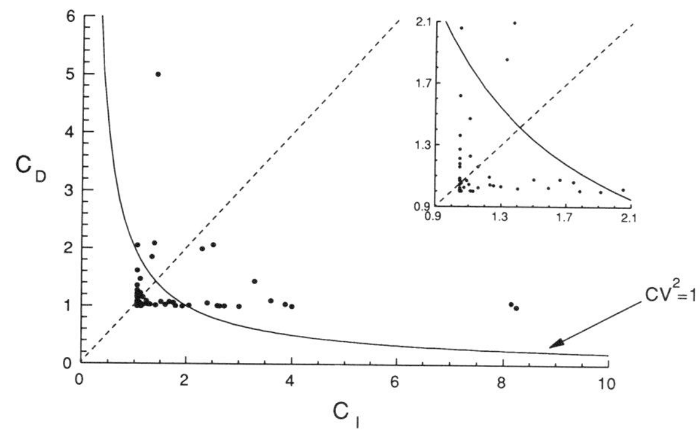

Figure 4. Host density dependent heterogeneity (C_D) plotted against host density independent heterogeneity (C_I). The points correspond to the 65 examples in tables 2 and 3. The solid curve is the contour where $CV^2 = 1$. The dashed line separates points for which $C_I > C_D$ from those for which $C_D > C_I$. The inset magnifies the cluster of points nearest to the origin. See text for further details. (After Pacala *et al.* 1990.)

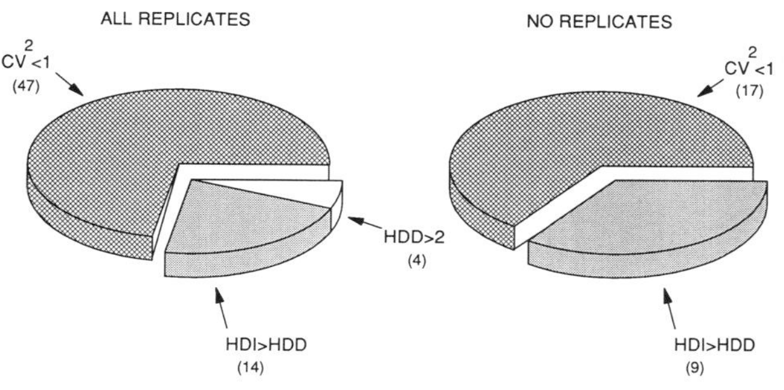

Figure 5. Pie diagrams distinguishing between examples in table 3 where $CV^2 < 1$ from those $CV^2 > 1$. (Left): all 65 data sets; (right) only the 26 data sets marked by asterisks which include no spatial or temporal replicates. Where $CV^2 > 1$, the shaded segment corresponds to host density independent heterogeneity predominating and the unshaded segment to where host density dependent heterogeneity is the more important.

sufficient to stabilize the populations. Interestingly, in 14 of these 18 cases $C_I > C_D$ and, furthermore, in each case $C_I > 2$, indicating that the level of heterogeneity in C_I alone is sufficient to make $CV^2 > 1$. In the remaining 47 examples in figure 4 where $CV^2 < 1$, the majority (see inset) CV^2 show heterogeneity having no appreciable affect on population regulation. In eight cases, however, $CV^2 > 0.6$, at which level heterogeneity could be promoting stability to some degree. Figure 5 further summarizes these results and also shows that the pseudoreplication does not appear to be biasing the different categories since the same qualitative picture emerges from the reduced data set of 26 different species pairs.

Thus, contrary to the popular view, this analysis suggests that density independent spatial patterns of parasitism (e.g. figure 1*c*) may be more important in promoting population regulation than density dependent patterns. In the six cases where HDD is important (i.e. C_D contributes substantially to the $CV^2 > 1$), the density dependence is inverse in only one case (data set 55; $\mu = -1.97$). This ratio is smaller than expected from the overall frequencies of significant direct and inverse density dependent patterns summarised in figure 6.

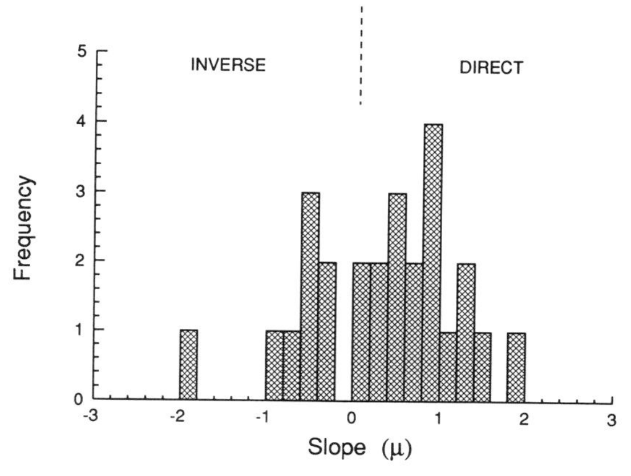

Figure 6. Frequency distribution of estimated values of μ from table 3 which are significantly different from $\mu = 0$ ($p < 0.05$).

CONCLUSIONS

Although, the discrete-time models discussed in this paper show that the '$CV^2 > 1$ rule' may apply across a wide variety of host–parasitoid interactions, several fundamental assumptions about the biology of the interactions have been made which do not apply to all host–parasitoid systems.

(i) The assumption of coupled, synchronized interactions restricts our analysis to parasitoids that are effectively specialists on the one host species. The dynamics of generalist parasitoids and their hosts can be very different (Hassell & May 1986; Latto & Hassell 1988) and will require a separate treatment.

(ii) Having discrete host and parasitoid generations does not permit appreciable overlapping of host and parasitoid generations which would be better represented in continuous time. Whether or not comparable stability criteria exist for the affects of heterogeneity in such continuous interactions has yet to be determined.

(iii) The extent to which the parasitoids encounter hosts at random within patches, as assumed in our models, and the importance of any deviations from this, will depend in part on the size of the patches relative to the foraging area of the parasitoids. This introduces important questions on the scales at which HDD and HDI heterogeneity exert their main effects. For example, any covariance between parasitoid distribution and local host densities per patch is likely to be scale-dependent, since it depends critically on what a foraging parasitoid recognizes as a patch (Waage 1979). Likewise, HDI heterogeneity depends on any differences in the attractiveness of patches independent of the host density that they contain, as well as any Poisson and other 'errors' in the parasitoids' decision making. Both of these are likely to be strongly influenced by the scale of patchiness being examined.

(iv) By neglecting interference between parasitoids or competition among hosts, the models in this paper focus on situations where interactions between host and parasitoid populations are of predominant im-

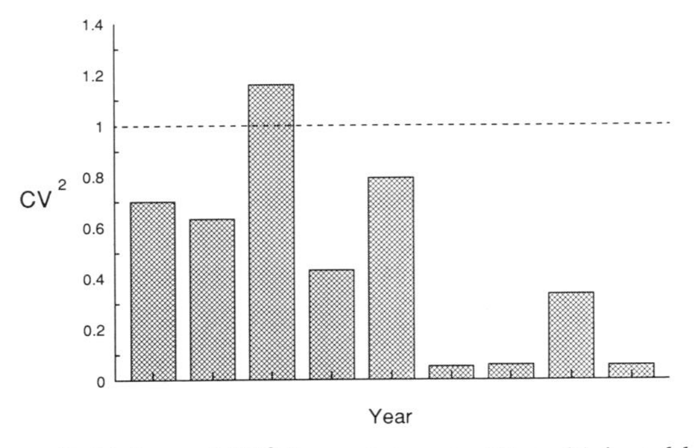

Figure 7. Values of CV^2 from data sets 23 to 31 in table 3. These are annual replicates between 1980 and 1989. (From T. H. Jones & M. P. Hassell, unpublished data.)

portance for the overall dynamics. Any additional density-dependent and density-independent factors that markedly influence fecundity or survival of the hosts and parasitoids will also affect dynamics and change the overall stability conditions. In neglecting these, our emphasis is primarily on understanding the extent to which one factor, the distribution of parasitism amongst hosts, can contribute to population regulation.

(v) Many of the examples in tables 2 and 3 come from single-generation studies with no temporal replication. We can thus only predict effects on dynamics assuming that the estimated values of CV^2 are typical for each interaction. In fact, those studies with some temporal replication show considerable variation in CV^2 from generation to generation, as shown by the example in figure 7. Procedures are now needed for evaluating how fluctuating CV^2 values affect population dynamics.

An often contentious issue in ecology has concerned the relevance of general models in understanding the dynamics of natural systems in the field. This study shows how relatively simple models of host–parasitoid systems can profitably be applied to field data on levels of parasitism in a patchy environment. Such heterogeneity has often been regarded as a complicating factor in population dynamics, and one that rapidly leads to analytical intractability. Clearly, this need not necessarily be so. The $CV^2 > 1$ rule explains the consequences of heterogeneity for population dynamics in terms of a simple description of the heterogeneity itself. The rule gives a rough prediction of the effects of heterogeneity and also identifies the kinds of heterogeneity that contribute to population regulation.

APPENDIX

In this appendix we derive simplified expressions for the stability conditions (2*a*) and (2*b*). Let

$$f(N_t, P_t) = E[n\,\mathrm{e}^{-P_t g(n)\epsilon}], \qquad \text{(A 1)}$$

where E ( ) stands for expectation (average across all patches), n is the relative abundance of hosts in a patch (local abundance divided by mean abundance), $g(n)$ describes the tendency of local parasitoid abundance to change with n, and ϵ is a unit-mean random variable that governs spatial variance in local parasitoid abundance which is unrelated to local host density. Also, the term $\mathrm{e}^{-P_t g(n)\epsilon}$ is the probability of surviving in a patch if the parasitoids in the patch search randomly (see the explanation of model I in the text). In a pure HDD model, ϵ is a constant and equal to one, and in a pure HDI model, $g(n)$ is constant. Hassell *et al.* (1990) review the many previous studies that employ special cases of (A 1).

Further, let $\bar{g} = E(g(n))$ and $\bar{g}' = E(ng(n))$, and observe that:

$$-\partial f(N^*, P^*)/\partial P_t = E(ng(n)\,\epsilon\,\mathrm{e}^{-P^* g(n)\epsilon}). \qquad \text{(A 2)}$$

The covariance between local parasitoid and unparasitized host density at the end of a growing season is:

$$\begin{aligned}\begin{bmatrix}\text{end-of-season}\\ \text{covariance}\end{bmatrix} &= E\left[\begin{pmatrix}\text{parasitoids}\\ \text{in patch } i\end{pmatrix}\begin{pmatrix}\text{surviving}\\ \text{hosts in patch } i\end{pmatrix}\right] - E\begin{bmatrix}\text{parasitoids}\\ \text{in patch } i\end{bmatrix} E\begin{bmatrix}\text{surviving}\\ \text{hosts in patch } i\end{bmatrix},\\ &= E[(P^* g(n_i)\,\epsilon)(N^* n_i\,\mathrm{e}^{-P^* g(n)\epsilon})] - E[P^* g(n_i)\,\epsilon]E[N^* n_i\,\mathrm{e}^{-P^* g(n)\epsilon}],\\ &= P^* N^*\{E[n_i\, g(n_i)\,\epsilon\,\mathrm{e}^{-P^* g(n)\epsilon}] - E[g(n_i)\,\epsilon]\,E[n_i\,\mathrm{e}^{-P^* g(n)\epsilon}]\}.\end{aligned}$$

Of the three expectations in the above equation, the first is given by (A 2), the second is $\bar{g}$ and the third is given by (A 1). Thus

$$[\text{end-of-season covariance}] = P^* N^*[\partial f(N^*, P^*)/\partial P_t - \bar{g} f(N^*, P^*)].$$

At equilibrium, the mean parasitoid abundance per patch is simply $P^*\bar{g}$, the mean end-of-season abundance of unparasitized hosts is $f(N^*, P^*)\,N^*$, and $f(N^*, P^*) = 1/\lambda$. Thus, C_e, the end-of-season covariance divided by the product of mean parasitoid and unparasitized host abundance ($N^*P^*\bar{g}/\lambda$) is:

$$C_e = [-\lambda/\bar{g}\cdot\partial f(N^*, P^*)/\partial P_t] - 1. \qquad \text{(A 3)}$$

Similarly, the equilibrium covariance at the beginning of the season is:

$$\begin{aligned}[\text{beginning-of-season covariance}] &= E(N^* n P^* g(n)\,\epsilon) - N^* P^* \bar{g},\\ &= N^* P^*[E(ng(n)\,\epsilon) - \bar{g}],\end{aligned}$$

and so the scaled covariance C_b is:

$$C_b = \bar{g}'/\bar{g} - 1. \qquad \text{(A 4)}$$

By using (A 3) and (A 4), we may write condition (2*a*) as:

$$C_e + 1/C_b + 1 < \lambda - 1/\lambda P^* \bar{g}' = Z(\lambda). \qquad \text{(A 5)}$$

Now, consider the equilibrium equation from (1*a*):

$$1/\lambda = E[n\,\mathrm{e}^{-P^* g(n)\epsilon}]. \qquad \text{(A 6)}$$

It is straightforward to show from (A 6) that P^* increases monotonically from zero to infinity as λ increases from one to infinity. As $\lambda \to 1$, (A 6) approaches:

$$1/\lambda = E(n(1 - P^* g(n)\,\epsilon)),$$

which reduces to $1/\lambda = 1 - P^*\bar{g}'$.

Thus, as $\lambda \rightarrow 1$, condition (A 5) approaches:

$$C_e + 1 < 1/\lambda(C_b + 1),$$

and as $\lambda \rightarrow \infty$, (A 5) approaches:

$$C_e < -1.$$

Turning now to condition (2 *b*), we first observe from the equilibrium equation from (1 *b*) that:

$$P^* = N^*(1 - 1/\lambda).$$

Thus (2 *b*) may be written as:

$$P^*(-\partial f(N^*, P^*)/\partial P_t) > N^* \cdot \partial f(N^*, P^*)/\partial N_t.$$

If $P_t = e^X$ and $N_t = e^Y$, then

$$\partial f(N^*, P^*)/\partial X = P^* \, \partial f(N^*, P^*)/\partial P_t,$$

and

$$\partial f(N^*, P^*)/\partial Y = N^* \, \partial f(N^*, P^*)/\partial N_t,$$

and so condition (2 *b*) may be written:

$$-\partial f(N^*, P^*)/\partial X > \partial f(N^*, P^*)/\partial Y.$$

Because $\partial f/\partial X < 0$ (host survival decreases as parasitoid density increases) this condition is always satisfied unless $\partial f/\partial Y > 0$ (an increase in host abundance increases host survival). Although $\partial f/\partial Y$ may indeed be positive in some cases involving parasitoid satiation, in the majority of biologically plausible models there is either no density dependence affecting the host ($\partial f/\partial Y = 0$) or ($\partial f/\partial Y < 0$). As a result, condition (2 *a*) is usually necessary and sufficient for stability.

In summary, therefore, condition (2 *a*) is equivalent to:

$$C_e + 1 < (C_b + 1)\, Z(\lambda),$$

and condition (2 *b*) is equivalent to:

$$-\partial f(N^*, P^*)/\partial X > \partial f(N^*, P^*)/\partial Y.$$

REFERENCES

Anderson, R. M. & May, R. M. 1984 Spatial, temporal and genetic heterogeneity in host populations and the design of immunization programmes. *IMA J. Math. appl. Med. Biol.* **1**, 233–266.

Atkinson, W. D. & Shorrocks, B. 1981 Competition on a divided and ephemeral resource: a simulation model. *J. Anim. Ecol.* **50**, 461–471.

Auslander, D., Oster, G. & Huffaker, C. B. 1974 Dynamics of interacting populations. *J. Franklin Inst.* **297**, 345–376.

Brown, M. W. & Cameron, E. A. 1979 Effects of dispalure and egg mass size on parasitism by the gypsy moth egg parasite, Ooencyrtus kuwani. *Envir. Entomol.* **8**, 77–80.

Casas, J. 1988 Analysis of searching movements of a leafminer parasitoid in a structured environment. *Physiol. Entomol.* **13**, 373–380.

Casas, J. 1989 Foraging behaviour of a leafminer parasitoid in the field. *Ecol. Entomol.* **14**, 257–265.

Chesson, P. L. & Murdoch, W. W. 1986 Aggregation of risk: relationships among host-parasitoid models. *Am. Nat.* **127**, 696–715.

Comins, H. N. & Hassell, M. P. 1987 The dynamics of predation and competition in patchy environments. *Theor. Popul. Biol.* **31**. 393–421.

Crawley, M. J. 1983 *Herbivory: the dynamics of animal–plant interactions*. Oxford. Blackwell Scientific Publications.

Ehler, L. E. 1986 Distribution of progeny in two ineffective parasites of a gall midge (Diptera: Cecidomyiidae). *Envir. Entomol.* **15**, 1268–1271.

Elliott, J. M. 1983 The responses of the aquatic parasitoid *Agriotypus armatus* (Hymenoptera: Agriotypidae) to the spatial distribution and density of its caddis host *Silo pallipes* (Trichoptera: Goeridae). *J. Anim. Ecol.* **52**, 315–330.

Free, C. A., Beddington, J. R. & Lawton, J. H. 1977 On the inadequacy of simple models of mutual interference for parasitism and predation. *J. Anim. Ecol.* **46**, 543–554.

Freeman, B. E. & Parnell, J. R. 1983 Mortality of *Sceliphron assimile* Dahlbom (Sphecidae) caused by the eulophid *Melittobia chalybii* Ashmead. *J. Anim. Ecol.* **42**, 779–784.

Godfray, H. C. J. & Hassell, M. P. 1987 Natural enemies can cause discrete generations in tropical insects. *Nature, Lond.* **327**, 144–147.

Godfray, H. C. J. & Hassell, M. P. 1989 Discrete and continuous insect populations in tropical environments. *J. Anim. Ecol.* **58**, 153–174.

Godfray, H. C. J. & Hassell, M. P. 1990 Encapsulation and host-parasitoid population dynamics. In *Parasitism: co-existence or conflict?* (ed. C. Toft). Oxford University Press.

Godfray, H. C. J. & Pacala, S. 1991 Aggregation and the population dynamics of parasitoids and predators. *Am. Nat.* **136**. (In the press.)

Gurney, W. S. C. & Nisbet, R. M. 1985 Fluctuation periodicity, generation separation, and the expression of larval competition. *Theor. Popul. Biol.* **28**, 150–180.

Gurney, W. S. C., Nisbet, R. M. & Lawton, J. H. 1983 The systematic formulation of tractable single-species population models. *J. Anim. Ecol.* **52**, 479–496.

Hails, R. 1988 The ecology of *Andricus quercuscalicis* and its natural enemies. Ph.D. thesis, University of London.

Hanski, I. 1981 Coexistence of competitors in patchy environment with and without predation. *Oikos* **37**, 306–312.

Hassell, M. P. 1978 *The dynamics of arthropod predator–prey systems*. (237 pages.) Princeton University Press.

Hassell, M. P. 1984 Parasitism in patchy environments: inverse density dependence can be stabilizing. *IMA J. Math. appl. Med. Biol.* **1**, 123–133.

Hassell, M. P. & May, R. M. 1973 Stability in insect host-parasite models. *J. Anim. Ecol.* **42**, 693–726.

Hassell, M. P. & May, R. M. 1974 Aggregation in predators and insect parasites and its effect on stability. *J. Anim. Ecol.* **43**, 567–594.

Hassell, M. P. & May, R. M. 1986 Generalist and specialist natural enemies in insect predator–prey interactions. *J. Anim. Ecol.* **55**, 923–940.

Hassell, M. P. & May, R. M. 1988 Spatial heterogeneity and the dynamics of parasitoid-host systems. *Ann. Zool. Fennici* **25**, 55–61.

Hassell, M. P., Pacala, S., May, R. M. & Chesson, P. L. 1990. The persistence of host-parasitoid associations in patchy environments. I. A general criterion. *Am. Nat.* **136**. (In the press.)

Heads, P. A. & Lawton, J. H. 1983 Studies on the natural enemy complex of the holly leaf-miner: the effects of scale on the detection of aggregative responses and the implications for biological control. *Oikos* **40**, 267–276.

Hirose, Y., Kimoto, H. & Hiemata, K. 1976 The effect of host aggregation in parasitism by *Trichogramma papilonis* Nagarkarti (Hymenoptera: Trichogrammatidae), an egg parasite of *Papilio xuthus* Linne (Lepidoptera: Papilionidae). *Appl. Entomol. Zool.* **11**, 116–125.

Hochberg, M. & Lawton, J. H. 1990 Competition between kingdoms. *Trends Ecol. Evol.* **5**. (In the press.)

Ives, A. R. & May, R. M. 1985 Competition within and between species in a patchy environment: relations between microscopic and macroscopic models. *J. Theor. Biol.* **115**, 65–92.

Ives, A. R. 1990 The effects of density-dependent and density-independent aggregation on model host-parasitoid systems. *Am. Nat.* **136**. (In the press.)

Jones, T. H. & Hassell, M. P. 1988 Patterns of parasitism by *Trybliographa rapae*, a cynipid parasitoid of the cabbage root fly, under laboratory and field conditions. *Ecol. Entomol.* **13**, 309–317.

Kidd, N. A. C. & Mayer, A. D. 1983 The effect of escape responses on the stability of insect host-parasite models. *J. Theor. Biol.* **104**, 275–287.

Latto, J. & Hassell, M. P. 1988 Generalist predators and the importance of spatial density dependence. *Oecologia* **77**, 375–377.

Lessells, C. M. 1985 Parasitoid foraging: should parasitism be density dependent. *J. Anim. Ecol.* **54**, 27–41.

Lotka, A. J. 1925 *Elements of physical biology*. Baltimore: Williams & Wilkins.

May, R. M. 1974 *Stability and complexity in model ecosystems*. Princeton University Press.

May, R. M. 1978 Host–parasitoid systems in patchy environments: a phenomenological model. *J. Anim. Ecol.* **47**, 833–843.

McClure, M. S. 1977 Parasitism of the scale insect, *Fiorinia externa* (Homoptera: Diaspididae) by *Aspidiotiphagus citrinus* (Hymenoptera: Eulophidae) in a hemlock forest: density dependence. *Envir. Entomol.* **6**, 551–555.

Munster-Svendsen, M. 1980 The distribution in time and space of parasitism in *Epinotia tedella* (Cl.) (Lepidoptera: Tortricidae). *Ecol. Entomol.* **5**, 373–383.

Murdoch, W. W., Nisbet, R. M., Blythe, S. P., Gurney, W. S. & Reeve, J. D. 1987 An invulnerable age class and stability in delay-differential parasitoid–host models. *Am. Nat.* **129**, 263–282.

Murdoch, W. W. & Oaten, A. 1975 Predation and population stability. *Adv. ecol. Res.* **9**, 1–131.

Murdoch, W. W. & Oaten, A. 1989 Aggregation by parasitoids and predators: effects on equilibrium and stability. *Am. Nat.* **134**, 288–310.

Murdoch, W. W., Reeve, J. D., Huffaker, C. E. & Kennett, C. E. 1984 Biological control of scale insects and ecological theory. *Am. Nat.* **123**, 371–392.

Myers, J. H. 1988 Can a general hypothesis explain population cycles of forest Lepidoptera? *Adv. ecol. Res.* **18**, 179–242.

Nicholson, A. J. & Bailey, V. A. 1935 The balance of animal populations. Part 1. *Proc. zool. Soc. Lond.* **1935**, 551–598.

Nisbet, R. M. & Gurney, W. S. C. 1983 The systematic formulation of population models for insects with dynamically varying instar duration. *Theor. Popul. Biol.* **23**, 114–135.

Pacala, S. & Hassell, M. P. 1990 The persistence of host-parasitoid associations in patchy environments. II. Evaluation of field data. *Am. Nat.* **136**. (In the press.)

Pacala, S., Hassell, M. P. & May, R. M. 1990 Host–parasitoid associations in patchy environments. *Nature, Lond.* **344**, 150–153.

Reeve, J. D. 1988 Environmental variability, migration, and persistence in host–parasitoid systems. *Am. Nat.* **132**, 810–836.

Smith, A. D. M. & Maelzer, D. A. 1986 Aggregation of parasitoids and density independence of parasitism in field populations of the wasp *Aphytis melinus* and its host, the red scale *Aonidiella aurantii*. *Ecol. Entomol.* **11**, 425–434.

Stiling, P. D. 1980 Competition and coexistence among *Eupteryx* leafhoppers (Hemiptera: Cicadellidae) occurring on stinging nettles (*Urtica dioica*). *J. Anim. Ecol.* **49**, 793–805.

Stiling, P. D. 1987 The frequency of density dependence in insect host-parasitoid systems. *Ecology* **68**, 844–856.

Strassman, J. E. 1981 Parasitoids, predators and group size in the paper wasp, *Polistes exclamans*. *Ecology* **62**, 1225–1233.

Talor, L. R., Woiwod, I. P. & Perry, J. N. 1980 Variance and the large scale spatial stability of aphids, moths and birds. *J. Anim. Ecol.* **49**, 831–854.

Thorarinsson, K. 1990 Biological control of the cottony-cushion scale: experimental tests of the spatial density dependence hypothesis. *Ecology* **71**, 635–644.

Trexler, J. C. 1985 Density-dependent parasitism by a eulophid parasitoid: Tests of an intragenerational hypothesis. *Oikos* **44**, 415–422.

Volterra, V. 1926 Variazioni e fluttuazioni del numero d'individui in specie animali conviventi. *Mem. Acad. Lincei* **2**, 31–113.

Waage, J. K. 1979 Foraging for patchily distributed hosts by the parasitoid, *Nemeritis canescens*. *J. Anim. Ecol.* **48**, 353–371.

Waage, J. K. 1983 Aggregation in field parasitoid populations foraging time allocation by a population of *Diadegma* (Hymenoptera: Ichneumonidae). *Ecol. Entomol.* **8**, 447–453.

Walde, S. J. & Murdoch, W. W. 1988 Spatial density dependence in parasitoids. *Ann. Rev. Entomol.* **33**, 441–466.

Weseloh, R. M. 1972 Influence of gypsy moth egg mass dimensions and microhabitat distribution on parasitization by *Ooencyrtus kuwanai*. *Ann. entomol. Soc. Am.* **65** 64–69.

Discussion

G. A. Tingley (*Imperial College of Science, London, U.K.*). We have seen that patch size is of importance in the observed spatial pattern. What is the likely sensitivity of this type of analysis to the selection of inappropriate patch size in field observations?

M. P. Hassell. As our estimates of CV^2 are weighted for host density per patch, part of the problem of spatial scale is avoided: heterogeneity is measured at the level of the individual hosts. However, scale remains important in a number of ways. First, our assumption that the exploitation of hosts within a pâtch is random is satisfactory for relatively small patch sizes, but could introduce significant error as patch size gets large. Second, is the very interesting problem of identifying the scales at which host density dependent and host density independent heterogeneity have their maximal effect. These need not necessarily occur at the same scale.

Complex dynamics in multispecies communities

H. C. J. GODFRAY[1] AND S. P. BLYTHE[2]

[1] *Department of Biology & Centre for Population Biology, Imperial College at Silwood Park, Ascot, Berks SL5 7PY, U.K.*
[2] *Department of Statistics and Modelling Science, University of Strathclyde, Glasgow G1 1XH, U.K.*

SUMMARY

Communities of living organisms have potentially very complex population dynamics. Two components of complexity are considered, the dimensionality of the attractor underlying the persistent dynamics, and the presence of chaos. The dimensionality of real biological communities is unknown while there is great controversy about the presence of chaos in population dynamics. The evidence for chaos, and changes in the popularity of chaos among empirical biologists, is reviewed. Two new techniques developed in the physical sciences, attractor reconstruction and the estimation of the correlation dimension, are described and examples of their use in biology discussed. Although these techniques offer exciting new prospects for investigating community dynamics, there are some major problems in using them in biology. These problems include the length of biological time series, the ubiquity of noise, transient behaviour, Darwinian evolution and problems in interpretation. These problems are discussed and it is concluded that the best prospects of applying these techniques are using data collected in laboratory microcosms.

INTRODUCTION

Biological communities are normally composed of large numbers of interacting species of plants, animals and microorganisms. One consequence of the size of natural communities, and also of the fact that biological systems are subject to time-lags and have spatial extent, is that the dynamics of biological communities are potentially exceedingly complicated. The question of the actual complexity of the dynamics of natural communities is one of the major problems of contemporary population ecology.

It is important to state immediately what is meant by complexity because the word has many different interpretations in the study of large systems. We discuss here two elements of complexity. The persistent behaviour of a system can be described by the properties of the attractor underlying the dynamics. The attractor can be visualized as a geometrical object in a space of appropriate dimensionality determined by the dimensionality of the attractor itself. We take as one element of the complexity of community dynamics the number of dimensions necessary to describe the community attractor. This measure is related to the number of separate equations that are required to describe the persistent behaviour of the system: low dimensionality implies few equations. (Note, however, these equations are not normally the same as the biological equations underpinning the dynamics of individual species.) If the dynamics of the community are chaotic, then the attractor has fractal geometry and, normally, non-integer dimension. We take as our second element of complexity the presence of chaos in the dynamics of a community. The two elements of complexity are potentially orthogonal: chaos may be described by either high- or low-dimensional attractors.

Any answer to the question of community complexity will depend on the scale of investigation. Consider a resource-consumer system where the resource renews itself on a very short timescale compared to that of the dynamics of the consumer. A coarse exploration of the dynamics of the system might just consider the consumer, assuming that the resource instantaneously reaches the equilibrium appropriate to current consumer density. A more detailed exploration of the dynamics might reveal dynamic behaviour caused by short lags in the response of the resource to changes in community abundance. Thus questions of community complexity are inextricably linked to both the scale of investigation and also to problems of model abstraction: the size of the subset of species in the community that need explicit consideration when describing the dynamics of a particular target species.

Throughout the late 1960s and 1970s, the dominant research programme on the dynamics of communities was based on the assumption that all members of the community fluctuated about stable equilibrium population densities. The behaviour of the system in the neighbourhood of the equilibrium could then be studied by linearization about the equilibrium value. The community was characterized by the community matrix, the elements of which, a_{ij}, described the marginal influence of change in the abundance of species i on species j at equilibrium; the matrix included intraspecific density-dependence $(i=j)$ (Levins 1968). Study of linearized community models failed to support the commonly held notion that complex biological systems would automatically be more stable (May 1973) and initiated a continuing debate about whether an association between stability and complexity might be found in biologically realistic community matrices. Studies on the community matrix also examined the question of model abstraction (MacArthur 1972; Levine 1976; Lawlor 1979; Schaffer 1981).

The community matrix approach assumes both that communities have stable equilibria and that the communities exist sufficiently close to the equilibrium that their dynamics may be encapsulated by linearization. It has been known since the work of the founders of population ecology that biological populations are capable of persistent cyclic dynamics and there is little disagreement among empirical biologists that at least some species are cyclic. The discovery that natural populations are potentially chaotic is much more recent, and the question of the existence of chaos in nature far from settled.

In the first section we review the evidence for chaos as an empirical phenomenon in population biology. We then describe a number of new techniques from applied mathematics and the physical sciences that potentially allow the resolution of the problem of the complexity of natural systems. We describe their application to biological data and particular problems that arise in using these techniques in biology.

CHAOS IN ECOLOGICAL SYSTEMS

Chaos was not discovered by mathematicians but by scientists such as meteorologists and biologists working in other disciplines (Gleick 1988). In biology, chaos was discovered through the study of single species population models in discrete generations with self regulation. As the nonlinearities in the self-regulation term increase, the dynamic behaviour of the population changes from a monotonic approach to a stable equilibrium, to an oscillatory approach to equilibrium and then to stable limit cycles which undergo period doubling until chaos ensues. Embedded within the chaotic region of parameter space are small regions where low frequency limit cycles may be observed though these, again through period doubling, merge back into chaos as the strength of the self-regulation is increased (May 1974, 1976; May & Oster 1976). Chaos was also observed in time-lagged, differential equation models of populations (May 1980) and in systems of ordinary differential equation models, though at least three species are required before chaos can be observed (Gilpin 1979).

The possibility that biological populations have chaotic dynamics prompted investigation to see whether the nonlinearities in the dynamics of real populations were sufficient to cause chaos. In what became a very influential paper, Hassell *et al.* (1976) estimated the parameters of a simple single-species, discrete-generation model capable of showing chaos, by using 28 sets of insect population data. The dynamics of the model was characterized by two parameters, one representing fecundity and the second representing the degree of nonlinearity in density dependence. As the nonlinearity increased, the dynamics of the model followed the familiar course of monotonic and oscillatory approaches to equilibrium, limit cycles and chaos. With two exceptions, all the data sets fell within the regions of stable equilibria. The two exceptions were the well-known outbreak pest, the Colorado Beetle (*Leptinotarsus quadrilineata*), which fell in the region of persistent cycles, and Nicholson's laboratory experiments with blowflies which fell within the region of chaos. We shall discuss Nicholson's experiments in more detail below.

In recent years, this study has attracted much criticism as being over simplistic. These criticisms are often unfair as the authors were scrupulous in cataloguing a long list of simplifications and assumptions inherent in their analysis. Perhaps the most important of these simplifications is the assumption that the dynamics of a species embedded in a complex community can be abstracted by fitting the population data to a simple single-species model. The problems of this approach were clearly pointed out by Hassell *et al.* (1976) who noted that their procedure suggested that the Larch Budworm (*Zeiraphera diniana*) should have stable population dynamics whereas there is good evidence that it shows approximately eight-year cycles driven by interactions with its foodplant or a natural enemy (Baltensweiler 1968).

A number of workers looked for evidence of chaos in laboratory systems, again by fitting data to simple population models. In particular, two studies on *Drosophila* (Thomas *et al.* 1980; Mueller & Ayala 1981) failed to find any evidence for chaos. The studies on *Drosophila* prompted speculation that natural selection might be responsible for the absence of chaotic dynamics in nature. Thomas *et al.* suggested that alleles that led to non-chaotic population dynamics might be favoured through group selection. In contrast, Mueller & Ayala showed that, at least under certain circumstances, a population showing non-equilibrium population dynamics could be invaded by alleles that, when common, promoted stable population dynamics.

By the early 1980s, a consensus had arisen among experimental population biologists that chaos was unlikely to be significant in natural populations. The twin planks of this consensus were the failure to predict chaotic dynamics by using models with parameters estimated from real data, and the conjecture that natural selection would promote non-chaotic dynamics. However, we believe a sea-change occurred around the middle of the decade and that in the past five years there has been renewed interest in chaos among empirical ecologists. There are perhaps two reasons for this shift in opinion.

First, the view that chaos only occurs in the presence of biologically unrealistic nonlinearities has been challenged by a series of population models in which chaos occurs with parameters well within the bounds of biological realism. For example, Bellows & Hassell (1988) predicted chaos by using a detailed age-structured host–parasitoid model which they parameterized by using experimental data. Prout & McChesney (1985) found that *Drosophila* females developing in crowded cultures had reduced fecundity as adults. This delayed density-dependence, which can lead to chaotic dynamics, had not been appreciated by earlier workers attempting to assess the likelihood of chaos in laboratory fruitfly systems. These, and similar results, indicated that chaos might become more likely as the number of interacting species, and the complexity of the interaction, increased.

The second factor that has revitalized interest in chaos comes not from biology but from applied mathematics and the physical sciences. A series of new techniques have been developed that allow the analysis of time series data to estimate properties of the attractors underlying dynamic systems. In the next section we review two methods, attractor reconstruction and the estimation of correlation dimensions. The responsibility for bringing these techniques to the attention of biologists, and for pioneering their use on biological data, is largely due to the tireless advocacy of Schaffer and his colleagues (see, for example, Schaffer 1984, 1985; Schaffer & Kot 1985, 1986). We omit discussion of one very new technique, nonlinear forecasting (Sugihara & May 1990*a*), as this is the subject of the next chapter (Sugihara & Grenfell, this symposium).

ANALYSIS OF TIME SERIES

(*a*) *Reconstructing attractors*

Consider data collected on the population densities of a predator and its prey with overlapping generations. The raw data might consist of two parallel time series. It is often more informative to plot the numbers of prey at any particular time against the number of predators at the same time, the resulting plot being known as a phase plot drawn in phase space. Suppose that the interaction between the predator and its prey results in stable equilibria for both species. The phase plot now shows the populations being attracted towards a point in phase space that represents the stable equilibria. If the results from a deterministic model of the interaction was plotted in phase space, the trajectory would approach the single equilibrium and remain there for ever. For real data, the trajectory would be continually displaced from the equilibrium point by stochastic events and the phase plot would resemble a ball of wool centred on the equilibrium value.

The point equilibrium in the above example is an example of a dynamic attractor, in this case a point attractor with zero dimension. Trajectories originating in an area of phase space called the basin of the attractor will flow towards the attractor and, in the absence of perturbation, never leave it. Now suppose that instead of a stable equilibrium, the predator–prey system settles into some regular cycle. When plotted in phase space, the trajectory will, after the transients have died out, converge onto some roughly oval orbit. The equilibrium dynamics is now determined by a periodic attractor, a one-dimensional object plotted in a two dimensional space. More complex periodic cycles can arise by 'period-doublings', leading to figure-of-eights and further twisted structures in phase space.

The above attractors are the standard features of Newtonian dynamics. Chaos is normally produced by objects known as strange attractors. For overlapping generations (ordinary differential equations), a chaotic attractor can only exist in a three-dimensional phase space. So, instead of a simple predator–prey interaction, consider a three-trophic level interaction.† The dynamics of the system may still be described by a classical attractor, either a point attractor for a stable equilibrium, a periodic attractor or even a two-dimensional attractor. However, chaotic behaviour on a strange attractor is now a possibility.

Figure 1 shows a strange attractor in three-dimensional phase space. After any transients have decayed, and again ignoring any stochastic perturbation, all population trajectories originating in the basin of attraction will come to lie on the surface of the attractor. To see why the attractor earns the name strange, consider a cloud of points lying close together on the attractor (see also Schaffer (1984) for a similar explanation). Now follow the fate of the points as they move around the surface of the attractor. This process is exactly analogous to trying to predict the future population densities of several populations that at the present moment have roughly similar population levels. As the cloud of points moves around the surface of the attractor, the cloud is first stretched on the surface of the attractor and then folded over on to itself, a process that happens once for every circuit of the orbit. This continual stretching and folding means that trajectories that started off in close proximity soon become separated. More technically, nearby points become homogenized on the surface of the attractor. The practical consequence of this homogenization is that populations that initially have similar population densities quickly diverge in densities. One of the hallmarks of chaos is extreme sensitivity to initial conditions; present population densities can only be used to predict future densities in the very short term. The pattern of the decay of predictive power with time is further discussed by Sugihara & Grenfell (this symposium).

The attractor in figure 1 appears to be a twisted diaphanous sheet suggesting a two-dimensional structure. However, the apparent two-dimensionality conceals greater complexity. Each time the sheet is folded back on itself, the two halves do not merge but retain their structure such that the sheet consists of an infinite number of separate layers: the ultimate *millefeuille*. The object is in fact a fractal: if a cross section of the flow is taken and examined under increasing magnification, greater and greater detail will be revealed and, in addition, the details will be self repeating. More specifically, the object is a type of Cantor Set, a set of disconnected points that poses problems for traditional

† The restriction of chaotic attractors in continuous systems to representation in three or more dimensions only partially precludes chaotic dynamics in two-species or even one-species systems. A two-species interaction will always have a zero or one dimensional attractor (and a one-species interaction a zero-dimensional attractor) unless extra degrees of freedom are supplied by the incorporation of a time lag or a spatial component into the interaction. More technically, the dimensionality of the system is equivalent to the number of ordinary differential equations (ODEs) necessary for its specification. A predator–prey interaction described by two differential equations can only display stable or periodic dynamics. However, a time-lagged differential equation, or a partial differential equation (incorporating spatial coordinates) both potentially require an infinite number ODEs for their specification. In practice, the attractors underlying many time-lagged and partial differential equations can be described in low-dimensional phase space.

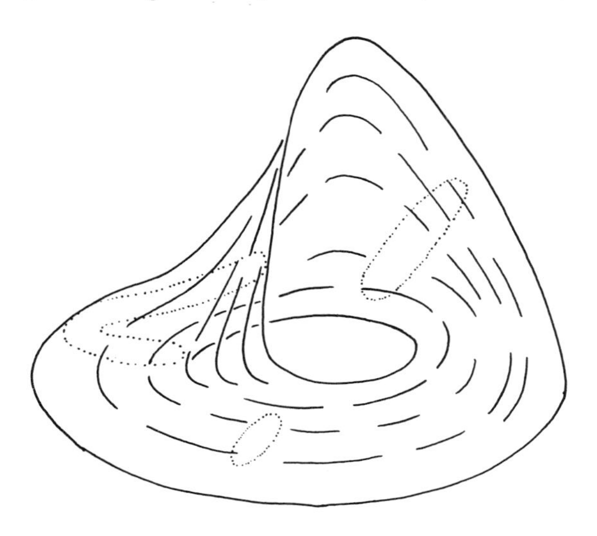

Figure 1. Schematic diagram of a strange attractor (the Rössler attractor). The dotted circle represents a collection of nearby trajectories, which as they orbit the attractor, are initially stretched and then folded over on themselves. As a result, nearby trajectories diverge and prediction is not possible far into the future.

concepts of dimensionality (see Sugihara & May (1990*b*) for a recent discussion of dimensionality and measure theory in biology). Such objects have fractional dimension; for example the object in figure 1 has a dimension of between one and two. It looks like a twisted sheet because it is nearly two-dimensional while the additional complexity of the fractal structure is represented by the fractional part of the dimensionality.

The rate of divergence of nearby trajectories on an attractor is best measured using Lyapunov exponents. Positive Lyapunov exponents indicate divergence in some direction and negative exponents convergence; a strange attractor must have at least one positive Lyapunov exponent. For an essentially two-dimensional attractor such as figure 1, Lyapunov exponents are calculated, as an average over the attractor, (1) in the direction of the flow, (2) perpendicular to the flow in the plane of the attractor and (3) perpendicular to both the flow and the plane of the attractor. In the direction of flow, nearby points are part of the same trajectory and so neither diverge or converge: the Lyapunov exponent is zero. Nearby trajectories in the plane of the attractor diverge (due to stretching and folding) and are associated with positive Lyapunov exponents while those perpendicular to the plane of the attractor, and which are pulled down onto the surface of the attractor, converge and are associated with a negative Lyapunov exponent. A strange attractor that exists in three-dimensional space must have one positive and one negative Lyapunov exponent. Higher dimensional strange attractors may have different combinations of positive and negative Lyapunov exponents and this provides a partial taxonomy of the attractors.

The characteristic shape of strange attractors suggests a way of identifying chaos in time series. Plot the time series in phase space and then examine the flow on the attractor to look for stretching and folding. However, to construct the phase plot one needs to know the population densities for each species in the system. A biologist typically does not have this data and frequently does not even know the number of species for which data is required. A solution to this problem was provided by Packard *et al.* (1980) and Takens (1981) who proved that the properties of the attractor governing the system can be obtained from a single time series by plotting trajectories in a phase space constructed from lagged coordinates. For example, a three-dimensional plot might be obtained by giving each point $s(t)$ in the series the coordinates (x, y, z), where $x = s(t)$, $y = s(t-\tau)$ and $z = s(t-2\tau)$ where τ is a suitably chosen time lag.

An immediate question is the number of dimensions that are necessary in order to draw the attractor. Takens (1981) and Packard *et al.* (1980) proved that for a system of m equations, a sufficient number of dimensions is $2m+1$. This provides rather cold comfort, especially for high m. However, examination of physical systems suggests that the dynamics of at least some complex systems are governed by low dimensional attractors. For example, a famous non-equilibrium chemical reaction, the Belousov–Zhabotinskii reaction, involving possibly over 25 chemical compounds, has been shown to be described by an attractor with a dimension near two (Roux *et al.* 1983). If complex systems of interacting species also have low-dimensional attractors, and if sufficient data is at hand, plotting time series in lagged coordinate space should reveal the shape of the attractor.

Supposing the structure of the attractor has been convincingly recreated in phase space, further analysis is possible by associating the dynamics on the attractor with a one-dimensional return map. Return maps are most familiar in biology as a means of analysing populations with discrete generations. The size of the population in generation $t+1$ is plotted against the size of the population in generation t, the resulting line is often called a Moran curve or Ricker curve. Return maps are also useful for visualizing the onset of chaos in populations with discrete generations (May & Oster 1976). Return maps can be generated from smooth flows by sectioning the attractor in a plane perpendicular to the flow (in general, the intersection of a multidimensional flow with a lower dimensional object perpendicular to the flow is known as a Poincaré section). In the case of a nearly two dimensional object such as the attractor in figure 1, the Poincaré section is taken using a plane and the intersection of the plane with the attractor is a set of points that almost form a line (figure 2). The next stage is actually to fit a line to the series of points and to measure the distance of each point along the line. Each time the population completes a whole circuit of the attractor, it will intersect the Poincaré section once. Thus each point on the line is associated with a position in the time series. A return map is constructed by plotting the position of each point along the line against the point that preceded it.

An examination of the one-dimensional return map can reveal much about the behaviour of the whole attractor. For example, suppose the return map is

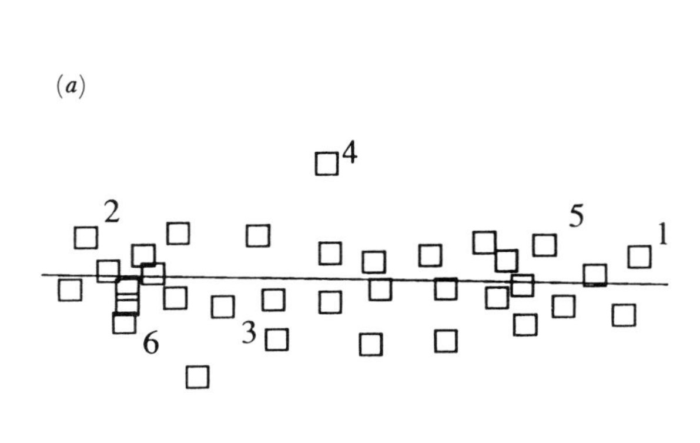

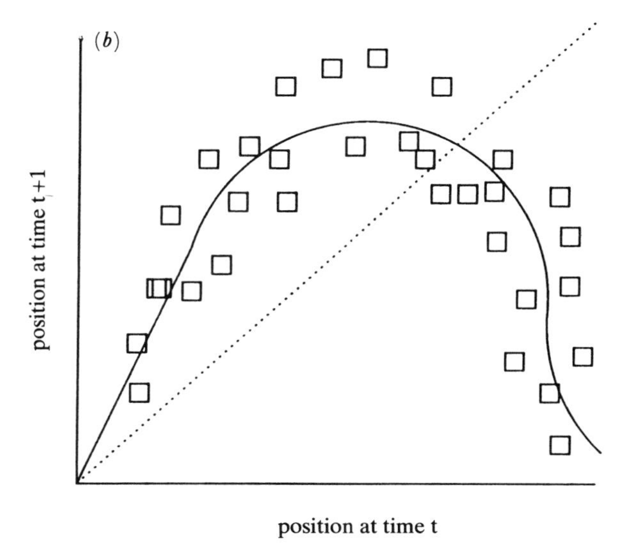

Figure 2. The intersection of a strange attractor, such as that in figure 1, and a plane perpendicular to the flow is a Poincaré section. As explained in the text, the Poincaré section is approximately a straight line (*a*). Successive intersections of the orbit with the section (numbered) can be used to construct a one-dimensional return map (*b*).

unimodal, an examination of the shape of the curve will show whether the population will behave chaotically and also reveal the way chaos arises through a series of bifurcations leading to period doubling. Note, an examination of a Ricker curve reveals exactly the same thing for a single-species population in discrete generations. In addition, the map can be used to estimate Lyapunov exponents.

Finally, we point out that chaotic attractors arise in a number of ways apart from through period doubling (Schaffer *et al.* 1988). If interacting populations are affected by some external cyclic influence, for example daily or annual environmental influences, the attractor may appear like the surface of a hollow doughnut or torus. Here, instead of stretching or folding, clouds of nearby points are stretched and then overlain on each other as they move around the torus. If the attractor is drawn in phase space (as above) and then a Poincaré section taken, the intersection is a circle rather than a line, a circular map rather than a one-dimensional map is then studied. Various complications arise in the presence of environmental forcing. For example, the dynamics may become phase-locked: exhibiting fluctuations of the same frequency as the environmental fluctuations, or of a harmonic of that frequency.

(*b*) *Estimating dimensions*

Ideally, one would like to be able to measure the number and magnitude of Lyapunov exponents directly from the data. This would provide information both on the dimensionality of the attractor and also on the presence of chaos. In practice, estimating Lyapunov exponents is computationally difficult and demanding of data, especially when the number of Lyapunov exponents is unknown. An alternative to the calculation of Lyapunov exponents is the estimation of entropy measures which provide some of the same information.

Consider an arbitrarily small box in the phase space containing an attractor. In two dimensions, the box is a square; in three dimensions a cube and in higher dimensions a hypercube. An entropy or information measure gives a value for the uncertainty of the future behaviour of the system, based on the knowledge of the box containing the system trajectory at the present time.

The uncertainty can be partitioned into different components corresponding to the dimensions of phase space. For example, consider the strange attractor shown in figure 1. Knowledge of the presence of the trajectory at a particular point gives us different information about the future position of the system (1) in the direction perpendicular to the plane of the attractor, no information, (2) along the flow of the trajectory, perfect information, and (3) across the fractal surface of the attractor, an intermediate amount of information, the quality depending on the rate of trajectory divergence. Adding up the partial dimensions in each direction provides an overall information dimension that is related to the fractal dimension of the attractor (Grassberger 1986*a*).

The calculation of information measures poses many of the same problems as the calculation of Lyapunov exponents. However, it is usually simpler to calculate an overall information dimension. A number of different dimensions may be calculated depending on the particular definition of entropy employed. A useful family of entropies is the order-α generalized Renyi information.

$$S^{(\alpha)}(\epsilon) = \frac{1}{1-\alpha} \log [\sum_i p_i^{\alpha}]. \qquad (1)$$

Where α is a parameter, and p_i is the probability that the ith box of side ϵ contains a trajectory, summed over the non-empty boxes. As the p_i are in a sense the 'weights' of each box, $S^{(\alpha)}(\epsilon)$ is related to the mass distribution of trajectories over the attractor, at resolution ϵ. The information dimensions $D^{(\alpha)}$ are calculated from

$$D^{(\alpha)} = \lim_{\epsilon \to 0} \left[\frac{S^{(\alpha)}(\epsilon)}{\log(1/\epsilon)} \right]. \qquad (2)$$

The sequence $D^{(\alpha)}$ is always a decreasing function for increasing alpha. There is considerable confusion in the terminology applied to these dimensions, which we will try not to add to here. $D^{(0)}$ is the fractal dimension, as $S^{(0)}(\epsilon)$ is the log of the number of non-empty ϵ-boxes. $D^{(1)}$ (the limit as $\alpha \to 1$) is confusingly called the information dimension, as $S^{(1)}(\epsilon)$ reduces to the more familiar Shannon information measure for entropy. $D^{(2)}$ is known as the correlation dimension or exponent, and as Grassberger (1986*a*, p. 305) says, '[it] is the easiest generalized dimension to estimate, even if it is not the most interesting'. Most practical calculations of 'dimension' of an attractor have involved $D^{(2)}$, either as a quantity in its own right, or as a lower bound to the fractal dimension $D^{(0)}$.

To calculate the correlation dimensions from experimental data, one embeds a time series in spaces of different dimensions using lagged coordinates (see previous section). Then, in each dimension, one calculates the correlation integral, $C(x)$, the proportion of points that are separated by a distance less than a threshold, x. When x is very small, all, or nearly all, points will be separated by a distance greater than x and the value of $C(x)$ will be around zero. When x is very large, all points will be separated by a distance less than x and consequently $C(x) = 1$. Grassberger & Procaccia (1983*a*, *b*) showed that for intermediate values of x,

$$C(x) \approx x^{d_C}, \qquad (3)$$

or equivalently

$$\ln(C(x)) \approx d_C \ln(x). \qquad (4)$$

The value of d_C calculated using this method asymptotes at the value of the correlation dimension $D^{(2)}$ as the embedding dimension increases.

The size of the interval on x for which equation (3) holds (the scaling region) depends both on the embedding dimension and on the noisiness of the data. The scaling region decreases in size as the embedding dimension increases setting an upper limit on the dimensionality of the attractor that can be detected for finite data sets. Noise also reduces the size of the scaling region, typically by obscuring the correlation for low values of x.

We show the calculation of correlation dimensions by using an example from a model with chaotic behaviour. Hochberg *et al.* (1990) studied the three-species interaction between an insect and its specific parasitoid and pathogen. The pathogen is contracted by eating infected food and also exists in a protected stage that allows persistence of the disease in the temporary absence of the host. Parasitoid attack is independent of the presence of the disease and may be either random or clumped. For different parameter values, host-pathogen, host-parasitoid or three species interactions may be stable. The model is phrased as a

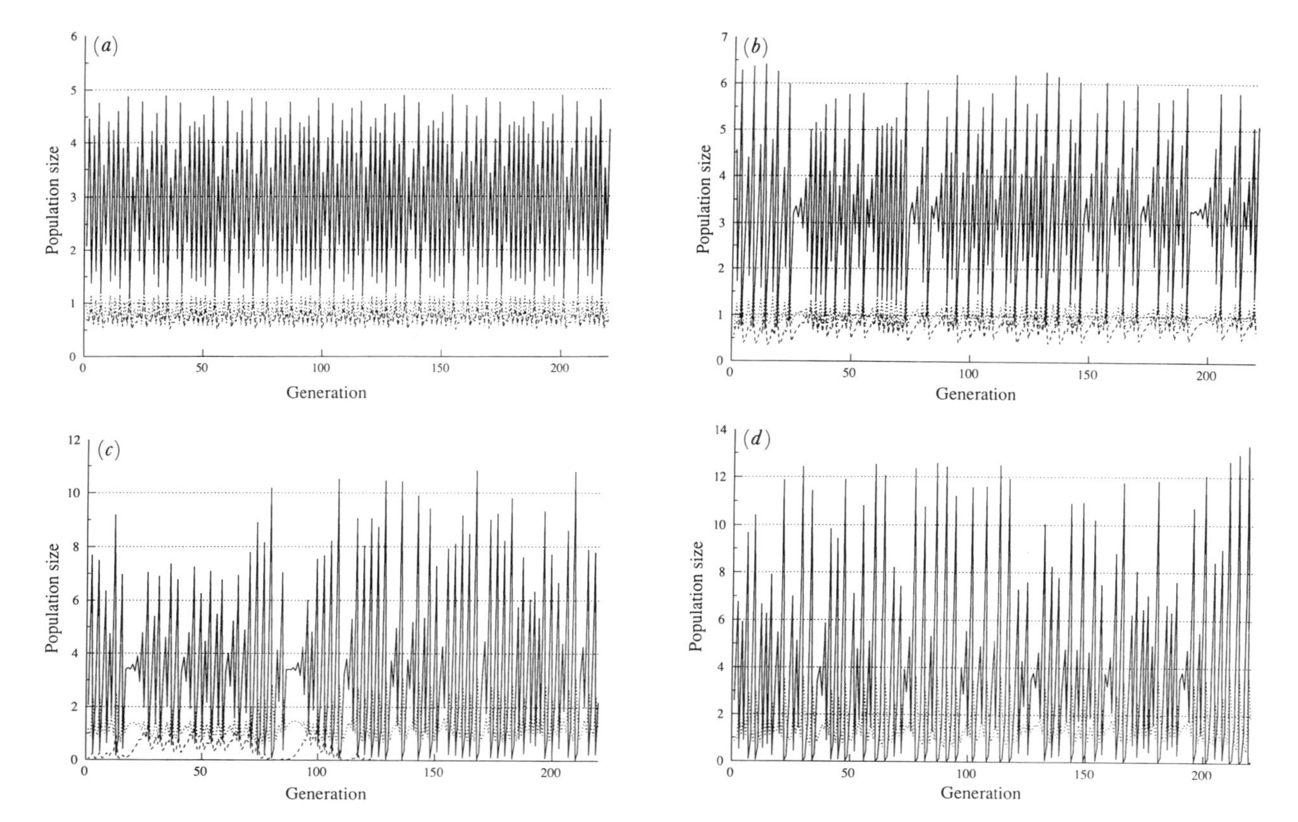

Figure 3. An example of host–parasitoid–pathogen dynamics from a model of Hochberg, Hassell & May (1990). All parameters are constant except host fecundity. Unbroken line, the host; dashed line, the parasitoid; dotted line, the pathogen. (*a*) Fecundity = 6; (*b*) fecundity = 8; (*c*) fecundity = 10; (*d*) fecundity = 12.

system of three difference equations with a time step equal to the host and parasitoid generation times. Each host generation, a pathogen epidemic occurs that determines the number of infectious particles in the next generation.

The dynamic behaviour of the system is strongly influenced by host fecundity. For one particular parameter set, we chart the change in system dynamics as host fecundity increases from 2 to 12. When the fecundity is two, a stable three species equilibrium is found. As fecundity increases stable limit cycles are found which increase in period until the system becomes chaotic. As fecundity is further increased within the chaotic region, the apparent 'randomness' of the population trajectories also increases (figure 3). At a fecundity of about 10, the parasitoid becomes extinct though the remaining host–pathogen interaction is also chaotic.

We ran the model for 500 generations, which appeared to eliminate transient behaviour and then used the results for another 500 generations to estimate the correlation dimension of the attractor in the chaotic region. Stable equilibria (zero-dimensional) and periodic behaviour (one-dimensional) give way to dynamic behaviour governed by attractors of between 1 and 1.5 (figure 4). When the parasitoid drops out of the interaction, the correlation dimension falls from about 1.4 to 1.1 though the non-integer dimensionality confirms that the interaction is still chaotic.

EXAMPLES OF USE

(a) *Laboratory data: Nicholson's blowflies*

During the 1950s the Australian entomologist A. J. Nicholson performed a series of cage experiments using populations of the sheep blowfly *Lucilia cuprina* (Wied.), under a variety of food-supply regimes, to investigate the process of density dependence in population dynamics (Nicholson 1954, 1957). In one of these experiments, Nicholson (1957) produced a striking pattern of variation in adult blowfly numbers with time (Figure 5), starting with irregular cyclic behaviour, and changing after about one year to a pattern of large and erratic fluctuations superimposed upon an increasing average population size.

There has been much debate on the population

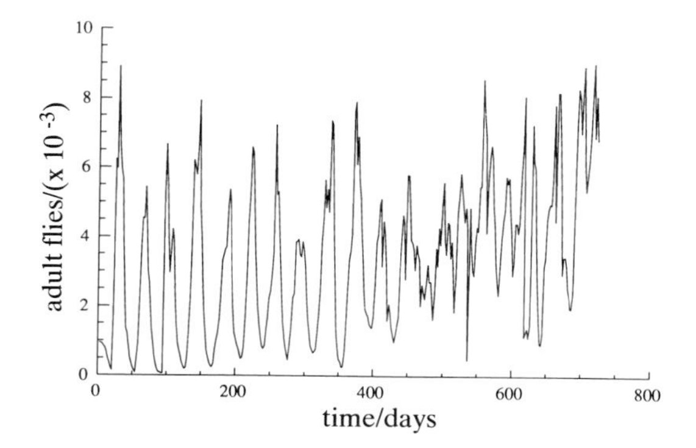

Figure 5. Changes in blowfly numbers in Nicholson's long-term laboratory cage experiment.

dynamics underlying Nicholson's results. As mentioned above, Hassall *et al.* (1976) suggested that the dynamics were chaotic after using the data to estimate the parameters of a single-species population model in discrete generations. More detailed stage-structured models have been fitted to the same experimental data by a number of groups (Oster & Takahashi 1974; Gurney *et al.* 1980; Readshaw & Cuff 1980; Brillinger *et al.* 1980; Stokes *et al.* 1988). There is general agreement that at least the earlier part of the data can be explained by perturbed limit cycles (rather than quasi-cycles, see below). However, there is disagreement about whether the perturbed limit cycles might better be explained by a model with a chaotic component, and whether the data set becomes chaotic latterly.

The results of the estimated correlation dimension for this data set are shown in figure 6 (Blythe & Stokes 1988). The correlation dimension increases with the embedding dimension; though there is some suggestion that the correlation dimension is beginning to stabilize at a value of around 5, there is only weak evidence of a low-dimensional attractor. Two possible explanations for the poor performance of the technique are the relatively small data set, and the possibility that the system was subject to natural selection over the course of the experiment. We return to this last point below.

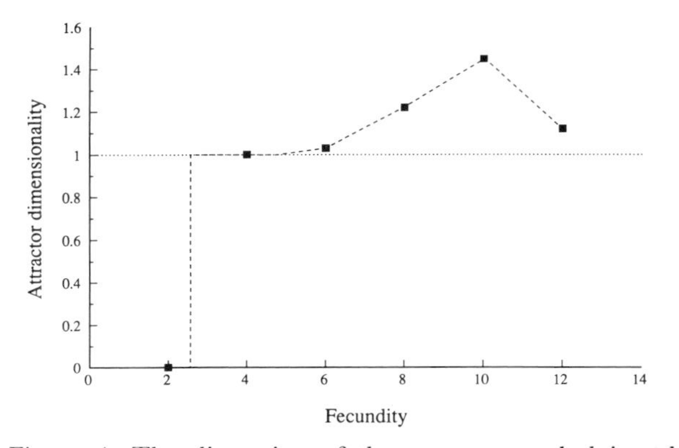

Figure 4. The dimension of the attractor underlying the host–parasitoid–pathogen dynamics illustrated in figure 3 for different values of the fecundity of the host.

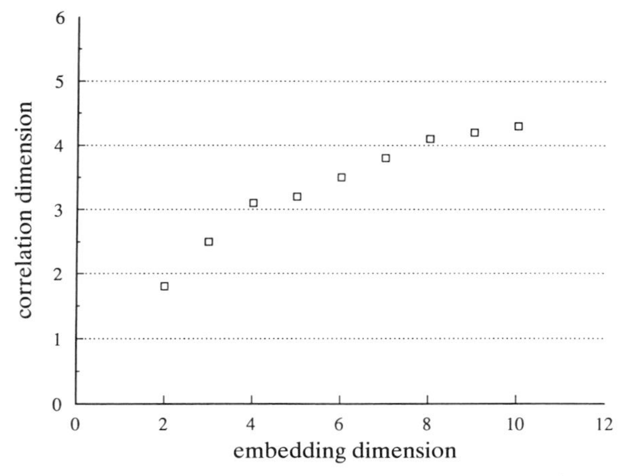

Figure 6. Estimating the correlation dimension for the blowfly population in Figure 5. The graph shows the estimated value of the correlation dimension, d_c, as a function of the embedding dimension.

(*b*) *Field data: lynx and measles attractors*

The technique of reconstructing an attractor from a lagged time series, followed by inspection of the form of the attractor for the presence of stretching and folding, has been applied to data on the Canadian Lynx and on measles epidemics by Schaffer (1984) and Schaffer & Kot (1985), respectively. In both cases the time series are strongly cyclic, in the case of the lynx with a period of approximately 9–10 years and in the case of the measles epidemics with a period of 2–3 years.

The data on lynx populations was obtained indirectly from the returns of fur trappers in the Canadian arctic between 1735 and 1934. This famous data set was first studied by Elton & Nicholson (1942) and by many subsequent workers. Schaffer analysed data from 1821 to 1913, a period of sustained 9–10 year oscillations. By using a time lag of three years and an embedding dimension of three, the attractor appeared as an essentially two-dimensional sheet, twisted into a cone. Poincaré sections were nearly one-dimensional and revealed some evidence of stretching and folding. The sections appeared to be describable by a unimodal map though the data was not of sufficient quality for the dynamic properties of the whole system to be precisely reconstructed from the map. Schaffer concludes that the most likely explanations for the data are that they are twice periodic or slightly chaotic, the chaos possibly caused by an interaction between an underlying twice-periodic orbit and noise.

Data on measles was obtained from medical health records from New York & Baltimore spanning the period 1928 until the advent of vaccination in 1963. Explanations for the presence of recurrent measles epidemics involve seasonal differences in the rate of transmission or random events leading to temporary extinction of the disease (see, for example, May & Anderson 1979). Schaffer & Kot reconstructed the attractor in three dimensions using a three month lag. Again the attractor appeared cone-like; essentially a two-dimensional object in three-dimensional space. Poincaré sectioning suggested stretching and folding while the properties of a one-dimensional return map constructed from the sections also suggested chaos (for example, a positive Lyapunov exponent was calculated from the return map). Finally, the correlation dimension was non-integer (fractal), again suggesting chaos. Sugihara & May (1990*a*; see also Sugihara & Grenfell, this symposium) also analyse this data set and conclude that there is evidence of chaos.

(*c*) *Field data: dimension of plankton dynamics*

We have attempted to apply the correlation dimension technique to a classic long-term data set, the Continuous Plankton Record (H. C. J. Godfray *et al.* unpublished data). Since 1948, regular monthly surveys of plankton have been carried out by using continuous plankton recorders towed behind merchant ships and weather ships in the North Sea and Atlantic (Colebrook 1960, 1975). Plankton are collected at a standard depth of 10 m and identified to as low a taxonomic level as possible. Each data point represents the estimated density for a particular group of plankton in an area of sea.

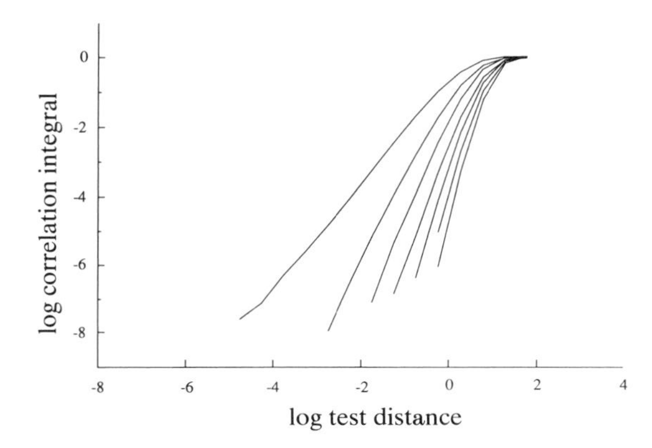

Figure 7. An attempt to estimate d_c from plankton data collected as part of the Continuous Plankton Record (the copepod *Calanus finmarchicus* from area C2 in the North Sea, see Colebrook, 1960). The slope of the graph of the correlation integral plotted against the test distance fails to asymptote as the embedding dimension increases.

From the large number of possible data sets, we selected 20 time series using as criteria those data sets with the smallest numbers of missing values and real zeros. When missing values occurred (not more than 10 in any data set), we estimated replacement values using linear interpolation. The data sets we used were chiefly groupings of copepods and euphausiaceans from the southern part of the North Sea (see H. C. J. Godfray *et al.* (unpublished data) for full details of data and localities). There is evidence of a long-term decline in the abundance of plankton in the area (see, for example, Colebrook 1978) and we have used the logarithm of the monthly differences as our raw data. In embedding the time series using lagged coordinates, we used a lag of seven months (for some data sets we explored other lags though our results were identical). Initially, we explored the data sets using traditional time series analysis. The majority (though not all) of the data sets showed annual cycles while four-yearly and four-monthly cycles were observed in some sets.

An example of the relation between the correlation integral and the test distance for the plankton data is shown in figure 7. The correlation integral increases with the test distance and there is no scaling interval that allows the estimation of the correlation dimension for any but the lowest embedding dimension. This result is typical of all the 20 data sets we explored. Thus there may be no low-dimensional attractor underlying the dynamics of these North Sea plankton. However, it is also possible that even the (relatively) outstanding quality of data of the Continuous Plankton Record is insufficient to allow the application of this technique.

BIOLOGICAL PROBLEMS

The techniques discussed above were chiefly developed with applications in the physical sciences in mind. Using these techniques in the biological sciences presents a number of problems, some peculiar to, and some exacerbated by, the need to work with living organisms.

(*a*) *Size of data base*

Attempts to reconstruct the shape of attractors and to estimate correlation dimensions demand long time series of data. Very few biological data sets are sufficiently long for these purposes. The exact length requirement will obviously depend on the quality of the data though little information about the structure of a strange attractor is likely to be obtained without information on 20 orbits, at the very least.

The lack of long term data on long-lived plants and animals is not surprising; few biologists embark on studies destined to occupy a substantial fraction of their own lives. It is perhaps more surprising that there is so little data on short-lived animals such as plankton that can be easily cultured in the laboratory for many generations.

The problem of small data sets is, of course, not unique to the biological sciences. The search for a global climatic attractor provides a cautionary tale about the dangers of over-interpreting short data sets. It has been suggested that the ratio of oxygen isotopes in cores from the seabed are correlated with climatic conditions at the time of sediment deposition. Nicolis & Nicolis (1984) examined a time-series of 184 oxygen-isotope data points. Unfortunately, the 184 data points were unevenly spaced over time, which presents obvious difficulties if the data set is to be embedded by using lagged coordinates. The solution adopted by Nicolis & Nicolis was to interpolate the data, increasing the length to just under 500 data points. Application of the Grassberger & Procaccia algorithm suggested an attractor with a dimension of 3.1. However, as was pointed out by Grassberger (1986*b*; see also Schaffer *et al.* (1988)), the process of interpolation introduces spurious correlation between data points and artefactually low estimates of attractor dimensionality. Unfortunately, the length of the data set is too short either to make a strong positive or negative statement about a potential attractor.

(*b*) *Noise*

Biologically data is typically much more noisy than data from the physical sciences and this can obfuscate underlying deterministic patterns. The effect of noise can be explored in model systems by artificially generating variability in parameters and observing the consequences on the detection of chaos and estimation of dimensionality (see, for example, Schaffer 1984). Most such experiments have involved uncorrelated, or white noise. Typically, longer data sets are required to overcome moderate levels of noise. A more serious problem, that has received less attention, is correlated or coloured noise. This is a greater problem as it results in correlation between data points that may influence the estimation of dimensionality.

Several workers have pointed out that moderate amounts of noise may actually help reveal underlying dynamics (Nisbet & Gurney 1982; Schaffer 1985). Consider a deterministic system whose dynamics are described by a unimodal return map. If the system has a point attractor then the addition of noise will result in a cloud of points about the stable point. However, if the system has a periodic solution then the addition of noise will lead to the system wondering over the surface of the map, revealing its structure in more detail. Of course, a chaotic system reveals its own map and extra noise is unlikely to increase the resolution.

Periodic attractors can also be transformed into strange attractors by the addition of noise. For example, the logistic map is said to become chaotic when the fecundity exceeds a certain value. In fact, the chaotic region contains an infinite number of periodic solutions. However, even the presence of very small amounts of noise results in most of the region being chaotic.

(*c*) *Transients*

The structure of an attractor can only be reconstructed if the trajectory of the system through phase space lies on the attractor. If the study is initiated when the system is some way from the attractor, the estimation of attractor properties will be impeded by the presence of transients. This problem is of course present in many dynamical experiments, but it may be particularly acute in population dynamics: laboratory populations and communities are frequently set up with little knowledge of the consequences of different starting conditions.

Blythe & Stokes (1988) examined some of the consequences of initial transients using a second-order ordinary differential equation,

$$\mathrm{d}^2x/\mathrm{d}t^2 + 2b\,\mathrm{d}x/\mathrm{d}t + \omega^2 x = 0, \quad x|_{t=0} = 1, \quad \mathrm{d}x/\mathrm{d}t|_{t=0} = 0. \tag{5}$$

This has a stable equilibrium at $x = 0$, approached via damped oscillations when $\omega^2 > b^2$ (figure 8*a*). The system thus has a point attractor with zero dimensions. However, if the correlation dimension technique is applied to data collected during the period of oscillatory transience, a dimension of about 0.9 is estimated. Thus an uncritical reading of the results would suggest a periodic attractor rather than a point attractor, possibly with a slight chaotic component.

When analysing data from the field, or from long-term laboratory experiments, any initial transients may be expected to be absent or at least to die away. However, a system that displays an oscillatory approach to equilibrium, and that is continually subject to external stochastic perturbations, may show a mixture of apparently cyclic and noisy dynamic behaviour that appears superficially very similar to chaos. Such behaviour has been called quasi-cycles by Nisbet *et al.* (1977) and we digress slightly to discuss how quasi-cycles may be characterized (see also Nisbet & Gurney (1982)).

Quasi-cycles can be described by a quantity the coherence number n_c, the number of cycles taken for the amplitude of the oscillation to decay by a factor e^{-1}. The coherence number equals $\omega/(2\pi\mu)$ where ω is the angular frequency of the oscillatory transients which decay at a rate proportional to $e^{-\mu t}$. Quasi-cycles are typically characterized by bursts of approximately $3n_c$ rough cycles, interspersed by noise. Quasi-cycles can be

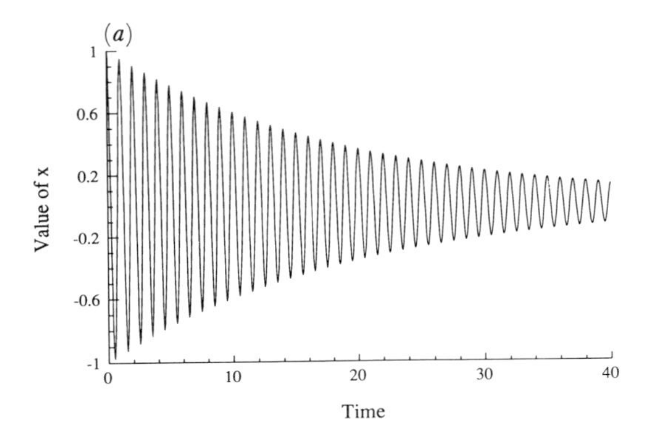

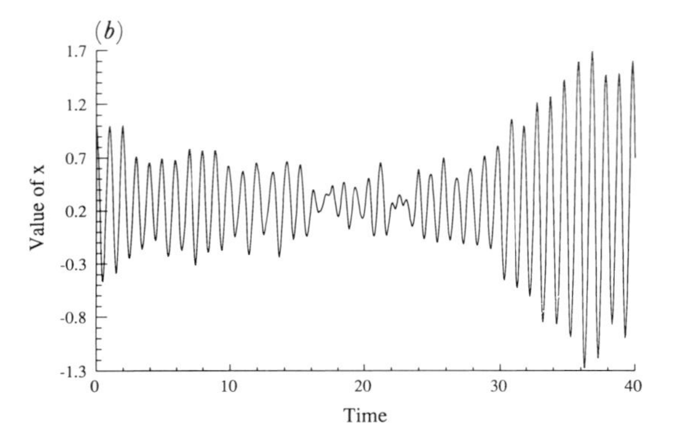

Figure 8. (*a*) The oscillatory approach (to a stable equilibrium) of equation (5); (*b*) quasi-cycles around a stable equilibrium of equation (7). In both cases $w = 2\pi$.

generated by random (white) noise when the transfer function $T(f)$ of the system has a sharp peak. The transfer function relates the amplitude of an external perturbation, at frequency f, to the consequent fluctuations at the same frequency in the population under study. If a population has a 'flat' transfer function then all external perturbations are faithfully mirrored by the population. If the transfer function has a sharp peak, say at frequency $\hat{f}$, then only perturbations of that frequency cause oscillations in the population. White noise contains fluctuations of all frequencies, including $\hat{f}$, and is thus able to cause well-characterized cycles in populations with a sharp transfer function. In general, if the range of frequencies present in the external noise is described by the spectral density s_n, the equivalent variance in the population fluctuations is

$$\sigma_x^2 = s_n \int_{-\infty}^{+\infty} |T(f)|^2 \, df. \tag{6}$$

The quantity within the integral is the modulus of the transfer function (in general a complex number) and represents the ratio of the amplitude of the environmental fluctuation and the response of the population.

To explore the influence of quasi-cycles on the calculation of dimensions, consider first a non-homogenous extension of equation (5)

$$d^2x/dt^2 + 2b\, dx/dt + \omega^2 x = \gamma(t), \quad x|_{t=0} = 1, \quad dx/dt|_{t=0} = 0. \tag{7}$$

Where $\gamma(t)$ is a source of Gaussian random noise with zero mean and spectral density s_n. Solutions of equation (7) exhibit coherent quasi-cycles (figure 8*b*). Blythe & Stokes (1988) calculated the correlation dimension for such a trajectory, embedding the time series in 2–11-dimensional space. No convergent scaling regions were found and so no estimate of attractor dimensionality was possible. Thus the quasi-cycles obscure the presence of a simple point attractor though the correlation dimension distinguishes quasi-cycles from, at least, low-dimensional chaos.

It may easily be shown that the variance in the fluctuations of equation (7), σ_x^2, is

$$\sigma_x^2 \approx s_n/2b\omega^2. \tag{8}$$

As $\omega = 2\pi$ in the example in figure 8, we would expect to need a large spectral density s_n in order to produce significant variance in the population fluctuations. Also, for that example, $n_c \approx 20$, so that bursts of up to 60 reasonably well-formed cycles would be expected during the coherent phase.

A second, and biologically more plausible, example of quasi-cycles can be obtained from the time-delayed logistic equation with a randomly perturbed parameter. Here, the coherence number is lower and the sensitivity of population variance to noise spectral density somewhat greater. In scaled (dimensionless) form with X the population variable, the time-delayed logistic is

$$\frac{dX_t}{dt} = aX_t[1-(1+\gamma(t))\, X_{t-1}], \quad x_t|_{t \leqslant 0} = 1, \tag{9}$$

where a is scaled fecundity. The scaled carrying capacity = 1 and we assume is subject to environmental perturbation described by $\gamma(t)$ with mean = 0 and variance s_n. When $\gamma(t) = 0$ ($S_n = 0$), equation (9) has a stable point for $a < \frac{1}{2}\pi$, and regular cycles (period initially ≈ 4) if $a > \frac{1}{2}\pi$. We assume that the population is at carrying capacity as an arbitrary initial history.

The quasi-cycle behaviour of equation (7) was examined by Nisbet *et al.* (1977) and Nisbet & Gurney (1982). If we choose a value of $a = 1.4$, then we have $n_c \approx 2.8$, and a sharp peak in the transfer function at scaled angular frequency 1.70, so that we might expect up to 8 or 9 rough cycles per coherent burst. The relation between s_n and σ_x^2 (equation 4) is approximately linear, with a proportionality constant of order unity (Nisbet *et al.* 1977), so that we may expect much less regularity in the population fluctuations obtained from equation (7) in comparison with equation (5) (figure 9). The pattern of irregular quasi-cyclic bursts is not atypical of population data, and it is certainly not clear, upon inspection only, whether we are seeing chaos, noisy limit cycles, or quasi-cycles.

We attempted to estimate the attractor dimension of the data in figure 9 by embedding the time series in embedding dimensions of 2–8 (we used a longer data run than that illustrated in the figure). We found no evidence for the gradient reaching a limiting value as the embedding dimension increases. If the dimension

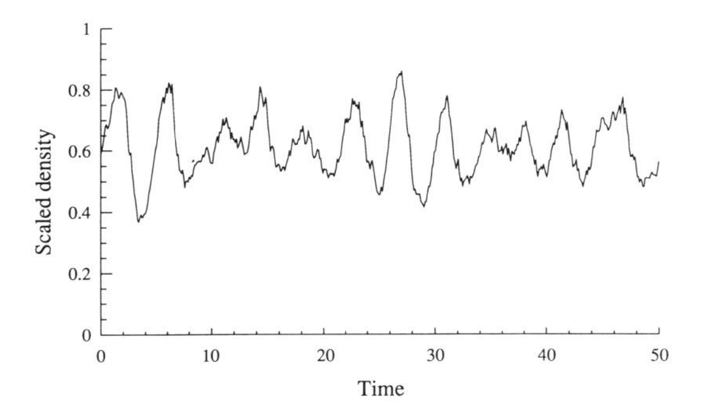

Figure 9. The dynamics of the time-delayed logistic equation with stochastic variation in carrying capacity (equation 9), $a = 1.4$, spectral density $= 1.5$.

were all the information available to us, we would be inclined to regard the data in figure 9 as essentially random. But spectral analysis of the data reveals a very strong periodic elements, clearly contradicting the notion of a noise process.

In conclusion, the calculation of correlation dimensions is fraught with difficulty in the presence of noise and transients, sometimes giving false positives, as in the case of a simple transient from initial conditions, and sometimes giving false negatives, as in the case of coherent quasi-cycles.

(*d*) *Evolution*

One view of organic evolution is that animals and plants spend most of their time at evolutionary equilibrium. From time to time they are confronted by new challenges, either from the biotic or abiotic environments, which cause relatively fast shifts to new equilibria through the action of natural selection. The alternative view is symbolized by van Valen's Red Queen Hypothesis (1973). Just as Lewis Carroll's Red Queen had to keep running to remain still, so all organisms have to continually evolve in the face of ever changing selection pressures. In van Valen's view, selection resulting in change in one organism in a community alters the selection experienced by other organisms, which in turn feeds back to cause more change in the first organism.

A more concrete example of possible red queen evolution has been suggested by Hamilton (1980) to occur between parasites and their hosts. If certain host genotypes are more susceptible to certain parasite genotypes then, under certain circumstances frequency dependent selection can lead to persistent cycles in host and parasite gene frequencies. The maintenance of such cycles requires continual, often quite heavy, mortality from disease in the host. As was pointed out by May & Anderson (1983), the genetic interactions between host and parasite will have important consequences on the population dynamics.

How might evolution affect the detection of dimensionality and chaos? If individuals from wild populations are introduced into laboratory microcosms, the novel conditions they experience may lead to selection on life history parameters (Blythe & Stokes 1988). As a result, the dynamic properties of the system may change. The Nicholson blowfly experiments (see above) are a plausible example of evolution occurring over the course of a laboratory experiment. In a sense, this form of evolutionary change is analogous to a transient as after the population has adapted to the new conditions in the laboratory, the dynamic behaviour should be unaffected by evolution. However, when interpreting results from the asymptoting dynamics, it must be borne in mind that the observed dynamics in the laboratory may differ from those in the field. It is also possible that the invariable simplicity of laboratory ecosystems in comparison with their natural counterparts might lead to consistent biases in the estimation of such properties as dimensionality.

Red Queen evolution may have a number of consequences for the population dynamics of a system. First, if the course of evolution was relatively slow compared to the length of the time series available for analysis, the form of an attractor may change over the course of the sampling period. If the shape of the attractor was reconstructed by embedding, then early and later trajectories might be physically separated in phase space. Similarly, such changes in the properties of an attractor would complicate estimation of dimensionality. A more interesting consequence occurs if the rate of evolution is relatively fast compared to the length of the time series. In these circumstances, population densities and gene frequencies may change on similar time scales and a reconstruction of the dynamics of the system would incorporate both population and genetic variables. The dimensionality of the system would be influenced by the degree of independence of population dynamics and population genetics.

(*e*) *Interpretation*

The attempts both to reconstruct attractors, and to estimate correlation dimensions, involve a degree of subjective pattern recognition. For example, the identification of regions of stretching and folding has to be made by eye. Taking Poincaré sections and estimating one-dimensional return maps increases objectivity though fitting curves to one-dimensional return maps with few data points results in large standard errors on parameters and consequent ambiguity in predicted dynamics. Similarly, when estimating correlation dimensions, the size of the scaling region has to be chosen subjectively.

Biologists, brought up on a diet of strict hypothesis testing, are typically more suspicious than physical scientists of conclusions drawn from the visual inspection of geometrical objects. Thus Berryman & Millstein (1989), unfairly in our view, dismiss the attempts of Schaffer and his colleagues to reconstruct attractors as providing evidence for chaos that is 'more illusory than scientific': the case for at least a chaotic measles attractor is now very strong (Schaffer 1985; Sugihara & May 1990*a*). In part, the problems of subjectivity can only be resolved through the use of better data. However, we also believe that biologists should adopt a more robust approach to scientific hypothesis testing.

CONCLUSIONS

The new techniques from applied mathematics and the physical sciences offer very exciting prospects of new approaches to some of the most interesting questions in population and community ecology. Nevertheless, there are formidable problems in applying these techniques in ecology: very few data sets are of sufficient quality and length to be analysed in this way. Further effort is needed by applied mathematicians to develop techniques better suited to the biological, as opposed to the physical sciences.

We believe that the best prospects of investigating community dynamics in biology lie with the study of experimental microcosms of short-lived organisms, for example, freshwater zoo-plankton. Obviously, the study of experimental communities is second-best to the study of natural communities. However, if it is not possible to understand the dynamics of a simplified community, under controlled conditions, there is very little prospect of understanding the dynamics of real communities. We strongly agree with a recent remark of Kareiva's (1989) 'ecologists gave up bottle experiments too soon'.

REFERENCES

Baltensweiler, W. 1968 The cyclic population dynamics of the grey larch tortrix, *Zeiraphera griseana* Hubner (= *Semasia diniana* Quenec) (Lepidoptera: Tortricidae). In *Insect abundance* (Ed. T. R. E. Southwood), (*Symp. R. ent. Soc. Lond.* **4**), pp. 88–97.

Bellows, T. S. & Hassell, M. P. 1988 The dynamics of age-structured host-parasitoid interactions. *J. Anim. Ecol.* **57**, 259–268.

Berryman, A. A. & Millstein, J. A. 1989 Are ecological systems chaotic—and if not, why not? *Trends Ecol. Evol.* **4**, 26–28.

Blythe, S. P. & Stokes, T. K. 1988 Biological attractors, transients and evolution. In *Ecodynamics* (ed. W. Wolff, C.-J. Soeder & F. R. Drepper), pp. 309–318. Berlin: Springer–Verlag.

Brillinger, D. R., Guckenheimer, J., Guttorp, P. & Oster, G. 1980 Empirical modelling of population time series data: the case of age and density dependent vital rates. *Lect. Math. Life Sci.* **13**, 65–90.

Colebrook, J. M. 1960 Continuous plankton records: methods of analysis 1950–1959. *Bull. mar. Ecol.* **5**, 51–64.

Colebrook, J. M. 1975 The continuous plankton recorder survey: automatic data processing methods. *Bull. mar. Ecol.* **8**, 123–142.

Colebrook, J. M. 1978 Continuous plankton records: zooplankton and environment, north-east Atlantic and North Sea, 1948–1975. *Oceanol. Acta* **1**, 9–23.

Elton, C. & Nicholson, M. J. 1942 The ten-year cycle in numbers of the lynx in Canada. *J. Anim. Ecol.* **11**, 215–244.

Gilpin, M. E. 1979 Spinal chaos in a predator–prey model. *Am. Nat.* **113**, 306–308.

Gleick, J. 1988 *Chaos*. New York: Viking.

Grassberger, P. 1986*a* Estimating the fractal dimension and entropies of strange attractors. In *Chaos* (ed. A. V. Holden), pp. 291–312. Manchester University Press.

Grassberger, P. 1986*b* Do climatic attractors exist? *Nature, Lond.* **323**, 609–612.

Grassberger, P. & Procaccia, I. 1983*a* Characterization of strange attractors. *Physics Review Letters* **50**, 346–349.

Grassberger, P. & Procaccia, I. 1983*b* Measuring the strangeness of strange attractors. *Physica* **9D**, 189–208.

Gurney, W. S. C., Blythe, S. P. & Nisbet, R. M. 1980 Nicholson's blowflies revisited. *Nature, Lond.* **187**, 17–21.

Hamilton, W. D. 1980 Sex versus non-sex versus parasite. *Oikos* **35**, 282–290.

Hassell, M. P., Lawton, J. H. & May, R. M. 1976 Patterns of dynamical behaviour in single species populations. *J. Anim. Ecol.* **45**, 471–486.

Hochberg, M. E., Hassell, M. P. & May, R. M. 1900 The dynamics of host-parasitoid-pathogen interactions. *Am. Nat.* **135**, 74–94.

Kareiva, P. 1989 Renewing the dialogue between theory and experiments in population ecology. In *Perspectives in ecological theory* (ed. J. Roughgarden, R. M. May and S. A. Levin), pp. 68–88. Princeton University Press.

Lawlor, L. R. 1979 Direct and indirect effects of n-species competition. *Oecologia* **43**, 355–364.

Levine, S. H. 1976 Competition interactions in ecosystems. *Am. Nat.* **110**, 903–910.

Levins, R. 1968 *Evolution in changing environments*. Princeton University Press.

MacArthur, R. H. 1972 *Geographical ecology*. New York: Harper & Row.

May, R. M. 1973 *Stability and complexity of model ecosystems*. Princeton University Press.

May, R. M. 1974 Biological populations with non-overlapping generations: stable points, stable cycles and chaos. *Science, Wash.* **186**, 645–647.

May, R. M. 1976 Simple mathematical models with very complicated dynamics. *Nature, Lond.* **261**, 459–467.

May, R. M. 1980 Nonlinear phenomena in ecology and epidemiology. *Ann. N.Y. Acad. Sci.* **357**, 267–281.

May, R. M. & Anderson, R. M. 1979 Population biology of infectious diseases Part II. *Nature* **280**, 455–461.

May, R. M. & Anderson, R. M. 1983 Epidemiology and genetics in the coevolution of parasites and hosts. *Proc. R. Soc.* B **321**, 565–607.

May, R. M. & Oster, G. F. 1976 Bifurcations and dynamic complexity in simple ecological models. *Am. Nat.* **110**, 573–599.

Mueller, L. D. & Ayala, F. J. 1981 Dynamics of single-species population growth: Stability or chaos. *Ecology* **62**, 1148–1154.

Nicholson, A. J. 1954 An outline of the dynamics of natural populations. *Aust. J. Zool.* **2**, 9–65.

Nicholson, A. J. 1957 The self adjustment of populations to change. *Cold Spring Harb. Symp. quant. Biol.* **22**, 153–173.

Nicolis, C. & Nicolis, G. 1985 Is there a global climatic attractor? *Nature, Lond.* **311**, 529–532.

Nisbet, R. M. & Gurney, W. S. C. 1982 *Modelling fluctuating populations*. London: Wiley.

Nisbet, R. M. & Gurney, W. S. C. 1976 Population dynamics in a periodically varying environment. *J. Theor. Biol.* **56**, 459–475.

Oster, G. & Takahashi, Y. 1974 Models for age-specific interactions in a periodic environment. *Ecol. Monogr.* **44**, 483–501.

Packard, N. H., Crutchfield, J. P., Farmer, J. D. & Shaw, R. S. 1980 Geometry from a time series. *Phys. Rev. Lett.* **45**, 712–716.

Prout, T. & McChesney, F. 1985 Competition among immatures affects their adult fertility: population dynamics. *Am. Nat.* **126**, 521–558.

Readshaw, J. L. & Cuff, W. R. 1980 A model of Nicholson's Blowfly cycles and its relevance to predation theory. *J. Anim. Ecol.* **49**, 1005–1010.

Roux, J.-C., Rossi, A., Bachelart, S. & Vidal, C. 1981 Experimental observations of complex dynamical behaviour during a chemical reaction. *Physics* **2D**, 395–403.

Schaffer, W. M. 1981 Ecological abstraction: the consequence of reduced dimensionality in ecological models. *Ecol. Monogr.* **51**, 383–401.

Schaffer, W. M. 1984 Stretching and folding in lynx fur returns: evidence for a strange attractor in nature? *Am. Nat.* **124**, 798–820.

Schaffer, W. M. 1985 Order and chaos in ecological systems. *Ecology* **66**, 93–106.

Schaffer, W. M. & Kot, M. 1985 Nearly one dimensional dynamics in an epidemic. *J. Theor. Biol.* **112**, 403–427.

Schaffer, W. M. & Kot, M. 1986 Chaos in ecological systems: the coals that Newcastle forgot. *Trends Ecol. Evol.* **1**, 58–63.

Schaffer, W. M., Truty, G. L. & Fulmer, S. L. 1988 *Dynamical software, users guide and introduction to chaotic systems.* (vol. I and II.) Tucson, Arizona: Dynamical Systems Inc.

Stokes, T. K., Gurney, W. S. C., Nisbet, R. M. & Blythe, S. P. 1988 Parameter evolution in a laboratory insect population. *Theor. Popul. Biol.* **33**, 248–265.

Sugihara, G. & May, R. M. 1990*a* Nonlinear forecasting as a way of distinguishing chaos from measurement error in time series. *Nature, Lond.* **344**, 734–741.

Sugihara, G. & May, R. M. 1990*b* Applications of fractals in ecology. *Trends Ecol. and Evol.* **5**, 79–86.

Takens, F. 1981 Detecting strange attractors in turbulence. In *Dynamical systems and turbulence* (ed. D. A. Rand & L. S. Young), pp. 366–381. Berlin: Springer–Verlag.

Thomas, W. R., Pomerantz, M. J. & Gilpin, M. E. 1980 Chaos, asymmetric growth and group selection for dynamical stability. *Ecology* **61**, 1312–1320.

van Valen, L. 1973 A new evolutionary law. *Evol. Theor.* **1**, 1–30.

Discussion

M. Williamson (*Department of Biology, University of York, U.K.*). When testing time series of biotic entities for chaos is there not a possibly confusing factor? It is one of the triumphs of nonlinear science to show that there is chaos in the physical environment on a variety of timescales from days (as in the weather) to millions of years (as in planetary orbits). If Dr Godfray shows chaos in a biological time series, may he not merely be showing chaos in the physical environment affecting the biological entity? Is he not trying to determine whether there is chaos in the population dynamics of that entity?

H. C. J. Godfray. One of the key issues here is the relative scales of the physical and biological processes. When experimentally studying biological populations, one is not normally concerned with climatic attractors or planetary motion with characteristic periods of very many generations. However, meteorological and tidal patterns may be of a scale likely to influence experimental results. Normally, it is far easier to measure the physical rather than the biological dynamics and so identification of the 'driving chaos' may be far simpler than the identification of chaos in the population dynamics. Of course, microcosm studies have the advantage of being able to control for external physical processes.

S. P. Blythe. Additional insight into this problem may be gained by modelling: consideration of forcing functions may be a good way forward in this area, as the study of complicated dynamics in these circumstances dates back to Cartwright (1948, *Journal of the Institute of Electrical Engineers* **95**, pp. 223) and many results are available. Note also that a chaotic forcing term is not needed to induce chaos: a periodic function will often do the trick (see, for example, Marcus *et al.* (1984), *FEBS Letters* 172, pp. 235). There is much scope (and need) for analysis in this area.

J. N. Perry (*Institute of Arable Crops Research, Statistics Department, Rothamsted Experimental Station, Harpendon, Herts. U.K.*). Dr Godfray refers to the study of Hassell *et al.* (1976) where an intrinsic growth parameter, λ, and a density-dependence parameter, β, were estimated from data. Since λ is calculated after allowing for all density-independent mortality, and since we now know that density dependence may often be present but undetected (and therefore underestimated), these estimates of λ may be biased downwards, i.e. underestimates. However, λ and β are clearly positively correlated; hence calculated β values may consequently also be underestimates. Therefore some data points in the figure might require amendment by a shift away from the origin, towards the chaotic region. I wonder to what extent some populations would exhibit chaos under Hassell *et al.*'s model, were all the density-dependence allowed for?

H. C. J. Godfray. This is an interesting additional caveat to the Hassell *et al.* study. However, my feeling is that 15 years on from this study we need more sophisticated techniques for detecting chaos: either more realistic underlying models, or techniques that do not make *a priori* assumptions about underlying mechanisms.

Distinguishing error from chaos in ecological time series

GEORGE SUGIHARA[1], BRYAN GRENFELL[2] AND ROBERT M. MAY[3]

[1] *Scripps Institution of Oceanography, University of California, San Diego, La Jolla, CA 92093, U.S.A.*
[2] *Department of Zoology, Cambridge University, Cambridge, U.K.*
[3] *Department of Zoology, Oxford University, Oxford, OX1 3PS, U.K. And Imperial College, Exhibition Road, London SW7 2AZ, U.K.*

SUMMARY

Over the years, there has been much discussion about the relative importance of environmental and biological factors in regulating natural populations. Often it is thought that environmental factors are associated with stochastic fluctuations in population density, and biological ones with deterministic regulation. We revisit these ideas in the light of recent work on chaos and nonlinear systems. We show that completely deterministic regulatory factors can lead to apparently random fluctuations in population density, and we then develop a new method (that can be applied to limited data sets) to make practical distinctions between apparently noisy dynamics produced by low-dimensional chaos and population variation that in fact derives from random (high-dimensional)noise, such as environmental stochasticity or sampling error.

To show its practical use, the method is first applied to models where the dynamics are known. We then apply the method to several sets of real data, including newly analysed data on the incidence of measles in the United Kingdom. Here the additional problems of secular trends and spatial effects are explored. In particular, we find that on a city-by-city scale measles exhibits low-dimensional chaos (as has previously been found for measles in New York City), whereas on a larger, country-wide scale the dynamics appear as a noisy two-year cycle. In addition to shedding light on the basic dynamics of some nonlinear biological systems, this work dramatizes how the scale on which data is collected and analysed can affect the conclusions drawn.

1. INTRODUCTION

The classical debate between the biotic and climatic schools has divided opinion over the relative importance of deterministic versus stochastic forces in controlling ecological populations (Sinclair 1989). This long-standing debate over random versus determined variation has begun to take on new meaning with recent interest in chaos and nonlinear dynamics, and with the ever-increasing demonstrations of the applicability of these ideas to real data. Until recently, one would have viewed a time series such as the one shown in figure 1 *a* and concluded that the ecologically important information here rested in the smooth fitted line. There is, however, a change of view occurring in dynamics, similar to the change that fractals is bringing to geometry and the study of spatial pattern, which suggests that the most interesting things may be found in the irregularities rather than in the smoothed pattern (Lorenz 1969; Takens 1981; Schaffer & Kot 1986; Sugihara & May 1990). Although initially it appears that incorporating such detail into the population dynamics debate may further cloud the problem, we shall argue that the end result is not new difficulty, but the prospect of a new clarity and simplicity.

The paper is divided into four sections. In the first, we discuss the limitations of traditional approaches to analysing deterministic influences on population dynamics. The second section outlines the light which recent advances in nonlinear dynamics theory shed on these problems. In particular, it summarizes a new method for distinguishing noise from low dimensional determinism in ecological time series, based on their internal predictability (Sugihara & May 1990). The third section applies this method to time series of notified case reports for childhood diseases in developed countries. Because they are often relatively long, and reflect comparatively simple host–microparasite population interactions, these series are among the best ecological candidates for applying nonlinear methods (Schaffer & Kot 1986; Schaffer *et al.* 1988). After reviewing previous work in this area, we present a new analysis of the dynamics of measles in English cities, which has significant implications for the question of spatial scaling in ecological systems and the concept of stationarity as defined in traditional time series analysis. The final section draws these conclusions together and suggests lines for future work.

2. TRADITIONAL APPROACHES

The classical approaches to analysing population

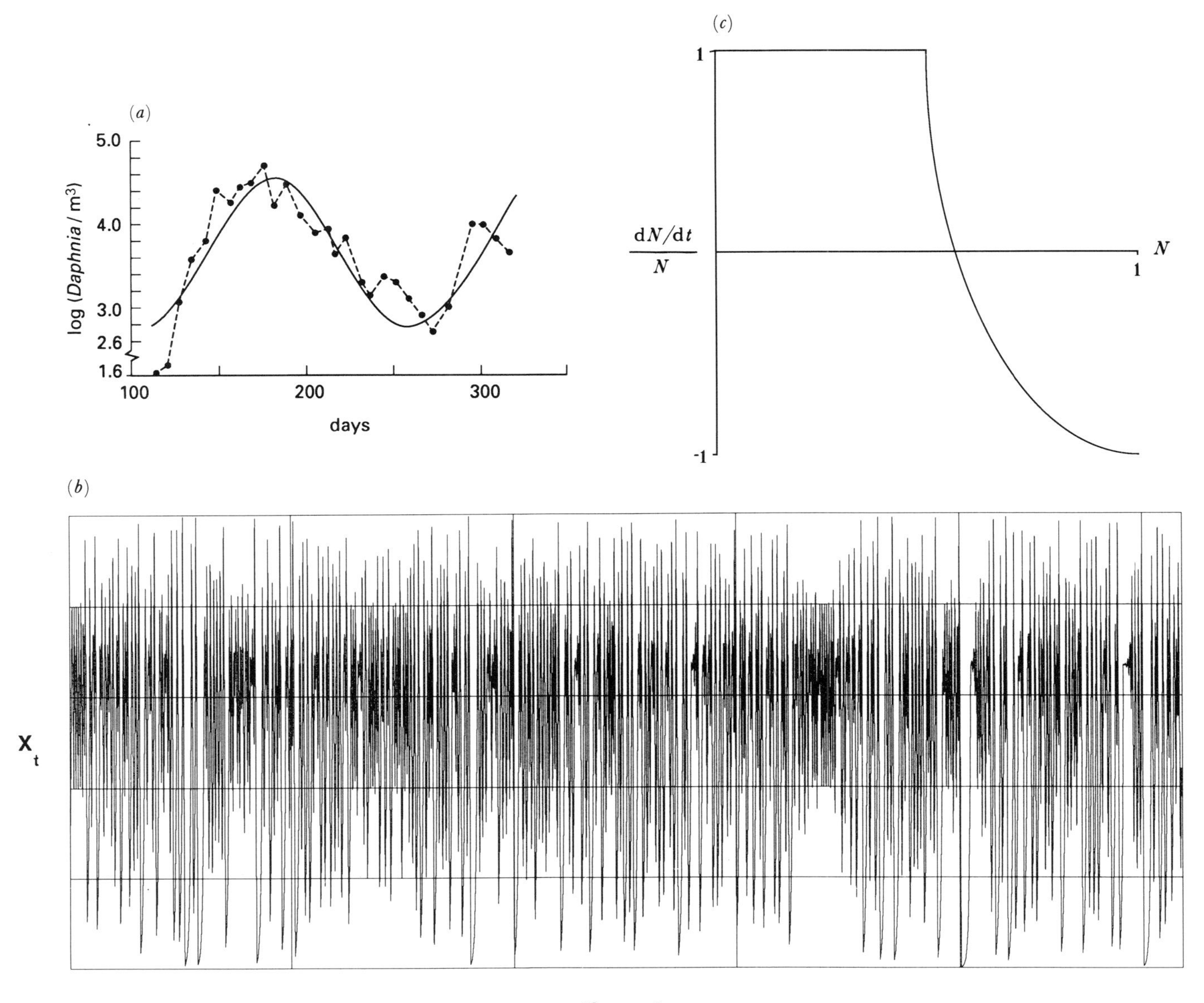

Figure 1. (*a*) An example of population dynamics for *Daphnia* and edible algae from Lake Washington (Murdoch & McCaughley 1985). The dashed line joins the observations and the smooth solid line is a sine wave fitted to the seasonal pattern. (*b*) Time series of 1000 points for the chaotic tent map: $x_{t+1} = 2x_t$ if $x_t < 0.5$; $x_{t+1} = 2\ (1-x_t)$ if $x_t > 0.5$. These data are in many ways indistinguishable from white noise. (*c*) Key factor analysis applied to the data in figure 1*b*. Although the signature appears to be an example of density vague population regulation (no density dependence at low densities with severe control at high densities), it was derived from a low-dimensional chaotic process.

dynamics are key factor analysis of observed data (to distinguish density-independent and nonlinear density-dependent influences) and the exploration of laboratory and mathematical models that reduce the complexity of real systems to a single or a few factors (Sinclair 1989). Although these methods have been highly successful in the main, recent developments in nonlinear theory identify a number of problems, which we summarize as follows.

(*a*) *Limitations of key factor analysis*

Consider the model time series shown in figure 1*b*, which appears to be stochastic. If this were a time series for a real population, one might easily conclude that it represents a population that is being buffeted by frequent random shocks with apparent occasional returns to a quasi-equilibrium. Indeed, as shown in figure 1*c*, a key factor analysis based on these data would lead one to the opinion that this is a classic example of what some would call 'density-vague control', i.e. no regulation at low densities with control occurring only at high densities. Yet these data do not represent density-vagueness at all, but are an example of simple chaotic dynamics that were generated from the deterministic tent map (Sugihara & May 1990). Thus, in this case, an interpretation of these data as arising from external unpredictibilities would have been incorrect. Conventional approaches would have mis-identified what was in fact simple chaotic dynamics

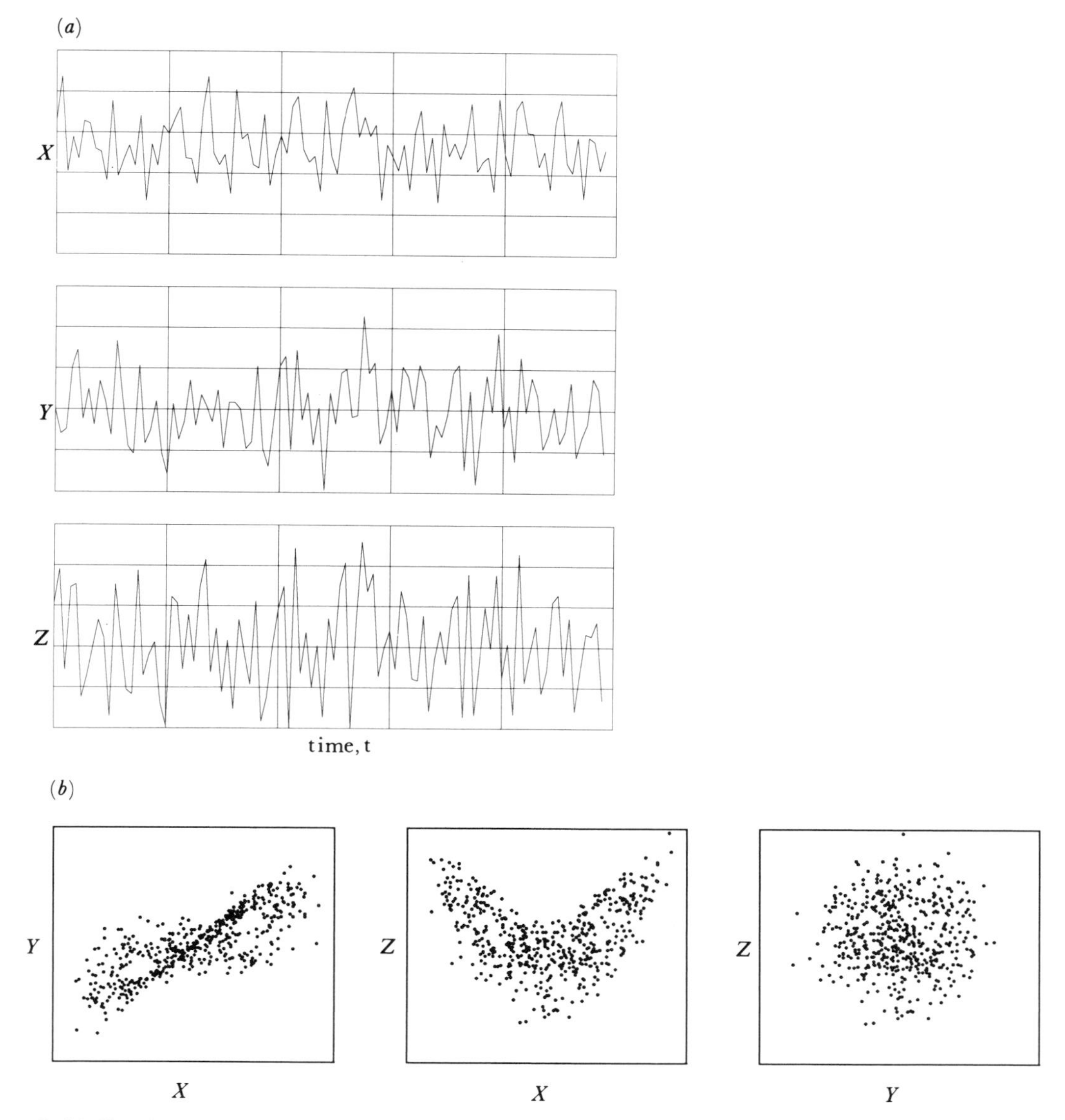

Figure 2. (*a*) Simultaneous time series for the three variables of the Lorenz system: X, Y and Z as functions of time. Each time-series represents the projection onto one axis of the Lorenz attractor (the so-called butterfly attractor) as it is embedded in three dimensions. (*b*) The time series in figure 2*a* were randomly sampled at 500 simultaneous times, and the correlations between all pairs of variables for the Lorenz attractor are shown here. The correlation between X and Y has $r = 0.636$; that between X and Z has $r = 0.001$; and Y and Z have $r = 0.000$. That is, although Y and Z are deterministically coupled they show no correlation.

(i.e., low dimensional chaos) as random variation within a density-vague envelope.

(*b*) *Limitations of the single factor approach*

To complicate matters further (as discussed here by Godfray & Blythe (1990)), the dynamics of real populations can only be properly understood if they are considered in their ecological context, that is in a way that recognizes the inherent complexity or multidimensionality of the problem. Populations do not exist singly, but are embedded in a dynamic web of other species and environmental forces. While we are consciously willing to acknowledge this fact, it is important to realize how such complexity could blur any relations that one could ever hope to see in a typical single-factor analysis (regressing one explanatory variable against another: a mainstay of ecological research) such as the key factor approach.

This can be shown with the following simple example. The three time series shown in figure 2*a* were generated from the three variables that describe the well known Lorenz equations (Lorenz 1969). The system is completely determined by these three variables. Suppose now that we do not know that these time series came from a Lorenz system. We only know that they are parallel measurements from some system. Typically when one is trying to understand a phenomenon one looks for patterns, in this case correlations between the time series. If we randomly sample these series at simultaneous times and look for pairwise correlations we get the result shown in figure 2*b*. While there is a significant relation between X and Y, there is no significant linear relation between X and Z

(although there is a significant parabolic relation), and no relation of any kind between Y and Z. Thus even though Y and Z are deterministically coupled they appear completely unrelated. That is to say, lack of correlation between pairs of variables does not imply lack of causation.

We are faced, therefore, with two dilemmas that at first glance are not obviously related: (i) how to distinguish randomness from low dimensional determinism, and (ii) how to explicitly acknowledge higher dimensions. These two problems are not only central to population regulation debates, but are at the heart of nonlinear theory as shall be discussed in the next section.

3. NONLINEAR PERSPECTIVE

Suppose we have perfect knowledge about the internal dynamics of a population, so that without arbitrary disturbance everything is determined and nothing is random. That is, we know all the variables and the functions describing how they are related. The space whose axes consist of each of these dynamically coupled variables is known as the 'state space.' For example, if this is an experimental multispecies system where the only important variables are other species, then the axes of the state space would be the population sizes of each of the coupled species. All population variability and motion would be constrained to some sub-manifold of that space set by the relations (functions) among variables. If, after an arbitrary perturbation, the population returns to it this sub-manifold, it is called an 'attractor' (Takens 1981; Abraham & Shaw 1982; Schaffer & Kot 1985; Godfray & Blythe 1990).

The central concepts here, of state space and attractor, represent the scientific ideal of perfect deterministic knowledge. In practice, however, we do not know the state space. Rather, we might have time series for one or more of the variables. Thus, in practice, the state space and its attractor are a black box, and the time series are observables or outputs. Each of these time series can be viewed essentially as a projection onto one axis of the state space through time (e.g. figure 2*a* shows the three time series for the Lorenz attractor). That is, each time series is a view in one dimension of a process occurring in higher dimensions. Therefore, in a perfectly deterministic world, much of the complexity or apparent randomness in a time series will arise from a state space having a high number of dimensions, or alternatively (and possibly in conjunction with), an attractor with chaotic dynamics.

Let us focus on the case where we have only one time series. How can we get information about the state space and attractor that produced it? The standard method here is Taken's (1981) technique of using lagged coordinates to embed a time series in higher dimensions (Godfray & Blythe 1990). Although embeddings can be created from the original state-space time series themselves, here we shall consider the worst case where there is only one time series to work with; this is also the case where the lagged coordinate idea is most valuable. Again, the space created by such an embedding is not the original state-space, but a mock version of it: something we shall call a 'phase space'. As proved by Takens (1981), the phase space retains essential properties of the original state-space including the dimensionality of the attractor it contains. Moreover, as we shall see, a phase space can be used to make forecasts, properties of which provide practical distinctions between low dimensional determinism and noise, even with limited data of the kind encountered in population biology (Sugihara & May 1990).

Thus even though our ultimate goal is to understand the population in state-space (i.e., how many dimensions does it have? are the dynamics low dimensional chaos or simply noisy? etc.), we may only have a time series for one variable of this space (namely for the population itself whose dynamics we are trying to understand). To get around this problem we shall construct a phase space as a surrogate having the same topological characteristics as the original state-space (again, a higher dimensional embedding of the time series by using time-lagged coordinates). Information about the original attractor can then be gained by exploring the properties of the mock attractor in phase space.

(*a*) *Nonlinear forecasting to differentiate noise from chaos: basic ideas*

The method outlined here is discussed in fuller detail in Sugihara & May (1990) and is based on theories of short-range prediction proposed in general terms by Lorenz (1969) and others (Tong & Lim 1980; Priestly 1980; Farmer & Sidorowich 1989). The basic idea here (which is classical in prediction) is that, if indeed deterministic laws govern a system, then even if the dynamical behaviour is chaotic, the future may be reckoned from the behaviour of past values that are similar to the present. The key, however, is in knowing the dimensionality within which the past, present and future are embedded.

Suppose (as discussed by Godfray & Blythe) we have properly embedded a time series in an E-dimensional phase space so that each lagged sequence of data points, $Z_t = \{x_t, x_{t-\tau}, \ldots, x_{t-(E-1)\tau}\}$ is a point in this E-dimensional space. Here we usually choose $\tau = 1$, but the results do not appear to be too sensitive to the choice of τ, provided it is not too large. This is like taking an E-pronged fork whose tines are separated by a distance τ, and dragging it sideways along the time series; the vector of time-series values, Z_t, formed by where the tines land at each instant, describes another E-dimensional point, and the set of vectors $\{Z\}$ describes the attractor. In general, $E_{\min} \leqslant 2D+1$, where D is the attractor dimension. Each predictee Z_i is now to be regarded as an E-dimensional point, comprising the present value x_i and the $E-1$ previous values each separated by one lag time τ. We now locate all nearby E-dimensional points in the phase space and choose a minimal neighbourhood so that the predictee is contained within the smallest simplex formed from its $E+1$ closest neighbours; a simplex containing $E+1$ vertices is the smallest simplex that can contain an

E-dimensional point as an interior point. When a minimal bounding simplex cannot be found (for example, for boundary points), we use the $E+1$ nearest neighbours. To obtain a prediction, we simply project the domain of the simplex into its range (that is, we keep track of where the other points in the simplex end up at p time steps), and compute the weighted centre of mass of the simplex to get the predicted value. In other words, we follow the short term destiny of nearby points in the attractor to see where they end up after p time steps. This is a non-parametric method, and it should apply to any stationary or quasi-ergodic process, including chaos. It uses no previous information about the model used to generate the series, only the information in the time series output itself. Unless otherwise stated, we shall construct the phase space from the first half of the time series to make predictions on the second half.

(*b*) *Nonlinear forecasting to differentiate noise from chaos: examples from models*

Figure 3*a* shows an application of this method to the white-noise time series produced by taking first differences of the tent map series shown in figure 1*b*. It compares predicted against actual results two time steps into the future: a time step where there is no significant correlation between values. Notice again, the phase space constructed from the first half is used to predict the values in the second half. Thus, this time-series, which by standard statistical analysis is uncorrelated white noise (unpredictable), in fact becomes strongly predictable when embedded in higher dimensions.

Figure 3*b* shows how predictability, as measured by the standard correlation coefficient, declines as the prediction interval T_p (i.e., how far into the future one projects) increases. Such a decrease in the correlation coefficient with increasing prediction time is the hallmark of chaos (or equivalently, of the presence of a positive lyapunov exponent, with the magnitude of the exponent being related to the rate of decrease of ρ with T_p). This property is noteworthy because it suggests a simple way to distinguish between additive noise and multiplicative chaos: predictions with uncorrelated additive noise will appear to have a fixed amount of error, regardless of how far or close into the future one tries to predict, whereas predictions with multiplicative chaos tend to deteriorate as one tries to forecast further into the future. This can be seen in figure 3*c* where it is shown that the characteristic signature of ρ decreasing with T_p does not arise when the erratic time series is in fact a noisy limit cycle (here additive noise superimposed on a sine wave). With uncorrelated additive noise, such as sampling variation, the error remains constant as the simplex is projected further into the future. In contrast, the dashed line in figure 3*c* represents the correlation coefficient (ρ) against prediction time (T_p) relation for a chaotic sequence generated as the sum of two independent runs of first-differences of the tent map. Although the two time series here both look alike as sample functions of some random process, the characteristic signatures differentiate random noise in one instance from deterministic chaos in the other.

The predictions in figure 3*a–c* are based on an embedding dimension of $E = 3$. These results are, however, sensitive to the choice of E. This is shown in figure 3*d* where results are summarized for prediction accuracy (correlation between predicted and observed, here at $T_p = 1$) versus embedding dimension E. It may at first sight appear surprising that having potentially more information erodes the accuracy of the predictions, since for large E there are more data summarized in each E-dimensional point, and a higher dimensional simplex of neighbours for each predictee. Sugihara & May (1990) have suggested that this effect may be caused by contamination of nearby points in the higher dimensional embeddings with points whose earlier coordinates are close but whose recent (and more relevant) coordinates are distant. If this is so, this method may have additional application as a trial and error method of computing an upper bound on the embedding dimension, and thence on the attractor dimension.

So far, we have compared relations between ρ and T_p for chaotic time series with the corresponding relation for additive white noise. More problematic, however, is the comparison with ρ–T_p relations generated by coloured noise spectra where there are significant short-term correlations but no long term ones. As with chaos, such correlated noise can also lead to declining ρ–T_p curves. Although, in the limit, the shallow form of the decline in simple cases may distinguish correlated noise from a chaotic signature (Sugihara & May 1990; Farmer & Sidorowich 1989), in a practical sense, particularly with limited data of the kind available in population biology, such distinctions may be difficult to find. One practical solution to this dilemma, suggested by Sugihara & May (1990), is that coloured noise may tentatively be distinguished from deterministic chaos if in addition to an exponentially declining ρ–T_p curve, the correlation, ρ, between predicted and observed values obtained by nonlinear methods is significantly better than the correlation obtained by the best-fitting autoregressive linear model. That is, if a time series is chaotic it should have both a steeply declining ρ–T_p curve and more predictability under the nonlinear hypothesis (i.e. that it was produced by nonlinear mechanisms), than if one assumed it was produced by linear mechanisms (i.e. by the linear superposition of simple cycles of various period and amplitude).

Perhaps most germane to the biotic/climate issue is the possibility of noise entering multiplicatively as a disturbance to population numbers which is then fed back into the dynamics. An example would be noise entering in the form $X_{t+1} = F(X_t + \Theta_t)$, where F is assumed to have stable dynamics. Here, if F is nonlinear, a nonlinear predictor may still perform better. However, one can again expect a slower than exponential decline in the ρ–T_p curve. An exponential decline which arises from locally exponentially diverging trajectories may be taken as the operational definition of chaos. A simple way to use nonlinear forecasting to distinguish this possibility is by examin-

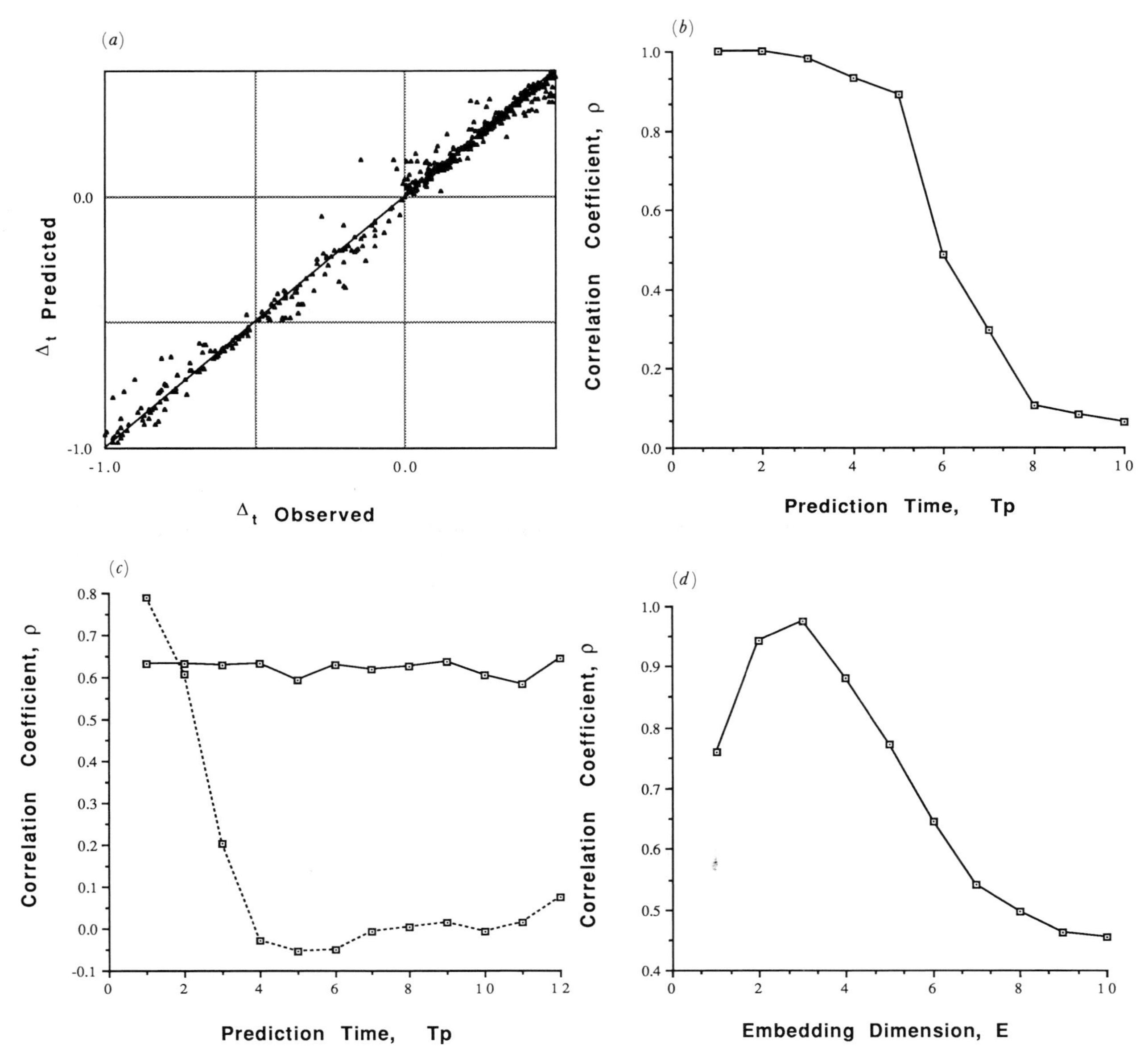

Figure 3. (*a*) Predicted values two steps into the future ($T_p = 2$) versus observed values for the white-noise time series produced by taking first differences of the tent map series shown in figure 1*b* (after Sugihara & May 1990). Specifically, the first 500 points in the series were used to generate a library of patterns, which were then used as a basis for making predictions for each of the second 500 points. As described in the text, the predictions were made using a simplex projection method (Sugihara & May 1990) with an embedding dimension and lag time of $E = 3$ and $\tau = 1$, respectively. Here the coefficient of correlation between predicted and actual values is $\rho = 0.997$ ($N = 500$). For comparison, we note that the corresponding correlation coefficient using an autoregressive linear model is $\rho = 0.04$. (*b*) Predictability measured by the standard correlation coefficient, ρ, as a function of how far into the future the forecast is made, T_p. The exponential decay in predictability with increasing prediction time, T_p, shown here is a characteristic of chaotic dynamics. (*c*) Additive noise (solid line) versus chaos (dashed line). The solid line shows that the correlation, ρ, between predicted and observed values for the case of additive noise (here white noise superimposed on a sine wave) does not decline as one tries to forecast further into the future. By contrast, the dashed line shows the declining signature characteristic of a chaotic sequence (here the sum of two separate tent map sequences). (*d*) Correlation coefficient, ρ, between predicted and observed values as a function of embedding dimension, E, for predictions one time step into the future ($T_p = 1$); like figures 3*a* and *b*, the figure is based on the time series shown in figure 1*b*.

ing whether the E-dimensional simplex tends to expand (chaos) or contract (noise fed through stable dynamics) when projected from its domain into its range; this idea will be developed further elsewhere.

4. APPLICATION: THE DYNAMICS OF CHILDHOOD MICROPARASITIC INFECTIONS

Because of their public health importance, the epidemiology of childhood viral diseases such as measles and chickenpox in developed countries is especially well documented (Nokes & Anderson 1986). In particular, the relatively long time series of case reports accumulated from disease notification programmes in Europe and the U.S.A. provide a mass of information about the characteristically oscillatory dynamics of these infections in human communities (Anderson *et al.* 1984). The dynamic origin of this recurrent epidemic behaviour has been extensively examined, both in terms of mathematical models (Bartlett 1957; Anderson & May 1985; May 1986; Hethcote & Levin 1989), and time-series analysis of epidemiological data (Anderson *et al.* 1984). The data for measles have received considerable attention recently, and have been the focus of a debate as to whether measles dynamics is simply a noisy limit cycle (Schwartz 1985), or low dimensional chaos superimposed on a seasonal cycle (Schaffer & Kot 1985, 1986; Schaffer *et al.* 1989). Much of this controversy has centred on Schaffer's pioneering analyses of case reports for New York, and so we begin by applying the prediction method to these data.

(*a*) *Measles and chickenpox in New York*

The methods described above have been used to analyse public health records of monthly changes in the reported incidence of measles and chickenpox in New York City (Sugihara & May 1990). The results are summarized in figure 4. The earlier arguments for chaos, based largely on qualitative judgements as to static properties of the attractor and model simulations (see, for example, Schaffer *et al.* 1989), were supported by the results of the forecasting analysis presented here. However, because prediction is a harder test of E-dimensional determinism than judgements as to the geometry of a putative attractor, we think this analysis constitutes the strongest evidence so far, for the measles attractor. Here we see a steeply declining ρ–T_p curve, with the characteristic signature of a chaotic process. The result is supported by the fact that the nonlinear predictor performs significantly better than the best linear predictor ($P < 0.0005$). An optimal embedding dimension of 5–8 is roughly consistent with our independent estimates of an attractor dimension of 2.5–3.5 by using the Grassberger–Procaccia (1983) algorithm (Schaffer & Kot (1985) also report an estimated attractor dimension of 2.5 for measles). Thus, we believe the apparent irregularity in the measles time series is not due to random effects (environmental shocks or measurement errors), but is generated by low dimensional chaos.

In contrast, the analysis for chickenpox (figure 4*b*) suggests that complexity here is not due to low dimensional chaos, but to noise (possibly high dimensional chaos), superimposed on a strong annual cycle. These data produce a flat ρ–T_p curve, similar to the additive noise case in figure 3*c*. Moreover, the best linear predictor was found to perform at least as well as our nonlinear predictor (the correlation between the predicted and observed monthly change in the numbers of chickenpox cases using the linear predictor was $\rho = 0.84$, and for the nonlinear predictor $\rho = 0.82$ where $N = 266$ and $T_p = 1$). Thus, there is no evidence that the irregularity in the chickenpox data is due to anything other than random noise.

As discussed in more detail elsewhere (Sugihara & May 1990), there are biological reasons to explain why measles may exhibit chaotic dynamics (essentially deriving from a roughly two-year 'inter-epidemic period' interacting with annual variations in transmissibility), while chickenpox (where infectiousness can recrudesce at older ages) exhibits only annual periodicities.

(*b*) *Measles in England and Wales*

We now extend the analysis to data on changes in the monthly incidence of measles in England and Wales (figure 5*a*). These and subsequent data were extracted from the Registrar General's Weekly Returns, for the period 1948 (when measles notifications began) to 1967 (just before the onset of mass measles vaccination in 1968 significantly altered the dynamics of the infection; Anderson *et al.* (1984)). As with earlier analyses (Sugihara & May 1990) we begin by transforming the data to first differences, partly to remove such linear trends as may exist and partly to increase the density of points in phase space. Because the time series here are very short ($N = 240$, roughly half the size of the New York series), to maximize the information content in estimating E we allow the library and predictions to span the full time period. However, to avoid circularity between our forecasts and the model, we sequentially exclude points from the library that are in the neighbourhood of each predictee (specifically the $(E-1)\tau$ points preceding and following each forecast). Similar but much noisier results for estimating the embedding dimension were obtained by using the standard protocol of the first half predicting the second half. The standard protocol was used for the ρ–T_p curves, where the pattern, though noisy, appeared to be more robust. The qualitative appearance of these curves was found to be much the same for all choices of library.

As shown in figure 5*b*, we obtain optimal embeddings at $E = 7$–10, which is similar to the range of values found for measles in New York ($E = 5$–8). However, unlike New York, it appears that the dynamics here are not produced by low dimensional chaos. The ρ–T_p curve (figure 5*c*) does not decay exponentially as it does for the measles incidence data from New York City, but rather has a flat appearance more reminiscent of the additive noise case we saw for chickenpox (figure 4*b*). This result is corroborated by the comparison with the optimal linear model. For predicted changes in measles frequency one month in advance ($N = 120$), the best linear autoregressive model gives the result $\rho = 0.797$, which is not significantly different from the correlation obtained with the nonlinear predictor, where $\rho = 0.790$. Thus, unlike measles incidence in New York City, measles in England and Wales appears not to be chaotic.

Thus we are faced with an apparent contradiction: why should measles in New York City be chaotic while the same disease in the United Kingdom is a simple

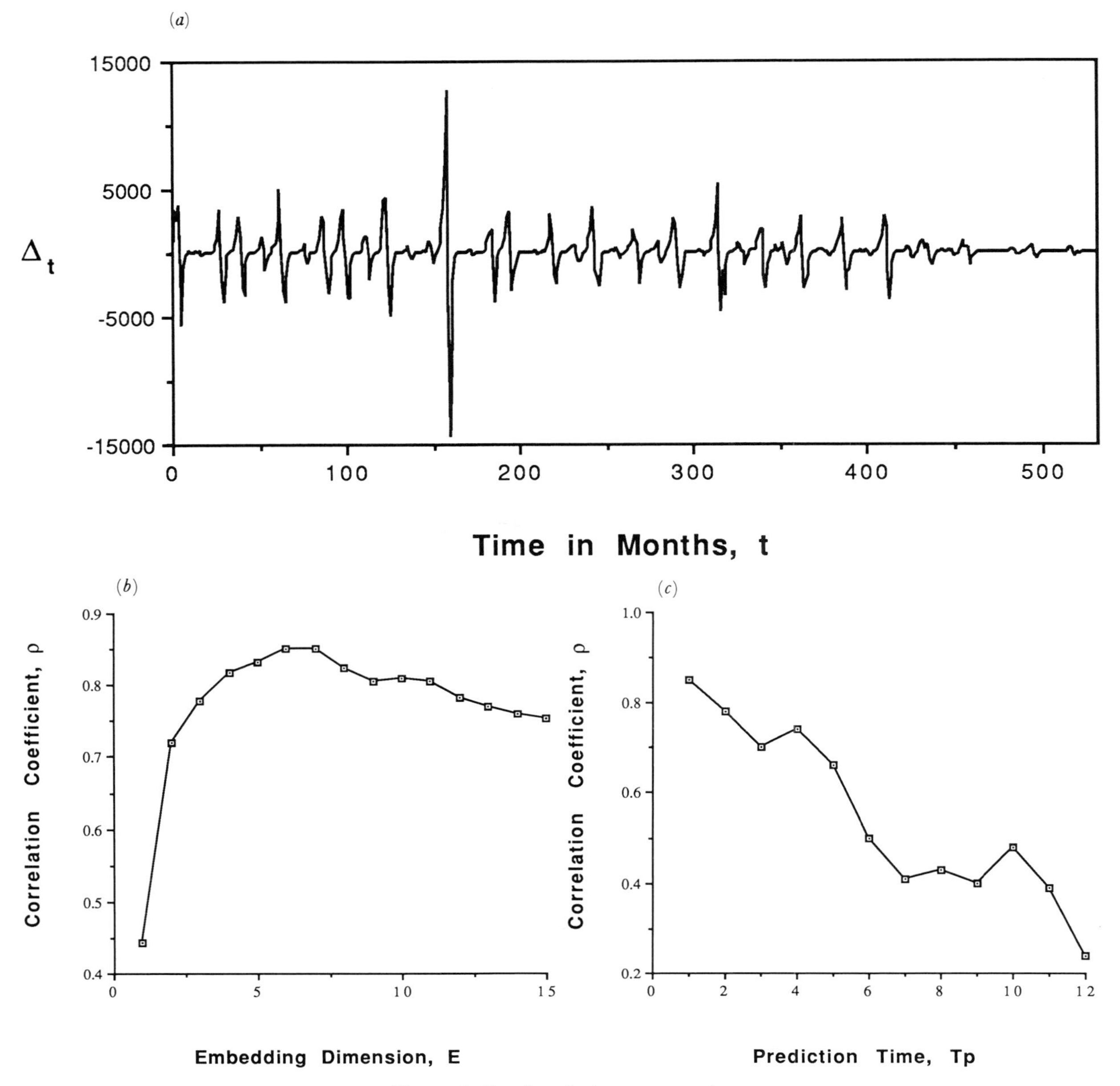

Figure 4. For description see opposite.

biennial cycle with additive noise? Could the contradiction be explained by differences in the population or spatial scales involved? For example, could individual cities in the U.K. be chaotic and nonlinear, but produce an emergent behaviour that appears linear when they are aggregated?

(*c*) *Spatial dynamics of measles*

The analysis of the spatial dynamics of measles has a distinguished pedigree in both biomathematics (Bartlett 1957) and spatial geography (Cliff & Haggett 1988). Before exploring the dynamics of measles in English cities, we clarify the dynamical implications of spatial heterogeneity with a simple model.

(i) *Scale dependence in models*

The sensitivity of ecological models to aggregation and scaling has been discussed in a number of different contexts (see, for example, Cohen 1979; Livdahl & Sugihara 1983; Sugihara *et al.* 1984; Ives & May 1985; Allen & Starr 1985; O'Neill *et al.* 1986; Sugihara *et al.* 1989). Here we test the theoretical possibility of emergent linearity from nonlinear parts with the following simple experiment.

We approximate measles dynamics within a single city as a chaotic logistic map ($X_{t=1} = aX_t(1-X_t)$), superimposed on a sine wave. We then investigate what happens as more of these 'sine+logistic' series are summed. In summing the series, we require the linear part (the sine waves) to be synchronized as the seasonal patterns in cities would be, but we allow the nonlinear parts to be independent to approximate spatial decoupling (or weak coupling by contagion). In effect, this is equivalent to averaging the output from independent logistic maps, and superimposing this net output on a sine wave. Because the dynamics in each city may not be perfectly identical, the logistic maps

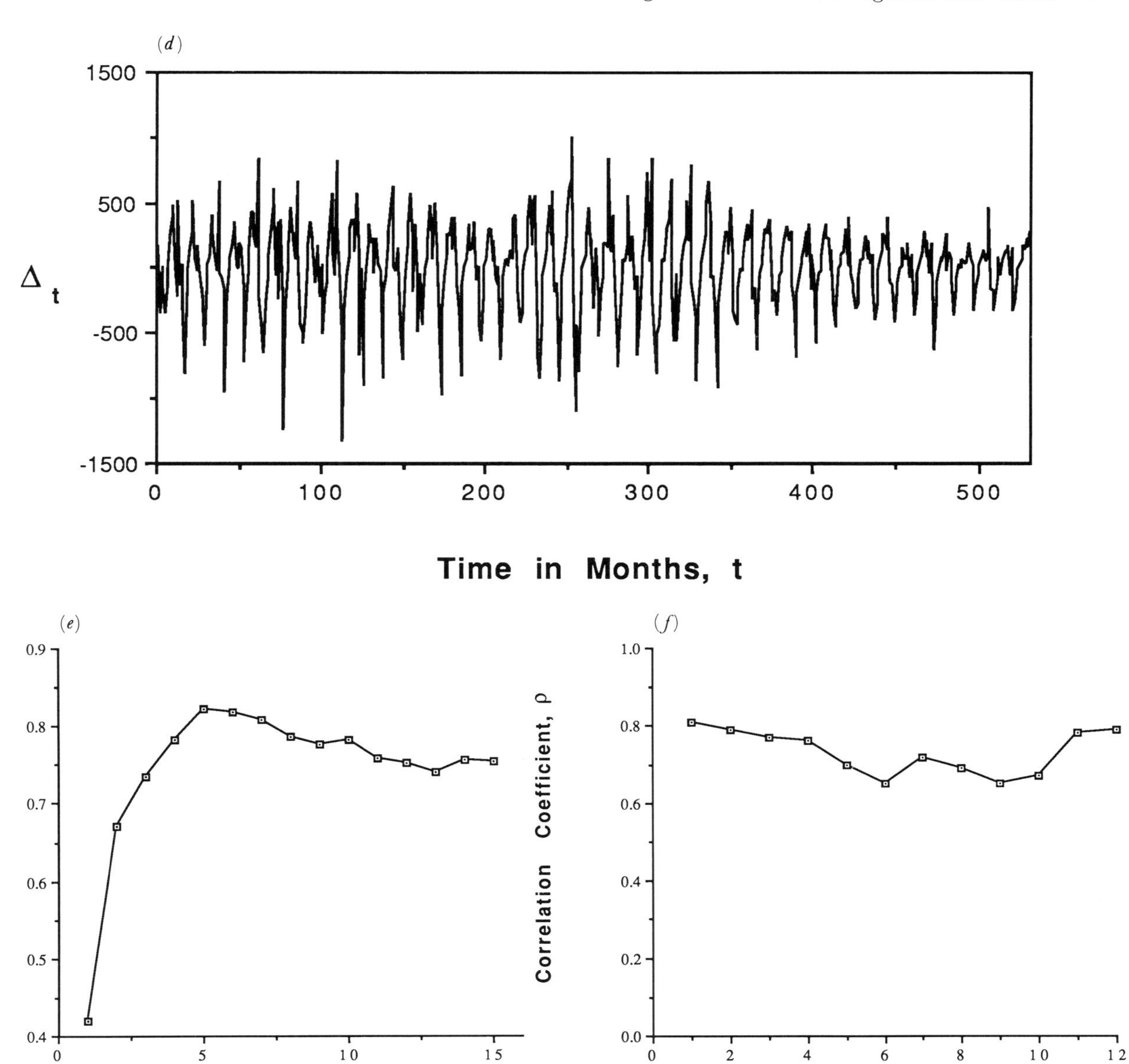

Figure 4. (*a*) Time series generated by taking first differences, $x_{t+1} - x_t$, of the monthly number of cases of measles reported in New York City between 1928 and 1972 (the first 532 points in the sequence shown here). After 1963, the introduction of immunization against measles had a qualitative effect on the dynamics of infection; this can be seen in the later part of the sequence shown here. (*b*) By using the methods described earlier, the first part of the measles time series (216 points, from 1928 to 1946) was used to construct a library, which was then used to predict forward from each point in the second part of the series (1946 to 1963). The correlation coefficient, ρ, between predicted and observed results is shown as a function of the embedding dimension, E, for predictions one time-step ahead, $T_p = 1$ (and $\tau = 1$). The figure suggests an optimal embedding dimension of $E \sim 5$–7. (*c*) Here ρ, between predicted and observed results for measles, is shown as a function of prediction interval T_p (for $E = 6$ and $\tau = 1$). The overall decline in prediction accuracy with increasing time into the future may be taken as indicative of chaotic dynamics, as distinct from uncorrelated noise. Figures 4*d*, *e*, *f* as for figures *a*, *b*, *c*, respectively, except now the data are for monthly case reports of chickenpox in New York City, from 1928 to 1972. Here, all 532 points are used in the analysis. Again figure 4*e* suggests an optimal embedding dimension $E \sim 5$–7. In marked contrast to figure *c*, *f* (calculated on the basis of $E = 5$ and $\tau = 1$) indicates pure additive noise, superimposed on a basic seasonal cycle. For a more detailed discussion of figure 4, see Sugihara & May (1990).

are given some variability by choosing the parameter a for each map uniformly in the interval (2.67, 3.67). However, similar results are obtained with independently initialised logistic maps having identical parameters.

Figure 6*a* shows how the ρ–T_p signature varies with increasing aggregation. As more independent logistic maps (cities) are folded into the picture the ρ–T_p signature becomes ever more shallow, giving much the appearance of the linear noise case. This is corroborated by figure 6*b* where the linear predictor tends to match the nonlinear predictor more closely as more

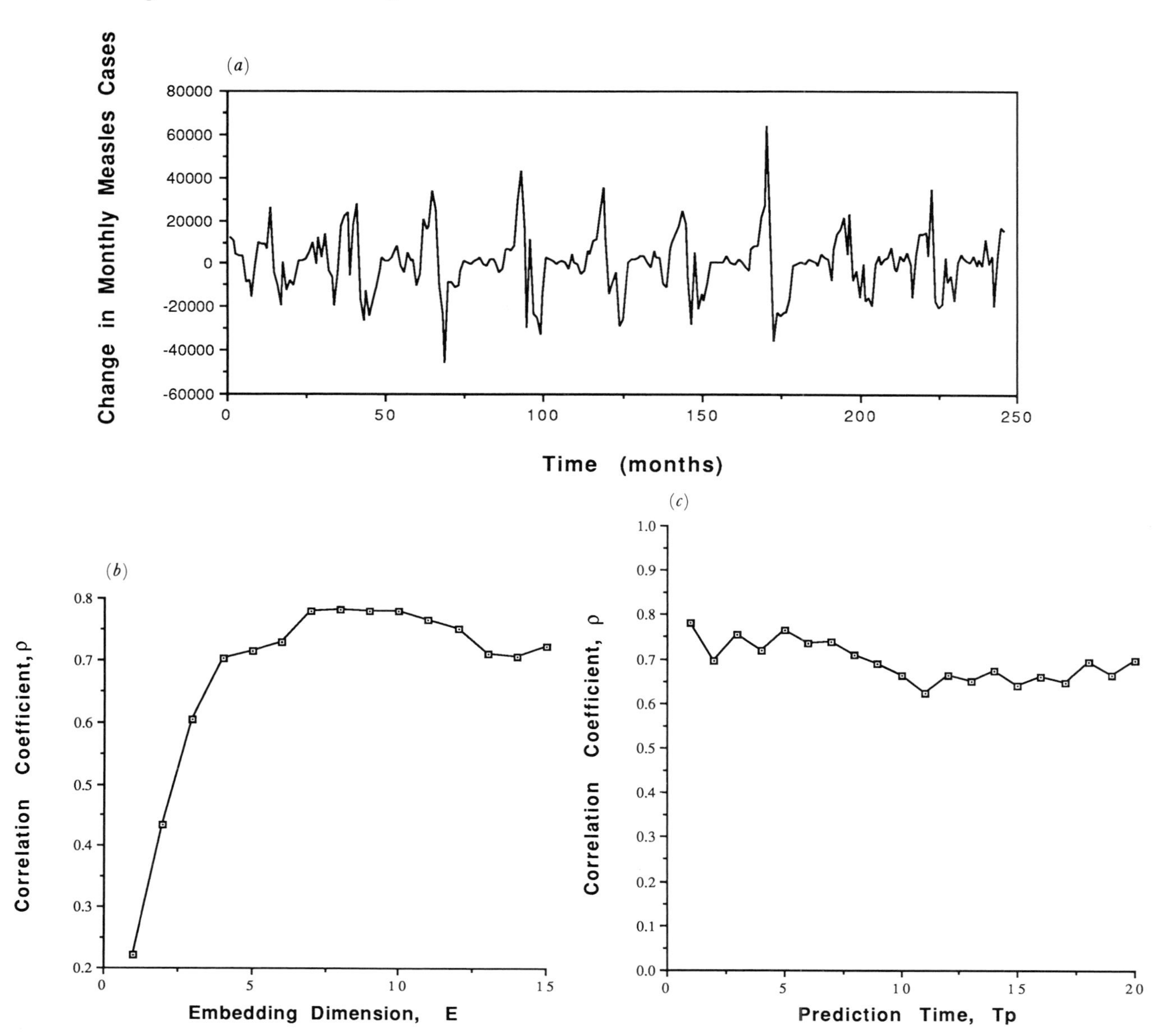

Figure 5 (*a*) Time-series of first differences in the number of cases of measles reported across England and Wales between 1948 and 1966. For data-source see Anderson *et al.* (1984). (*b*) Correlation coefficient or predictability, ρ, as a function of embedding dimension, E, for measles from England and Wales. Because of the low number of data points available ($N = 216$), we employed the whole series in forecasting in order to obtain the clearest estimate of optimal embedding (unbiased by the possibility of nonstationarity in the data). However, to ensure independence, the library used for each forecast was constructed to exclude points near the predictee in the time-series. A lag of $\tau = 2$ was used to embed these data. (*c*) Predictability, ρ, as a function of prediction time, T_p, for measles from England and Wales. Here the optimal parameters obtained above ($\tau = 2$ and $E = 8$) were used in forecasting; the library of patterns from the first half of the data was used to predict the second half of the data. The relatively flat pattern shown here is similar to the additive noise case seen for chickenpox in New York City. This figure shows that the large-scale aggregate behaviour of measles across England and Wales does not appear to be chaotic (in curious contrast to measles in New York City, figure 4*a*).

logistic series are summed. These trends are understandable in light of the following two facts. First, as more logistic maps are superimposed, the nonlinear signal becomes ever more complicated. High dimensional dynamics, chaotic or otherwise, are regarded as noise. Secondly, as more such series are superimposed, the amplitude of the nonlinear signal should decrease roughly as the square root of the number of independent chaotic logistic maps; this exposes more clearly the linear parts (seasonal sine wave) of the time series which are synchronized. Thus as more independent chaotic nonlinear series are aggregated, the nonlinear part should begin to resemble noise superimposed on a sine wave.

(ii) *Measles in English cities*

To test the applicability of these ideas to the observed patterns for measles in England and Wales, we have disaggregated the data, focusing on individual cities. The central question here is whether evidence for chaotic behaviour (which is not apparent in the countrywide analysis) emerges on a single-city scale.

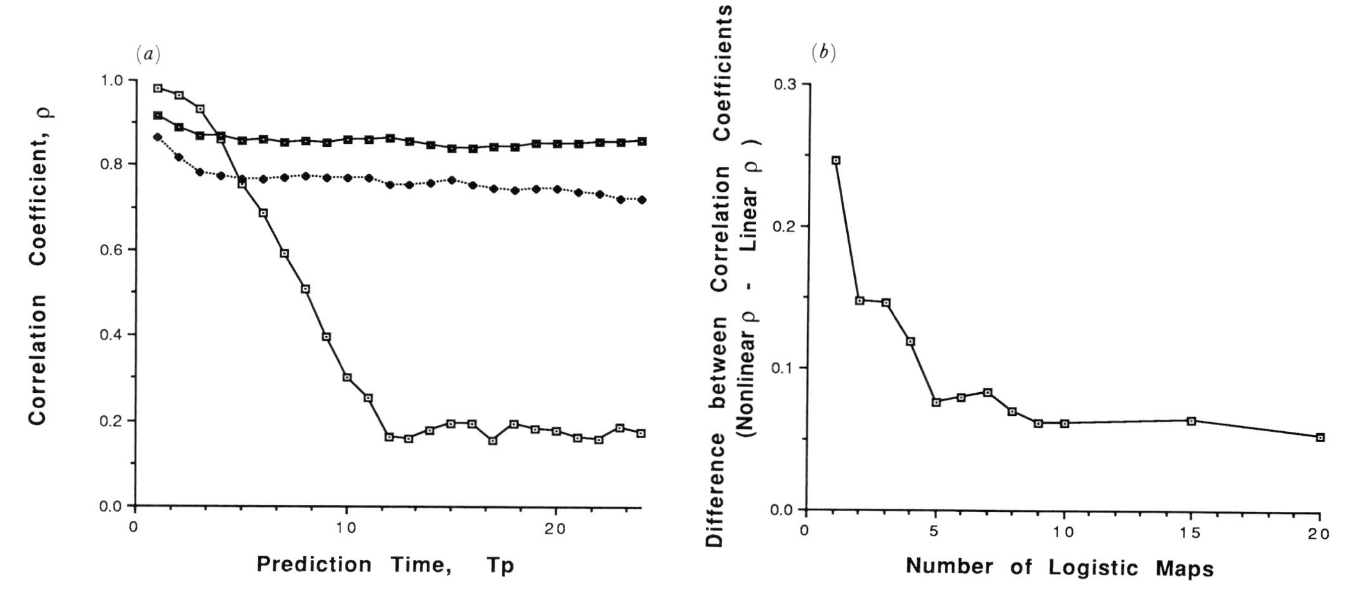

Figure 6. (*a*) Predictability, ρ, as a function of prediction time, T_p, at different levels of aggregation. Here we approximate measles dynamics in a single city as a chaotic logistic map superimposed on a sine curve, and investigate behaviour as more of these (sine+logistic) series are lumped (summed). The solid line with open boxes is for one (sine+logistic) ($\tau = 1$, $E = 3$), the dashed line with closed diamonds is for 10 (sine+logistic) series summed ($\tau = 1$, $E = 7$), and the solid line with solid boxes is for 20 series summed ($\tau = 1$, $E = 9$). The effect of such aggregation on the dynamics is to diminish the nonlinear chaotic portion of the signal, so that the ρ–T_p signature looks increasingly like the additive noise case. (*b*), the difference in predictability, ρ, between optimal linear autoregressive methods versus our nonlinear methods is shown, as a function of the number of (sine+logistic) maps that are lumped together. The maps are as described in figure 6*a*, and here $T_p = 1$ (and $E = 3$, $\tau = 1$). Note that the difference in ρ decreases with increasing aggregation.

Table 1. *Demographic summary for the seven English cities used in the spatial analysis of measles*

	distance (road miles)						
distance matrix	London	Birmingham	Liverpool	Manchester	Sheffield	Bristol	Newcastle
London	—	113	205	189	159	115	281
Birmingham	—	—	94	81	76	88	205
Liverpool	—	—	—	35	74	164	156
Manchester	—	—	—	—	38	164	132
Sheffield	—	—	—	—	—	164	128
Bristol	—	—	—	—	—	—	293
population[a] (thousands)	8282	1096	792	693	514	435	294

[a] Estimated population in 1960; from OPCS (1960).

We shall focus here on a representative sample of seven large English cities: table 1 lists the cities, along with their population sizes and a distance matrix, while figure 7*a* shows the associated measles time-series for the period 1948–67.

Figure 7*b* shows the embedding analyses for each of the seven cities (again, using the full data set ($N = 240$) to compute these correlations). All of the five most populous cities, London, Birmingham, Liverpool, Manchester and Sheffield, had optimal embeddings in a range similar to what was observed for New York ($E = 5$–8; Manchester, however, also had a peak at $E = 4$), and each had a local maximum at $E = 7$. Indeed these results seem to match the embedding results for New York better than those for the pooled data for Britain. On the other hand, the two least populous and most isolated cities appeared to require higher dimensional embeddings: for Bristol $E = 10$, and for Newcastle $E = 12$. Although one must be cautious not to overinterpret the specific figures obtained here, especially in light of the low number of data points involved ($N = 240$), it is interesting to note that both Bristol and Newcastle fall well below the population threshold believed necessary for the infection to remain endemic (Bartlett 1957). It is possible, therefore, that the higher dimensionality of the embedding here is because of the higher complexity coming from the required coupling to the outside world.

Figure 7*c* shows the ρ–T_p curves for the seven cities. The results here are not flat like the ones obtained with the aggregated data, but rather have a look very similar to the chaotic signature observed in New York

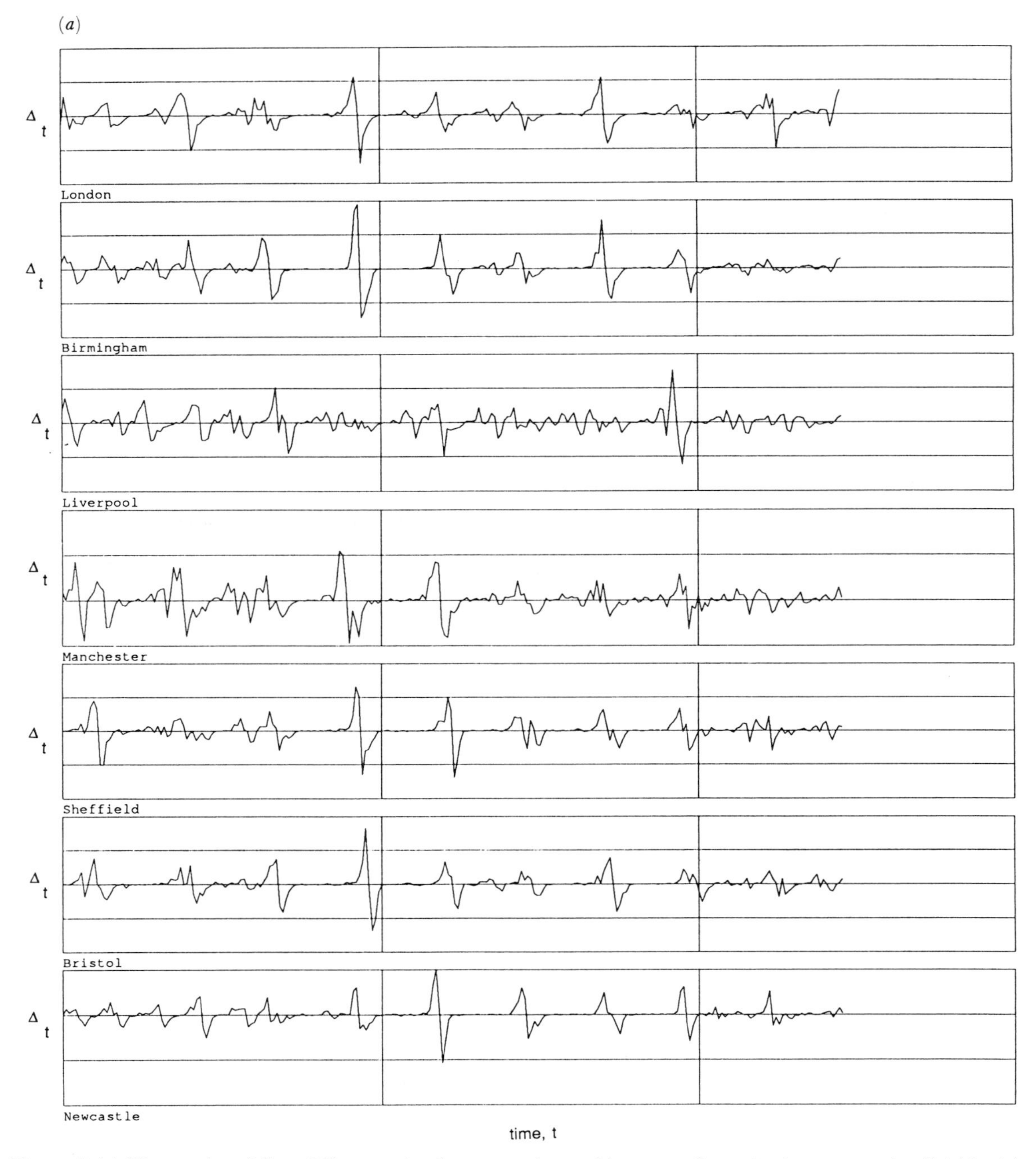

Figure 7. (*a*) Time-series of first differences in the reported monthly cases of measles in seven major British cities between 1948 and 1966 (arranged in order of city population size).

where predictability falls off steeply with increasing prediction time. Moreover, the results of the comparisons with the optimal linear predictor for each city shown in table 2 (ρ_{linear} against $\rho_{\text{nonlinear}}$ at $T_p = 1$, with the resulting p-level) firmly support the view that the dynamics are chaotic. Thus, it appears that scale considerations may help to resolve the apparent contradiction between the lumped analysis for measles in England and Wales, and the earlier analysis for New York City measles. In this regard, it is interesting that London, the most populous and geographically the largest of the British cities by almost an order of magnitude, appears to show the most gradual decline in its ρ–T_p curve.

These results show that the nonlinear dynamical features that are present in the individual cities of the U.K. are averaged out in the aggregate. Although a linear predictor worked well (at least as well as the nonlinear predictor) at forecasting changes in measles incidence country-wide, the greater success in predicting on a city-scale using a nonlinear predictor suggests that one might expect to produce better forecasts country-wide by combining the results of the component nonlinear predictors.

(*d*) ***Stationarity***

Finally, an issue that seldom appears in simple models but that is important when analysing data from the natural world, concerns the stationarity of the

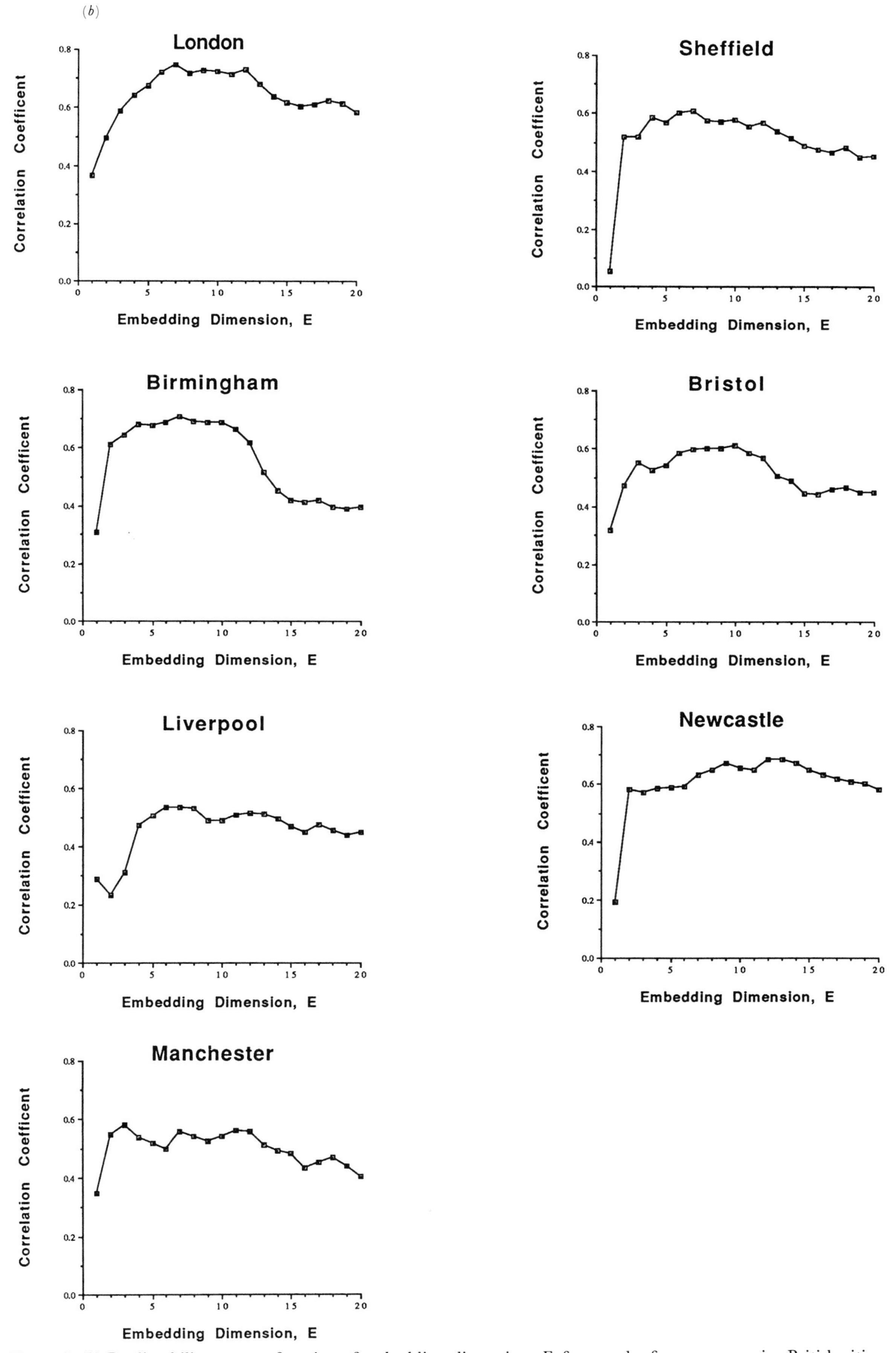

Figure 7. (*b*) Predictability, ρ, as a function of embedding dimension, E, for measles from seven major British cities. Clearest results were obtained by using $\tau = 1$ for Liverpool, Manchester, Sheffield and Newcastle, and $\tau = 2$ for London, Birmingham and Bristol.

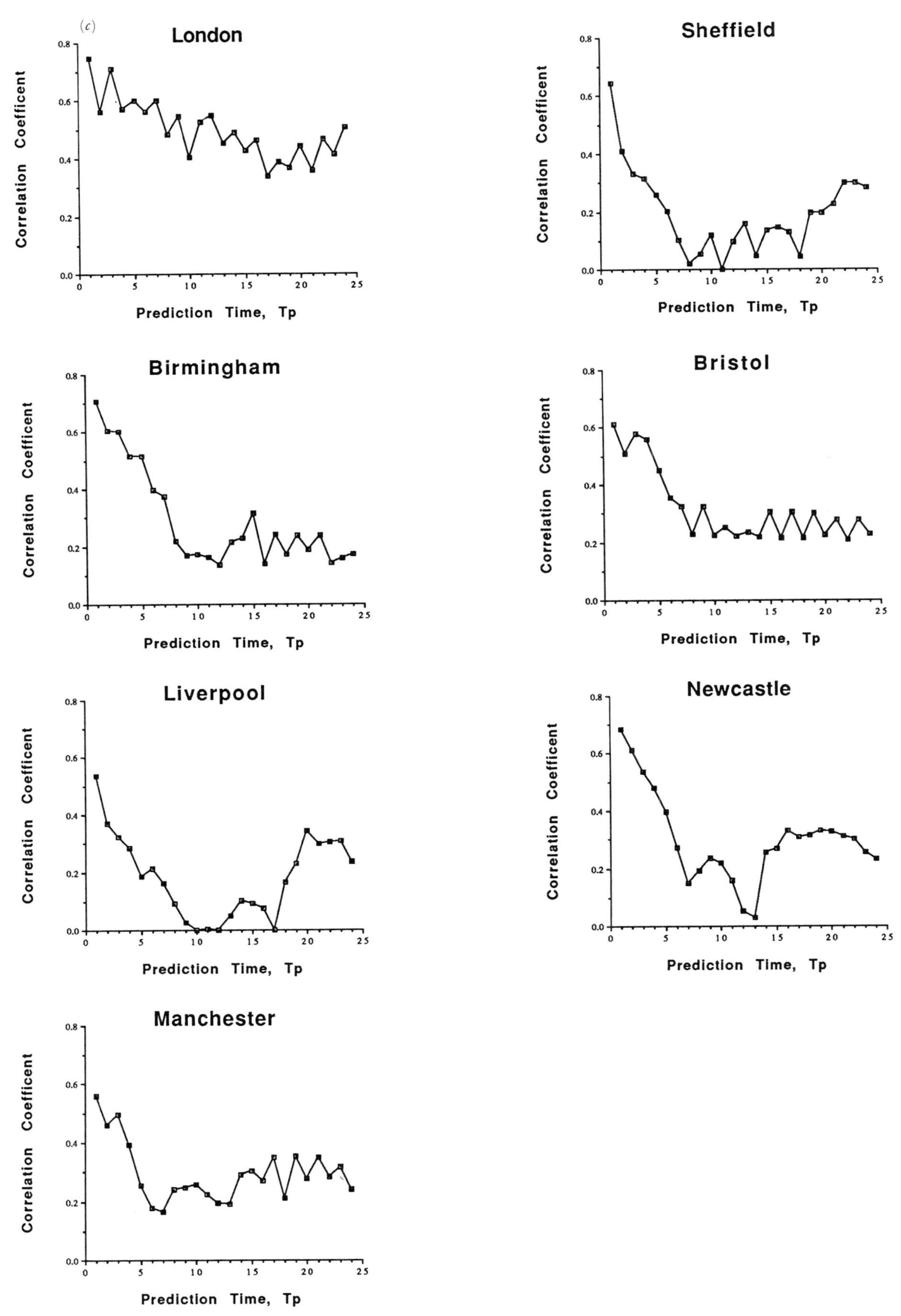

Figure 7.(*c*) Predictability, ρ, as a function of prediction interval, T_p, for measles from seven major British cities. The following parameters were used: London $\tau = 2$, $E = 7$; Birmingham $\tau = 2$, $E = 7$; Liverpool $\tau = 1$, $E = 7$; Manchester $\tau = 1$, $E = 7$, Sheffield $\tau = 1$, $E = 6$; Bristol $\tau = 2$, $E = 10$; Newcastle $\tau = 1$, $E = 12$. With the possible exception of London, all of the above cities show the characteristic decline in predictability with increasing prediction interval associated with chaotic dynamics, as seen in New York City (figure 4*a*).

Table 2. *Comparison between linear autoregressive methods and the nonlinear simplex predictor*

City	ρ_{linear}	$\rho_{nonlinear}$	Significance
London	0.63	0.76	$p < 0.001$
Birmingham	0.37	0.71	$p < 0.0005$
Liverpool	0.40	0.48	$p < 0.05$
Manchester	0.38	0.57	$p < 0.0005$
Sheffield	−0.02	0.64	$p < 0.0005$
Bristol	−0.01	0.67	$p \ll 0.0005$
Newcastle	−0.10	0.69	$p \ll 0.0005$

process generating the time-series. In all of the ρ–T_p analyses above, the first half of the series was used to construct a library of patterns that was then used to predict the second half of the time series. In the natural world, where parameters can undergo systematic changes over time, past patterns can be of dubious relevance to an altered present or an even more different future. An inspection of several of the time series in figure 7*a*, particularly those for Sheffield, Bristol and Newcastle, suggests that non-stationarity might indeed be a problem here.

One way to gauge whether secular trends might confound the forecasting results given above has been discussed by Sugihara & May (1990). Rather than using the first half of the time series to construct the library, and the second half to compute correlations between predictions and observations, we instead investigate what happens when the library and predicted halves are chosen in all combinations (table 3). That is, we use the first half to predict itself ($1 \rightarrow 1$) and then use it to predict the second half ($1 \rightarrow 2$), and the second half to predict itself ($2 \rightarrow 2$) and then use it to predict the first half ($2 \rightarrow 1$). We then compare the correlation coefficients obtained in each of these four cases. If the time-series shows a secular trend, we should find higher correlations when the library and predicted segments span the same time period, and lower ones when a library from one time span (e.g. first half) is used to forecast values from another time span (e.g. the second half of the series).

Table 3 shows the results of such an analysis for each of the English cities in figure 7. Although there is a certain amount of variation in the predictability of each of the reciprocal combinations, there is no systematic trend for higher correlations when the fitted half is used to predict on itself. This is most clearly evidenced in the summary statistics given at the bottom of table 3. Moreover, all reciprocal combinations gave similar ρ–T_p curves. Thus at least from a nonlinear perspective, these time series do not appear to contain secular trends.

On the other hand, when reciprocal pairings between fitted and predicted halves are made using the linear autoregressive approach, they can appear highly nonstationary (table 4). Notice that the linear autoregressive model keys on repeated patterns in one dimension (the time-series itself). Thus the time series for Sheffield, Bristol and Newcastle, for which there was almost no linear predictability between the first half and the second half, appear most clearly non-

Table 3. *Test of measles data for nonlinear stationarity* ($N = 120$, $T_p = 1$)

	library half → predicted half			
City	1 → 1	1 → 2	2 → 1	2 → 2
London	0.65	0.72	0.73	0.46
Birmingham	0.67	0.71	0.72	0.68
Liverpool	0.51	0.49	0.17	0.53
Manchester	0.48	0.54	0.48	0.27
Sheffield	0.67	0.44	0.47	0.44
Bristol	0.43	0.65	0.70	0.51
Newcastle	0.55	0.69	0.60	0.60

Table 4. *Test of measles data for linear stationarity* ($N = 120$, $T_p = 1$)

	fitted half → predicted half			
City	1 → 1	1 → 2	2 → 1	2 → 2
London	0.64	0.64	0.53	0.64
Birmingham	0.60	0.38	0.45	0.76
Liverpool	0.56	0.43	0.24	0.50
Manchester	0.64	0.38	0.42	0.52
Sheffield	0.60	−0.02	0.28	0.70
Bristol	0.73	−0.07	0.43	0.80
Newcastle	0.60	−0.10	0.35	0.57

stationary to the naked eye. None the less, when these time series are embedded in higher dimensions, the obvious secular trends disappear. The moral that emerges from this is that if a process is truly nonlinear, one needs to be careful in proclaiming nonstationarity based on linear criteria. A stationary process in higher dimensions may only appear to contain secular changes when viewed in one dimension.

5. DISCUSSION

Our preliminary analysis of the spatial dynamics of measles suggests two fruitful avenues for future work. First, the measles data for England and Wales are available on a much finer spatial scale than the crude city-by-city division examined here. In particular, a further subdivision of the London measles data would provide a much more refined test for the 'emergence' of chaos at smaller spatial scales. Secondly, we require more detailed mechanistic models, which allow explicitly for the impact of spatial heterogeneities in transmission on the dynamics of the host–parasite interaction (May 1986; May & Anderson 1984). As shown by the simple ('sine+logistic') spatial model considered above, the analysis of time series simulated from such models can provide important insights into the dynamics of the real system.

Two main points emerge from our paper, the first having to do with dynamical details and the second with general principles. First, growing understanding of deterministically chaotic systems suggests that apparently random time series may in fact be generated by deterministic mechanisms, and that techniques may be available to distinguish such low-dimensional chaos from externally imposed environmental noise or sampling error. Many of these techniques, however, require longer time series than are typically available in ecological or epidemiological contexts. We have outlined methods, based on the ability to make short-term (but not long-term) forecasts from deterministically chaotic data, but not from 'really noisy' data, that appear to work with relatively short runs of data. Application of these ideas to epidemiological time series for the incidence of measles and chickenpox suggests that fluctuations arise from deterministic chaos for measles and from sampling error or other sources of noise for chickenpox. This work sheds a wholly different light on earlier controversies as to whether populations are governed by environmental fluctuations or deterministic regulatory factors: with sufficient nonlinearity, deterministic factors also can give erratic fluctuations, but the different kinds of apparent randomness that arise from deterministic chaos versus external noise may be distinguished.

Secondly, the qualitative difference between the patterns seen for the incidence of measles aggregated over England and Wales, versus those seen in individual cities, provide a striking illustration of how the scale on which we collect and analyse data can affect our interpretation. Sufficiently aggregated, the England and Wales data for measles suggest a dynamical pattern of approximately two-year cycles with additive noise. Disaggregated city-by-city, a more detailed pattern of chaotic dynamics (and short-term prediction of the apparently erratic fluctuations) emerges. This clearly is a metaphor, albeit a very explicit metaphor, for a much larger range of issues in ecology.

REFERENCES

Abraham, R. H. & Shaw, C. D. 1982 *Dynamics: the geometry of behaviour* (*Vis. Math. Ser.* **1–4**). Santa Cruz: Ariel Press.

Allen, T. F. H. & Starr, T. B. 1985 *Hierarchy*. Chicago: Chicago University Press.

Anderson, R. M., Grenfell, B. T. & May, R. M. 1984 Oscillatory fluctuations in the incidence of infectious disease and the impact of vaccination: time series analysis. *J. Hyg. Camb.* **93**, 587–608.

Anderson, R. M. & May, R. M. 1985 Age-related changes in the rate of disease transmission: implications for the design of vaccination programmes. *J. Hyg., Camb.* **94**, 365–436.

Bartlett, M. S. 1957 Measles periodicity and community size. *J. R. Stat. Soc. Ser.* A, **120**, 48–70.

Cliff, A. D. & Haggett, P. 1988 *Atlas of disease distributions: analytic approaches to epidemiological data*. Oxford: Blackwell.

Cohen, J. E. 1979 Long-run growth rates of discrete multiplicative processes in markovian environments. *J. math. Analysis Applic.* **69**, 243–251.

Farmer, J. D. & Sidorowich, J. J. 1989 Exploiting chaos to predict the future and reduce noise. In *Evolution, learning and cognition* (ed. Y. C. Lee), pp. 277–304. New York: World Scientific Press.

Grassberger, P. & Procaccia, I. 1983 Measuring the strangeness of strange attractors. *Physica* **9**D, 189–208.

Hethcote, H. W. & Levin, S. A. 1989 Periodicity in epidemiological models. In *Applied mathematical ecology* (ed. S. A. Levin, T. G. Hallam & L. J. Gross), pp. 193–211. New York: Springer–Verlag.

Ives, A. R. & May, R. M. 1985 Competition within and between species in a patchy environment: relations between microscopic and macroscopic models. *J. theor. Biol.* **115**, 65–92.

Livdahl, T. & Sugihara, G. 1984 Nonlinear interactions of populations and the importance of estimating per capita rates of change. *J. Anim. Ecol.* **53**, 573–580.

Lorenz, E. N. 1969 Atmospheric predictability as revealed by naturally occurring analogues. *J. atmos. Sci.* **26**, 636–646.

May, R. M. 1986 Population biology of microparasitic infections. In *Mathematical ecology: an introduction* (ed. T. G. Hallam & S. A. Levin), pp. 405–442. New York: Springer–Verlag.

May, R. M. & Anderson, R. M. 1984 Spatial heterogeneity and the design of immunization programs. *Math. Biosci.* **72**, 83–111.

Murdoch, W. W. & McCaughley, E. 1985 Three distinct types of dynamic behaviour shown by a single planktonic system. *Nature, Lond.* **316**, 628–630.

Nokes, D. J. & Anderson, R. M. 1986 Rubella epidemiology in South East England. *J. Hyg., Camb.* **96**, 291–304.

O'Neill, R. V., DeAngelis, D. L., Waide, J. B. & Allen, T. F. H. 1986 *A hierarchical concept of ecosystems*. Princeton: Princeton University Press.

OPCS. 1960 *Registrar General's Annual review for England and Wales*. London: Office of Population Censuses and Surveys.

Priestley, M. B. 1980 State dependent models: a general approach to nonlinear time series analysis. *J. Time Ser. Anal.* **1**, 47–71.

Schaffer, W. M. & Kot, M. 1985 Nearly one dimensional dynamics in an epidemic. *J. theor Biol.* **112**, 403–427.

Schaffer, W. M. & Kot, M. 1986 Differential systems in ecology and epidemiology. In *Chaos* (ed. A. V. Holden), pp. 158–178. Princeton: Princeton University Press.

Schaffer, W. M., Olsen, L. F., Truty, G. L., Fulmer, S. L. & Graser, D. J. 1988 Periodic and chaotic dynamics in childhood infections. In *From chemical to biological organisation* (ed. M. Markus, S. C. Muller & G. Nicolis), pp. 331–347. New York: Springer-Verlag.

Schwartz, I. B. 1985 Multiple recurrent outbreaks and predictability in seasonally forced nonlinear epidemic models. *J. Math. Biol.* **21**, 347–361.

Sinclair, A. R. E. 1989 The regulation of animal populations. In *Ecological concepts* (ed. J. M. Cherrett) pp. 197–241. Oxford: Blackwell Scientific.

Sugihara, G. (rapporteur) *et al.* 1984 Ecosystems dynamics. In *Exploitation of marine communities* (ed. R. M. May), pp. 131–153. New York: Springer–Verlag.

Sugihara, G. & May, R. M. 1990 Nonlinear forecasting as a way of distinguishing chaos from measurement error in time series. *Nature, Lond.* **344**, 734–741.

Sugihara, G., Schoenly, K. & Trombla, A. 1989 Scale invariance in food web properties. *Science, Wash.* **245**, 48–52.

Takens, F. 1981 Detecting strange attractors in turbulence. *Lect. Notes Math.* **898**, 366–381.

Tong, H. & Lim, K. S. 1980 Threshold autoregression, limit cycles and cyclical data. *J. R. Stat. Soc.* B**42**, 245–292.

Discussion

P. Chesson (*Ohio State University, Columbus, Ohio, U.S.A.*). I am sure Professor Sugihara's technique will be a useful one for understanding chaotic systems, but I have concerns about the range of alternative causes of fluctuations in ecological time series that he has considered. He has spoken as if the only alternative to chaotic dynamics is measurement error. However, in nature stochastic factors affect population trajectories, not just their measurement. Thus one should expect populations to fluctuate (to have stochastic dynamics) as a consequence of these stochastic factors, quite apart from any apparent fluctuations that are introduced as an artefact of measurement error. It is quite likely that such stochastic factors will not be additive and will cause effects that are vastly different from the additive measurement error that Professor Sugihara considers.

His technique of distinguishing chaos from additive error will work when these two are the only alternatives, but will be incapable of distinguishing between stochastic dynamics and chaos. I accept Professor Sugihara's point that from some perspectives, high dimensional chaos and stochasticity are the same. It then appears that he intends his technique to distinguish between measurement error and dynamical uncertainty, whatever the cause of the latter, be it low-dimensional chaos or stochasticity.

H. M. PLATT (*The Natural History Museum, London*). As I understand it, Professor Sugihara became interested in the England and Wales measles data because, unlike those of New York, they seemed to be additive noise. However, when he backtracked to individual sets for cities he found the chaos patterns again. How does Professor Sugihara know that those data which he suggests display real additive noise patterns, such as the chicken pox set, are themselves not in fact assemblages of chaos patterns, which he may or may not be able to get at.

M. WILLIAMSON (*University of York, York, U.K.*). Is it not possible to get almost any shape of prediction curve by choosing various models for both dynamical chaos and for systems with measurement error?

G. SUGIHARA. As discussed more fully in the *Nature* paper (Sugihara & May 1990), it seems likely that a specific pattern of autocorrelated noise could be hand-tailored, to mimic any specified relation between ρ (correlation coefficient) and T_p prediction interval), such as that found for the chaotic test map. The converse is surely not true! Chaotic dynamical systems of low dimension will always show a systematic decline in with increasing T_p (at a characteristic prediction interval set by the Lyapunov exponent). We conjecture that, in general, such artificially designed patterns of autocorrelation will typically give flatter ρ–E (embedding dimension) relations than are found for simple time series generated by low-dimensional attractions (see Farmer & Sidorowich 1989; Sugihara & May 1990).

References

Farmer, J. D. & Sidorowich, J. J. 1989 Exploiting chaos to predict the future and reduce noise. In *Evolution, learning and cognition* (ed. Y. C. Lee), pp. 277–304. New York: World Scientific Press.

Sugihara, G. & May, R. M. 1990 Nonlinear forecasting as a way of distinguishing chaos from measurement error in time series. *Nature, Lond.* **344**, 734–741.

The interplay of population dynamics and the evolutionary process

JOSEPH TRAVIS

Department of Biological Science, Florida State University, Tallahassee, Florida 32306-2043, U.S.A.

SUMMARY

Long-term maintenance of genetic diversity is affected by ecological forces that are driven in turn by current levels of genetic variation. The strength of population regulation and the consequent patterns of population fluctuations determine the likelihood of genetic changes considered pivotal for rapid speciation. However, genetic diversity in the susceptibility to regulatory forces can reduce the magnitude of such fluctuations and minimize the likelihood of genetic revolutions. A group of populations that experiences local extinctions and recolonizations may hold lower levels of genetic diversity than in the absence of such extinctions, but local adaption, which provides enhanced genetic diversity, can reduce the likelihood of local extinctions. Tightly regulated populations experience different selection pressures than poorly regulated populations, although tighter regulation itself can evolve. When genotypic variation affects the outcome of interspecific interactions on a local scale, this effect, coupled with appropriate spatial variation, can enhance the resilience of the interactive system.

INTRODUCTION

In this paper several areas of interplay among population dynamics, ecological diversity, and the evolutionary process are explored. The primary goal is to review recent work that illuminates the many facets of that interplay. The secondary goals are to offer a synthesis of some diverse results and to focus attention on problems for which we lack sufficient information to draw general conclusions. I have tried to ensure that the literature cited can lead the general reader to a wide range of examples but that it can also lead the afficionado of specific topics to more detailed treatments.

ECOLOGICAL PROCESSES AND GENETIC DIVERSITY

(*a*) *Population fluctuations and genetic diversity*

When a population is reduced to a few individuals there are several population genetic consequences even in the absence of natural selection. The random sampling effects on gene frequencies that cancel one another in large populations can alter gene frequencies from the ancestral state. Similar sampling effects generate non-random combinations of the segregating alleles from many loci. Small numbers of individuals restrict the possibilities for accruing new genetic variation through mutation and promote inbreeding through the high levels of relatedness among subsequent progeny. The 'genetic size' of a population depends upon which of the above consequences are of interest because the same census number of individuals can represent different degrees of genetic 'smallness' for different consequences (Maruyama & Kimura 1980; Ewens 1982; Crow & Denniston 1988).

The measurement of 'genetic size', the effective population size, is the numerical size of an idealized population (random mating, no selection, equal contributions of all individuals to the next generation, etc.) that is suffering the genetic consequences of interest at the same rate or level as the real population under study. It is a surrogate number that encompasses the census number and other factors such as gender ratios, equality of contribution of progeny, breeding system, or mode of transmission of genes (Crow & Denniston 1988). Because these 'other factors' may change with density in natural populations the numerical relation between the census size and the effective size can be a complex one (see Travis & Mueller 1989).

Real populations fluctuate in their census numbers, so their effective sizes and consequent genetic properties fluctuate also. Periodic fluctuations in numbers generate matching periodic fluctuations in expected levels of genetic variation for many types of fluctuations (cyclical, stochastic, density dependent; see also the references in Motro & Thompson (1982)).

In many cases the genetic effects of one or more constrictions in numerical size linger for many generations after large numbers have been re-established (Nei *et al.* 1975; Chakraborty & Nei 1977; Motro & Thompson 1982; Maruyama & Fuerst 1985). Repeated constrictions in size reduce genetic variation dramatically when the cycle length of such repetitions is short and population growth rates are slow (Motro & Thompson 1982; see also Watterson (1989)).

The preceding theoretical results have inspired several hypotheses about specific empirical situations. Past bottlenecks have been postulated *a posteriori* to account for the cheetah's extraordinarily small amount of genetic variation for its current numerical size (see, for example, O'Brien *et al.* (1985); also Wayne *et al.*

(1986)). A more intriguing, *a priori* general prediction is that species experiencing regular dramatic fluctuations in population size with short cycle lengths should harbour less genetic variation than comparable species that do not have such fluctuations (Smith *et al.* 1975). This prediction fails for both cycling rodents (Gaines 1985) and birds (Gyllensten 1985) and the apparent reason for its failure shows the complexity of the interactions between ecological and genetic processes.

In cycling populations allele frequencies are altered during the low density phases because of genetic drift (Gaines & Whittam 1980; Bowen 1982; Gyllensten 1985). At high density phases there is increased gene flow among local subpopulations that reduces the levels of genetic differentiation among local populations and thereby increases dramatically the effective size of the entire ensemble (Bowen 1982; Plante *et al.* 1989). High emigration rates are regularly associated with the high density phase of cycling rodents (Krebs 1966; Gaines *et al.* 1979) and cycling grouse appear to migrate long distances (Gyllensten 1985). Thus changing movement patterns of individuals associated with local density fluctuations preclude lowering of the effective population size and concomitant genetic effects. This interpretation is supported by the observation that isolated vole populations on islands show a reduction in genetic variation (Kilpatrick & Crowell 1985).

The role of genetic interchange among subpopulations in overcoming the effects of local bottlenecks has been shown experimentally in plants (Polans & Allard 1989). A single generation of random mating restored the mean performance that had been lost due to inbreeding effects in a series of small isolates. If such interplay among local densities, migration rates and genetic structure were widespread then effects of periodic crashes in local populations on the genetic variation in the entire ensemble may be minimized. Present empirical data are insufficient for general conclusions. For example, insect emigration rates increase with increases in density (see, for example, Lawrence (1988) and references therein), but the extent to which such movement is translated into actual gene flow remains unclear.

There are circumstances in which fluctuations in population size can enhance the amount of additive genetic variation in a specific trait or set of traits. The additive variation is that portion of the genetic variation in a trait that is responsive to selection. When the population that enters a bottleneck has a high proportion of its total genetic variation for a trait contributed by epistatic effects (interactions among alleles at different loci) then the population can emerge from the bottleneck with more additive genetic variation than it had upon entering (Goodnight 1988). This occurs through the conversion of epistatic variance into additive variance. The most conducive situation for such conversion is an extended numerical constriction at a moderate effective size (Goodnight 1988). Traits likely to have the genetic prerequisites for this effect are those with either a long history of directional selection (reviewed in Travis & Mueller (1989)) or those strongly associated with mate recognition (Bryant & Meffert 1988). However, the absolute magnitude of epistatic variances in natural populations and the consequent magnitude of any potential enhancement of additive genetic variance is an unresolved empirical issue. The level of epistasis among individual loci appears small (Barker 1979; Hastings 1985) but interactions among groups of loci (blocks of genes on separate chromosomes, structural genes and their polygenic regulators, sex chromosomes and autosomes) often generate considerable epistatic variance (Kidwell 1969; Sved & Ayala 1970; Jones & Yamazaki 1974; Malpica & Vasallo 1980; Bao & Kallman 1982; Miyashita & Laurie-Ahlberg 1984; Kallman 1984).

Enhancement of genetic variation after a constriction in size plays a key role in two related areas. First, enhanced additive genetic variation allows further response to selection and enables a population to increase its overall level of adaptation to a specific selection pressure (Lande 1976, 1979). Second, the rearrangement of genetic variation after a constriction may allow a population to respond to selection dramatically and rapidly and thereby enhance the likelihood for rapid genetic differentiation among conspecific populations. This is the essential model for speciation via founder effects (Carson & Templeton 1984). Rapid differentiation via numerical constriction without such a conversion of genetic variation appears to be unlikely (Rouhani & Barton 1987; Charlesworth & Rouhani 1988) so this effect plays a central role in theories about the most profound stage in the evolutionary process.

(*b*) *The extinction and recolonization of subpopulations*

The numerical dynamics produced by the extinction and recolonization of independently fluctuating subpopulations is a subject of considerable debate with respect to its role in producing population stability (Kareiva, this symposium). There are two areas in which this ecological process has been a focus of population genetic debate. First, does such a process enhance the overall amount of genetic variation present in a species? Second, does this process affect the distribution of the extant genetic variation in a way that enhances the amount of differentiation among populations? The answers to these distinct questions bear on two larger contexts, the preservation of genetic variation in captive or managed populations and the creation of conditions conducive to a 'shifting balance' process of evolutionary change (Wade & McCauley 1988).

The first question has been addressed analytically by Ewens (1989) and via simulation by Maruyama & Kimura (1980), but with only limited success. If new colonists are drawn from only one randomly chosen subpopulation, if the extinction rate is not high, and if there is negligible gene flow among extant subpopulations, then a system with such an extinction–recolonization process will lose genetic variation more slowly than a comparable system without it

(Ewens 1989). Other models of the colonization process give different results (Maruyama & Kimura 1980; Ewens 1989) so no general answer is apparent.

The second question also has contingent answers (Slatkin 1977; Wade & McCauley 1988). When new populations are founded by colonists from many extant populations and if there is little gene flow among extant populations then this process impedes the stochastic differentiation of subpopulations. However, when colonists originate from a single subpopulation and the number of colonists is below the typical migration rate among extant subpopulations then this process enhances local differentiation.

The empirical data on this topic are few but the theoretical treatments offer unusually clear guidelines towards which data are most critical for producing different outcomes. The issues are reviewed thoughtfully by McCauley (1989) and include the origins of colonists, the rate of 'normal' interchange among extant populations, the typical numbers of colonists that found a new population, the extinction rate, and the assumption that extinction probabilities are independent of the size or age of the subpopulation. The increasing attention that ecologists are giving to such systems in the context of population regulation offers an excellent opportunity to acquire the demographic data that would enable us to understand whether extinction and recolonization contribute to the genetic structure of natural populations.

ECOLOGICAL DIVERSITY AND GENETIC PROCESSES

(*a*) *Overview*

In the previous sections, population dynamics were taken as 'given' and the genetic consequences of various patterns were reviewed. Diversity in dynamic patterns generates different levels and arrangements of genetic variation. However, natural populations do not fluctuate in a vacuum. Changes in numbers are produced by a variety of influences on mortality and natality rates and the phenotypic variants in a population may be affected differently by different limiting or regulating factors. In addition there are, invariably, fundamental changes in ecological relations that are associated with changes in population density. These two effects have consequences for both the action of natural selection and the patterns of population dynamics that result from evolutionary changes in vital rates. In the next two sections each type of effect is highlighted.

(*b*) *Multiple agents of population limitation and genetic heterogeneity in susceptibility to those agents*

Whether a particular population is regulated tightly or whether its growth is limited in more stochastic fashion there are usually several abiotic and biotic agents that perform the regulatory or limiting task function. Long-term studies of a variety of natural populations have revealed that only a subset of these several forces may act in any given year or be important for any given generation (Grant 1986; Marquis 1990). In this section the consequences of this observation for the intertwined problems of the net intensity of selection and the stability of population dynamics are examined. While the importance of questions about the stability of population dynamics will be self-evident to many readers, that of net selection intensities may not. Large intensities of selection are not compatible with long-term predictions from quantitative genetic models of evolution so the actual distribution of selection intensities in natural populations is an important, but unresolved, empirical issue (Endler 1986; Travis & Mueller 1989).

Studies of natural selection offer convincing evidence that there is invariably some level of genetic variation within a population in the vulnerability to any single agent of mortality or the susceptibility to any single factor that limits fertility (Manly 1985; Endler 1986; Travis 1989). In many cases variation is wide enough to produce large selection intensities. A variety of theoretical approaches can be used to show that such internal heterogeneity in vital rates (what the population geneticist would call variation in absolute fitness) can enhance the stability of the ensuing population dynamics (Wallace 1977; Lomnicki 1988; see Chesson (1989) for an analogous argument in community ecology). This effect has been explored extensively in host–parasite systems (May & Anderson 1983; Cohen & Newman 1989), inspired in part by the ubiquity of genetically based variation in host vulnerability (see, for example, Madhavi & Anderson 1985; Parker 1988; Alexander 1989). The effect is not as well appreciated in more general contexts.

Multiple agents of mortality, or more generally, multiple agents of population limitation, introduce complications. In some years or generations only one agent may be critical while in others a different agent or all of the agents may be critical. The qualitative consequences of this ecological diversity will vary with the correlation structure of vulnerability to different agents, but will also depend on the other sources of internal heterogeneity. Each situation is likely to be idiosyncratic, and I offer a detailed example to show how complexity can arise.

Larval anurans in temporary ponds face a number of mortality risks (complete descriptions of this ecological system can be found in Morin (1983); Wilbur (1984); Travis *et al.* (1987)). The major agents of mortality include predatory salamanders, predatory insects and drying of the pond. Each type of predator is limited by prey size such that rapidly growing tadpoles escape predation risk sooner than more slowly growing ones. The extensive variation in growth rate, in part genetically based, provides internal heterogeneity in vulnerability that affects population dynamics and generates directional selection for rapid growth. Hence vulnerability to each type of predator, salamander and insect, is positively correlated among genetic variants.

Patterns of microhabitat use as prey refugia complicate the situation. Salamanders search the water column for moving tadpoles and the prey have a refuge from them in vegetated or littered microhabitats (Morin 1986). The major insect predators, odonate naiads, occupy those very habitats and force tadpoles into the water column. The refuge from one predator is the locus of primary susceptibility to the other. Travis *et al.* (1985) manipulated the presence and absence of both predators in field enclosures with habitat structure and showed that when only one predator was present there was little differential mortality with respect to growth rate, presumably because of the availability of refuge. When both predators were present overall mortality rates increased, but not beyond the product of each predator's individual effect. However, the amount of differential mortality increased dramatically: relative fitness of slowly growing tadpoles when both predators were present was significantly below the expectation based on each predator's individual effect as an agent of selective mortality.

With respect to selection the effect of the positive correlation in vulnerability acts to increase the intensity of directional selection when both predators are present. When only one predator is present spatial refugia reduce the intensity of selection below what one would expect based on innate vulnerability. The intensity of selection will vary markedly from generation to generation although its direction will not. This result suggests that isolated observations of very strong selection intensities in nature may represent extreme values of a distribution of selection intensities caused by ecological variability. With respect to population dynamics the positive correlations caused by innate heterogeneity ought to be destabilizing but the negative correlations induced by spatial heterogeneity would be stabilizing. There is considerable variation in the importance of each type of predator from year to year, and thus the net effect cannot be determined from merely the signs of the correlations.

There are other patterns of covariance in vulnerability that have very different consequences. A negative covariance was described by Weis & Abrahamson (1986) in their studies of a gall-making insect. Larger galls are more vulnerable to bird predation but less so to parasitoid attack; smaller galls leave the enclosed insect more vulnerable to parasitoids but less susceptible to bird predation. In this situation the net selection would be optimizing if both agents of mortality acted simultaneously in every generation (and concomitant enhancement of stability in dynamics relative to no heterogeneity). If the importance of each enemy varies among years then the net result is a regime of fluctuating directional selection that will have the same long-term effect as optimizing selection (see Travis 1989).

In many cases genetic variation in the vulnerability to one ecological factor is uncorrelated with vulnerability to another. These cases include variation among plant genotypes in resistance to different species of herbivore (Simms & Rausher 1989; Marquis 1990), variation among herbivores in their success on different species of plant hosts (Rausher 1984; Via 1984), and variation in performance at different densities and competitive situations in plants (Khan *et al.* 1976; Shaw 1986; Miller & Schemske 1990) and insects (Bradshaw & Holzapfel 1989). In these cases there is always a significant selection pressure, but that pressure will differ qualitatively as the ensemble of critical factors varies from generation to generation. A pattern of uncorrelated resistance to multiple agents enhances stability of numerical dynamics above that expected from no heterogeneity but the net effect relative to other patterns of correlation would seem to depend on the exact distribution of the yearly importance of each agent.

(*b*) *Density-dependent selection and response*

No concept has captured the shared imagination of ecologists and geneticists as much as the contrasting patterns of evolution in populations with chronically high or low densities. There has been a considerable amount of verbal and mathematical theory on the subject (reviewed lucidly by Mueller (1988)), but the existing empirical data are highly variable and often paradoxical. In this section I highlight some of these contradictions and suggest that paradoxical results have their genesis in the complexity of the relation between population densities, specific ecological agents of mortality, and the expression of genetic variation. The interrelations between population density and specific regulating forces will not surprise ecologists. Conversely, the changing genetic expression with density will not surprise geneticists.

The crucial basis for density-dependent selection is that the appropriate measure of fitness changes from the intrinsic rate of increase of a genotype (the measure in unregulated, density-independent contexts) to a function of the carrying capacity of a genotype (assuming weak selection at the genetic level (see Charlesworth (1980) and references therein)). The demographic results of evolution at single loci in these contrasting conditions are a maximization of population growth rate under density-independent conditions or the maximization of the number of individuals in the age class in which numerical regulation occurs under density-dependent conditions. There may or may not be significant elevation of the overall carrying capacity under density-dependent evolution; theory points only to the numbers in the critical age class. The age structure of density-dependence will determine the demographic outcome of evolution.

The importance of age-structure is shown in a number of paradoxical results from studies of the evolution of vital rates and life histories under the contrasting regimes. A common expectation derived from verbal models of life-history evolution is that evolution in a tightly regulated population will delay the onset of reproduction and lower adult fecundity. This expectation is supported by mathematical theory but so is the counter expectation of earlier reproduction and high fecundity; the differences arise from different assumptions of how increased densities affect the age-structured vital rates (Charlesworth 1980; Iwasa & Teramoto 1980). Several empirical laboratory studies of different species and different husbandry regimes have produced nearly the full spectrum of mathematically derived predictions that, in many cases, contradict the simpler verbally derived predictions

(Luckinbill 1978; Taylor & Condra 1980; Mueller & Ayala 1981; Bergmans 1984; Bierbaum *et al.* 1989).

More variability in outcome is evident when one examines the specific phenotypic traits that evolve under these contrasting conditions. Evolution acts on the genetic covariance between trait variation and fitness variation (Crow & Nagylaki 1977) and it is reasonable to expect that different traits will covary with a rate of increase than with a carrying capacity. This expectation is the basis for the predictions made by many ecologists (see, for example, Pianka (1970)). Density-dependent covariances that allow evolution involve two components, the covariance of the phenotypic trait variation with fitness and the requisite expression of genetic variation for that phenotypic trait at that specific density. These patterns of covariance differ from species to species in natural populations and show the complexity of trait evolution under varying density régimes.

Larval anurans in temporary ponds again offer a simple illustration. At very low tadpole densities predatory salamanders have no appreciable impact because tadpole growth rates are high and encounter rates are low due to both contact probabilities and refugia (Wilbur 1984; Travis *et al.* 1985). Greater numerical impact of salamanders and their role as a selective force for rapid growth rate occur only at higher tadpole densities. At extremely high densities all tadpoles grow so slowly, due to effective food limitation, that prey can be driven extinct (Wilbur 1984). The density of tadpoles can also influence the expression of genetic variation in larval growth rate (figure 1); in the spadefoot toad more genetic variation is expressed at low densities than at high ones. In other species there is no such effect of density (Travis 1983). The net result is that effective selection for growth rate (that producing a genetic response) will occur only within certain regions of ecological parameter space with respect to population density and those regions differ even among species using the same habitat.

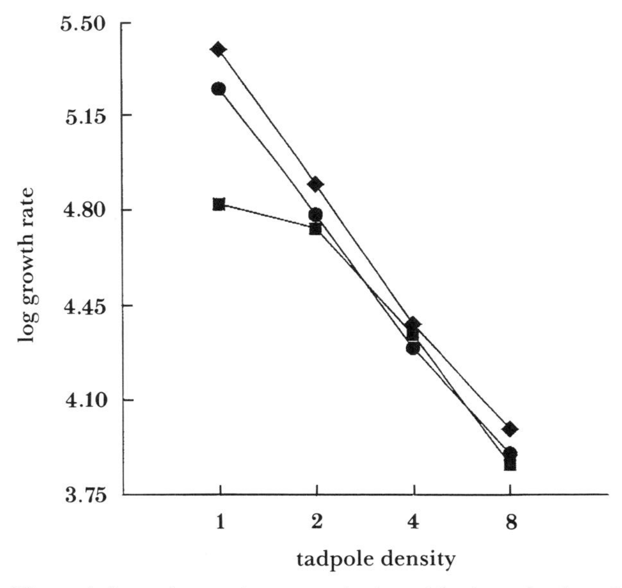

Figure 1. Larval growth rate on the logarithmic scale plotted against tadpole density (numbers of tadpoles per 500 ml water) for the eastern spadefoot toad, *Scaphiopus holbrooki*, of North America. Each symbol denotes the average value of eight replicates (density of 1) or four replicates (other densities). Different symbols denote different full sib families. Sibship differences exist at the lowest density but not at the others. This laboratory experiment is similar to those reported by Travis (1983).

These complexities are not restricted to viability selection in larval anurans. Conner (1989) pointed out that studies of sexual selection show all possible patterns of change in selection intensity with changes in population density. Traits that confer increased survival under high density conditions are irrelevant at low densities (Mueller & Sweet 1986). These patterns show that the frequency distribution of the critical ecological conditions must be known even to calculate the net direction and strength of selection on such traits (Gupta & Lewontin 1982). Density-dependent expression of trait variation is well known in laboratory populations of *Drosophila* (Clare & Luckinbill 1985; Luckinbill & Clare 1986; Bierbaum *et al.* 1989) and in some cases genetic variation exists in the sensitivity of trait expression to density effects (Khan *et al.* 1976; Marks 1982; cf. Travis 1983). Given these ecological and genetic complexities it may only be possible to make accurate predictions about density-dependent evolution from detailed models tailored to the ecology of specific organisms (Mueller (1988) shows this approach).

I owe a debt of gratitude to Professor M. P. Hassell and Professor R. M. May for provoking my thoughts on several of these topics. I thank Dr T. Miller, Dr D. Simberloff, Dr D. Strong and Dr A. Winn for discussions during the writing of this paper. My work has been supported by the National Science Foundation and during the writing of this paper I have been supported by BSR 88-18001.

REFERENCES

Alexander, H. M. 1989 An experimental field study of anther-smut disease of *Silene alba* caused by *Ustilago violacea*: genotypic variation and disease incidence. *Evolution* **43**, 835–847.

Barker, J. S. F. 1979 Interlocus interactions: a review of experimental evidence. *Theor. Popul. Biol.* **16**, 323–346.

Bao, I. Y. & Kallman, K. D. 1982 Genetic control of the hypothalamo-pituitary axis and the effect of hybridization on sexual maturation (*Xiphophorus*, Pisces, Poeciliidae). *J. exp. Zool.* **220**, 297–309.

Bergmans, M. 1984 Life history adaptation to demographic regime in labortory-cultured *Tisbe furcata* (Copepoda, Harpacticoida). *Evolution* **38**, 292–299.

Bierbaum, T. J., Mueller, L. D. & Ayala, F. J. 1989 Density-dependent evolution of life-history traits in *Drosophila melanogaster*. *Evolution* **43**, 382–392.

Bowen, B. S. 1982 Temporal dynamics of microgeographic structure of genetic variation in *Microtus californicus*. *J. Mammal.* **63**, 625–638.

Bradshaw, W. E. & Holzapfel, C. M. 1989 Life-historical consequences of density-dependent selection in the pitcher-plant mosquito, *Wyeomyia smithii*. *Am. Nat.* **133**, 869–887.

Bryant, E. H. & Meffert, L. M. 1988 Effect of an individual bottleneck on morphological integration in the housefly. *Evolution* **42**, 698–707.

Carson, H. L. & Templeton, A. R. 1984 Genetic revolutions

in relation to speciation phenomena: the founding of new populations. *A. Rev. Ecol. Syst.* **15**, 97–132.
Chakraborty, R. & Nei, M. 1977 Bottleneck effects on average heterozygosity and genetic distance with the stepwise mutation model. *Evolution* **31**, 347–356.
Charlesworth, B. 1980 *Evolution in age-structured populations.* Cambridge University Press.
Charlesworth, B. & Rouhani, S. 1988 The probability of peak shifts in a founder population. II. An additive polygenic trait. *Evolution* **42**, 1129–1145.
Chesson, P. L. 1989 A general model of the role of environmental variability in communities of competing species. *Lect. Math. Life Sci.* **20**, 97–123.
Clare, M. J. & Luckinbill, L. S. 1985 The effects of gene-environment interaction on the expression of longevity. *Heredity* **55**, 19–29.
Cohen, J. E. & Newman C. M. 1989 Host-parasite relations and random zero-sum games: the stabilizing effect of strategy diversification. *Am. Nat.* **133**, 533–552.
Conner, J. 1989 Density-dependent sexual selection in the fungus beetle, *Bolitotherus cornutus*. *Evolution* **43**, 1378–1386.
Crow, J. F. & Denniston, C. 1988 Inbreeding and variance effective population numbers. *Evolution* **42**, 482–495.
Crow, J. F. & Nagylaki, T. 1976 The rate of change of a character correlated with fitness. *Am. Nat.* **110**, 207–213.
Endler, J. A. 1986 *Natural selection in the wild.* New Jersey: Princeton University Press.
Ewens, W. J. 1982 On the concept of the effective population size. *Theor. Popul. Biol.* **21**, 373–378.
Ewens, W. J. 1989 The effective population size in the presence of catastrophes. In *Mathematical evolutionary theory* (ed. M. W. Feldman), pp. 9–25. New Jersey: Princeton University Press.
Gaines, M. S. 1985 Genetics. In *Biology of New World* Microtus (ed. R. H. Tamarin), pp. 845–883. Pennsylvania: American Society of Mammalogists.
Gaines, M. S., Vivas, A. M. & Baker, C. L. 1979 An experimental analysis of dispersal in fluctuating vole populations: Demographic parameters. *Ecology* **60**, 814–828.
Gaines, M. S. & Whittam, T. S. 1980 Genetic changes in fluctuating vole populations: selective vs. nonselective forces. *Genetics* **96**, 767–778.
Goodnight, C. J. 1988 Epistasis and the effect of founder events on the additive genetic variance. *Evolution* **42**, 441–454.
Grant, P. R. 1986 *Ecology and evolution of Darwin's finches.* New Jersey: Princeton University Press.
Gupta, A. P. & Lewontin, R. C. 1982 A study of reaction norms in natural populations of *Drosophila pseudoobscura*. *Evolution* **36**, 934–948.
Gyllensten, U. 1985 Temporal allozyme frequency changes in density fluctuating populations of willow grouse (*Lagopus lagopus* L.) *Evolution* **39**, 115–121.
Hastings, A. 1985 Multilocus population genetics with weak epistasis. I. Equilibrium properties of two-locus two allele models. *Genetics* **109**, 799–812.
Iwasa, Y. & Teramoto, E. 1980 A criterion of life history evolution based on density-dependent selection. *J. Theor. Biol.* **13**, 1–68.
Jones, J. S. & Yamazaki, T. 1974 Genetic background and the fitness of allozymes. *Genetics* **78**, 1185–1189.
Kallman, K. D. 1984 A new look at sex determination in poeciliid fishes. In *Evolutionary genetics of fishes* (ed. B. J. Turner), pp. 95–171. New York: Plenum Press.
Khan, M. A., Antonovics, J. & Bradshaw, A. D. 1976 Adaptation to heterogeneous environments. III. The inheritance of response to spacing in flax and linseed (*Linum usitatissimum*). *Aust. J. Agric. Res.* **27**, 649–659.
Kidwell, J. F. 1969 A chromosomal analysis of egg production and abdominal chaeta number in *Drosophila melanogaster*. *Can J. Genet. Cytol.* **11**, 547–557.
Kilpatrick, W. W. & Crowell, K. L. 1985 Genetic variation of the rock vole, *Microtus chrotorrhinus*. *J. Mammal.* **66**, 94–101.
Krebs, C. J. 1966 Demographic changes in fluctuating populations of *Microtus californicus*. *Ecol. Monogr.* **36**, 239–273.
Lande, R. 1976 Natural selection and random genetic drift in phenotypic evolution. *Evolution* **30**, 314–334.
Lande, R. 1979 Quantitative genetic analyses of multivariate evolution, applied to brain: body size allometry. *Evolution* **33**, 402–416.
Lawrence, W. S. 1988 Movement ecology of the red milkweed beetle in relation to population size and structure. *J. Anim. Ecol.* **57**, 21–35.
Lomnicki, A. 1988 *Population ecology of individuals.* New Jersey: Princeton University Press.
Luckinbill, L. S. 1978 r- and K-selection in experimental populations of *Escherichia coli*. *Science, Wash.* **202**, 1201–1203.
Luckinbill, L. S. & Clare, M. J. 1986 A density threshold for the expression of longevity in *Drosophila melanogaster*. *Heredity* **56**, 529–535.
McCauley, D. E. 1989 Extinction, colonization, and population structure: a study of a milkweed beetle. *Am. Nat.* **134**, 365–376.
Madhavi, R. & Anderson, R. M. 1985 Variability in the susceptibility of the fish host, *Poecilia reticulata*, to infection with *Gyrodactylus bullatarudis* (Monogenea). *Parasitology* **91**, 531–544.
Malpica, J. M. & Vasallo, J. M. 1980 A test for the selective origin of environmentally correlated allozyme patterns. *Nature, Lond.* **286**, 407–408.
Manly, B. F. J. 1985 *The statistics of natural selection.* New York: Chapman and Hall.
Marks, R. W. 1982 Genetic variability for density sensitivity of three components of fitness in *Drosophila melanogaster*. *Genetics* **101**, 301–316.
Marquis, R. J. 1990 Genotypic variation in leaf damage in *Piper arieianum* (Piperaceae) by a multispecies assemblage of herbivores. *Evolution* **44**, 104–120.
Maruyama, T. & Fuerst, P. A. 1985 Population bottlenecks and nonequilibrium models in population genetics. III. Genic homozygosity in populations which experience periodic bottlenecks. *Genetics* **111**, 691–703.
Maruyama, T. & Kimura, M. 1980 Genetic variability and effective population size when local extinctions and recolonization of subpopulations are frequent. *Proc. natn Acad. Sci.* **77**, 6710–6714.
May, R. M. & Anderson, R. M. 1983 Parasite-host coevolution. In *Coevolution* (ed. D. J. Futuyma & M. Slatkin), pp. 186–206. Massachussetts: Sinauer.
Miller, T. E. & Schemske, D. W. 1990 An experimental study of competitive performance in *Brassica rapa* (Cruciferae). *Am. J. Bot.* **77**. (In the press.)
Miyashita, N. & Laurie-Ahlberg, C. C. 1984 Genetical analysis of chromosomal interaction effects on the activities of the glucose 6-phosphate and 6-phosphogluconate dehydrogenases in *Drosophila melanogaster*. *Genetics* **106**, 655–668.
Morin, P. J. 1983 Predation, competition, and the composition of larval anuran guilds. *Ecol. Monogr.* **53**, 119–138.
Morin, P. J. 1986 Interactions between intraspecific competition and predation in an amphibian predator–prey system. *Ecology* **67**, 713–720.
Motro, U. & Thomson, G. 1982 On heterozygosity and the

effective size of populations subject to size changes. *Evolution* **36**, 1059–1066.
Mueller, L. D. 1988 Density-dependent population growth and natural selection in food-limited environments: the *Drosophila* model. *Am. Nat.* **132**, 786–809.
Mueller, L. D. & Ayala, F. J. 1981 Trade-off between r-selection and K-selection in *Drosophila* populations. *Proc. natn Acad. Sci.* U.S.A. **78**, 1303–1305.
Mueller, L. D. & Sweet, V. F. 1986 Density-dependent natural selection in *Drosophila*: evolution of pupation height. *Evolution* **40**, 1354–1356.
Nei, M., Maruyama, T. & Chakraborty, R. 1975 The bottleneck effect and genetic variability in populations. *Evolution* **29**, 1–10.
O'Brien, S. J., Roelke, M. E., Marker, L., Newman, A., Winkler, C. A., Meltzer, D., Colly, L., Evermann, J. F., Bush, M. & Wildt, D. 1985 A genetic basis for species vulnerability in the cheetah. *Science, Wash.* **227**, 1428–1434.
Parker, M. A. 1988 Polymorphism in disease resistance in the annual legume *Amphicarpaea bracteata*. *Heredity* **60**, 27–31.
Pianka, E. R. 1970 On r- and K-selection. *Am. Nat.* **104**, 592–596.
Plante, Y., Boag, P. T. & White, B. N. 1989 Microgeographic variation in mitochodrial DNA of meadow voles (*Microtus pennsylvanicus*) in relation to population density. *Evolution* **43**, 1522–1537.
Polans, N. O. & Allard, R. W. 1989 An experimental evaluation of the recovery potential of ryegrass populations from genetic stress resulting from restriction of population size. *Evolution* **43**, 1320–1324.
Rausher, M. D. 1984 Tradeoffs in performance on different hosts: evidence from within- and between-site variation in the beetle *Deloyala gutata*. *Evolution* **38**, 582–595.
Rouhani, S. & Barton, N. H. 1987 The probability of peak shifts in a founder population. *J. Theor. Biol.* **126**, 51–62.
Shaw, R. 1986 Response to density in a wild population of the perennial herb *Salvia lyrata*: variation among families. *Evolution* **40**, 492–505.
Simms, E. S. & Rausher, M. D. 1989 The evolution of resistance to herbivory in *Ipomoea purpurea*. II. Natural selection by insects and costs of resistance. *Evolution* **43**, 573–585.
Slatkin, M. 1977 Gene flow and genetic drift in a species subject to frequent local extinctions. *Theor. Popul. Biol.* **12**, 253–262.
Smith, M. H., Garten, C. T. & Ramsey, P. R. 1975 Genic heterozygosity and population dynamics in small mammals. In *Isozymes. IV. Genetics and evolution* (ed. C. L. Markert), pp. 85–102. New York: Academic Press.
Sved, J. A. & Ayala, F. J. 1970 A population cage test for heterosis in *Drosophila pseudoobscura*. *Genetics* **66**, 97–113.
Taylor, C. E. & Condra, C. 1980 r- and K-selection in *Drosophila pseudoobscura*. *Evolution* **34**, 1183–1193.
Travis, J. 1983 Variation in development patterns of larval anurans in temporary ponds. I. Persistent variation within a *Hyla gratiosa* population. *Evolution* **37**, 496–512.
Travis, J. 1989 The role of optimizing selection in natural populations. *A. Rev. Ecol. Syst.* **20**, 279–296.
Travis, J., Keen, W. H. & Juilianna, J. 1985 The effects of multiple factors on viability selection in *Hyla gratiosa* tadpoles. *Evolution* **39**, 1087–1099.
Travis, J., Emerson, S. B. & Blouin, M. 1987 A quantitative genetic analysis of larvel life history traits in *Hyla crucifer*. *Evolution* **41**, 145–156.
Travis, J. & Mueller, L. D. 1989 Blending ecology and genetics: Progress toward a unified population biology. In *Perspectives in ecological theory* (ed. J. Roughgarden, R. M May & S. A. Levin), pp. 101–124. New Jersey: Princeton University Press.
Via, S. 1984 The quantitative genetics of polyphagy in an insect herbivore. II. Genetic correlations in larval performance within and among host plants. *Evolution* **38**, 896–905.
Wade, M. J. & McCauley, D. E. 1988 Extinction and recolonization: their effects on the genetic differentiation of local populations. *Evolution* **42**, 995–1005.
Wallace, B. 1977 Automatic culling and population fitness. *Evol. Biol.* **10**, 265–276.
Watterson, G. A. 1989 The neutral alleles model with bottlenecks. In *Mathematical evolutionary theory* (ed. M. W. Feldman), pp. 26–40. New Jersey: Princeton University Press.
Wayne, R. K., Modi, W. S. & O'Brien, S. J. 1986 Morphological variability and asymmetry in the cheetah (*Acinonyx jubatus*), a genetically uniform species. *Evolution* **40**, 78–85.
Weis, A. E. & Abrahamson, W. G. 1986 Evolution of host-plant manipulation by gall makers: ecological and genetic factors in the *Solidago–Eurosta* system. *Am. Nat.* **127**, 681–695.
Wilbur, H. M. 1984 Complex life cycles and community organization in amphibians. In *A new ecology: novel approaches to interactive systems* (ed. P. W. Price, C. N. Slobodchikoff & W. S. Gaud), pp. 196–224. New York: John Wiley & Sons.

Discussion

J. J. D. Greenwood (*British Trust for Ornithology, Herts, U.K.*). The possibility that for small mammals the fluctuations in effective numbers may be even greater than those in actual numbers should not be taken as a broader generalization. In the snail *Cepaea nemoralis*, for example, there is evidence that movement is greater at lower population densities than at higher densities, which will tend to lessen the variation in effective numbers (Greenwood 1974, 1976). One can see why emigration rates may adaptively increase at high densities in cyclical populations. But in species in which populations vary spatially in accord with the local availability of resources one might expect individuals to move around more in places where densities are low, to harvest the scarce resources in those places. We need to investigate several representative cases before coming to any general conclusions.

References

Greenwood, J. J. D. 1974 Effective population numbers in the snail *Cepaea nemoralis*. *Evolution* **28**, 513–526.
Greenwood, J. J. D. 1976 Effective population number in *Cepaea*: a modification. *Evolution* **30**, 186.

The fossil record: a sampler of life's diversity

JAMES W. VALENTINE

Department of Geology, University of California, Santa Barbara, California 93106, U.S.A.

SUMMARY

The fossil record is adequate to determine the general patterns of diversity of genera and higher taxa across geological time, for most groups of organisms. The Linnaean hierarchy, in which most of the fossil groups have been classified, is ideally suited for such studies. Marine invertebrates are represented by three successive faunas that display increased diversities, but lower evolutionary turnovers; perhaps increasing specializations favoured lineages with higher extinction resistance. Tetrapods are also represented by three faunas that display increasing diversities and similar though more complex patterns of decreasing evolutionary turnovers. Tracheophytes have been placed in four Phanerozoic floras with generally increasing diversities, but by contrast with animals display increased species turnover with increasing diversity, perhaps in response to competitive requirements imposed by the successive origination of major clades.

1. INTRODUCTION

The only direct evidence of the diversity of life through geological time is contained in the fossil record. It is clear that the record is incomplete and biased, so that in interpreting past diversities these difficulties must be kept in mind and their effects minimized as far as possible. Evaluation of the nature of the fossil record, and of the stratigraphic record in which it is contained, have led to strategies of interpretation aimed at deriving the most information possible while minimizing the risks of overinterpretation. However, there are no precise ways of testing the extent to which these aims are realized, at least for the present. Instead, we must rely chiefly on the plausibility and coherence of our interpretations when viewed in the light of biological knowledge of today's world.

2. QUALITY OF THE FOSSIL RECORD

The completeness of the stratigraphic record varies with the resolution required (Schindell 1980; Sadler 1981). That is, the record is less complete if one wishes to resolve intervals of 1 Ma than if one can settle for intervals of 5 Ma; there is simply a far better chance of finding sediment in a long interval than in a short one. On the other hand, if one actually has a sedimentary column, the rocks in shorter timespans will tend to be more complete than those in longer spans. This is because in an average section, most stratigraphic gaps are of short durations, but there are nevertheless many gaps of intermediate duration and a few that are quite long. The longer the timespan studied, the more chance of encountering larger gaps, and it happens that they render the longer spans increasingly less complete. These relations have been quantified by Sadler (1981) for an average section. For example, an average 30-million-year section of shelf carbonates and detrital sediments (from which the vast bulk of Phanerozoic marine fossils come) is of the order of 33% complete for one million year resolution intervals. Non-marine deposits tend to be less complete and more variable. As many sediments are unfossiliferous, the fossil record is usually less complete than the stratigraphic record; some exceptions may occur in cases of condensed sections or of reworked fossils.

Usually there are many stratigraphic sections scattered across the world to represent some span of geological time such as an Epoch (a subdivision of a Period, such as the Cambrian Period) or an Age (a subdivision of an Epoch, averaging around 8 million years). If the stratigraphic gaps are random from section to section, then for resolution intervals as long as 1 Ma the chance of finding contemporaneous sediment in two average 30-million-year sections is on the order of 0.11, and in five sections, 0.004 (Sadler in Valentine *et al.* (1990)). Faunas that appear to be roughly contemporaneous, then, and that are common and widespread, must either represent times when deposition among sections was not random, as during extensive transgressions, or times when the gross composition of a fauna persisted over many 1 Ma intervals. There is evidence that numbers of transgressive episodes correlate globally, though how common this may be is still under debate. At any rate, to the extent that gaps and depositional episodes correlate widely, stratigraphic correlations will be better than in a random model but, on the other hand, the faunal record will contain more global gaps.

A single faunal locality can hardly represent the global biota of the times. Species diversity may be accommodated by the number of provinces in the biosphere, and/or by the number of communities in provinces, and/or by the number of species in communities. How many samples are required adequately to represent the global biota depends partly upon the biotic heterogeneity at provincial and community levels. The present day may represent nearly or quite the peak of biotic heterogeneity during the Phanerozoic. The accumulation of potential fossils at present, or of actual fossils in the near past, should provide a good test of fossil completeness. A study of the extent to which the durably skeletonized portion of the living fauna is represented in the fossil record of an entire marine shelf province has been based on the living Californian Province fauna as contrasted with its fossil record during the past million years (Valentine 1989*b*). The record proved surprisingly complete; about 85% of the living species were captured by the fossil record, and a single locality yielded over 50% owing to intermixture of species from a variety of shallow-water habitats. The implications are that most species with durable skeletons are recruited into the fossil record, and that the spottiness and gaps found in the record are chiefly because of subsequent loss of sediments via erosion, or of fossils via diagenesis. However, when rich fossil localities or locality clusters do occur there is a good chance that a large fraction of the easily fossilized fauna that was present is represented. Of course, the more localities in the more varied sediments, the greater the completeness of the fossil fauna is likely to be. As there are literally hundreds of rich fossil localities in the Phanerozoic record, there are likely to be many local faunas that are rather complete with respect to durably skeletonized forms. For diversity studies, this situation is far better than having, say, five times as many localities that average only one fifth as complete, for it provides us with data points that suggest the general mangitude of the standing faunal diversities, or at least of a rather constant fraction of the fauna (we hope). It is thus posible to extrapolate geographically between roughly contemporaneous localities and stratigraphically between successive faunal samples. To be sure, many of the finer-scale fluctuations are missed, but on the other hand the coherence of the data appears to indicate that the broad patterns revealed by palaeodiversity studies are real. This point has long been taken on faith, but studies leading to quantitative assessments of the record, that are improving steadily, suggest that the faith is by and large justified.

A final consideration is that the available diversity record does not simply deteriorate in a linear fashion with geological time, although it does seem certain that as the ages pass, more and more fossils are lost to erosion and diagenesis. At the same time, however, buried fossils, a sort of fossil 'bank', are constantly being brought into the realm of the collectible by erosion. Whether the available record of any given period is being improved or degraded, then, depends partly upon structural and erosional history. For example, Sadler (in Valentine *et al.* 1990) points out that because Cambrian cratonic and mioclinal rocks are being well-exposed now to erosion, because of the timing and extent of structural uplifts, the present is a better time to study Cambrian faunas than the faunas of some later Periods, such as the Carboniferous. Exposure of Cambrian rocks is not yet optimal; the future will probably be an even better time to be a Cambrian palaeontologist (say, in 20 Ma).

3. NATURE OF THE TAXA

Most research on fossil diversity patterns has used the taxa of the familiar Linnaean hierarchy as units of diversity. Such usage has been criticized as inappropriate by proponents of cladistic classification (Wiley 1981; Eldredge & Salthe 1984; Vrba & Eldredge 1984; Patterson & Smith 1987). However, Linnaean taxa seem on the contrary to be the most appropriate for diversity studies, a contention that requires a brief justification.

The Linnaean taxa are monophyletic clusters of species (or a single species) that share a polythetic assemblage of characters; higher and higher taxa are clusters of such clusters that share polythetic assemblages of characters. (Polythetic units have many defining features in common, but not unit has all such features, and no such feature is found in all units; they share a 'family resemblance'.) Linnaean taxa at any level have arisen from a species that branched from an ancestral taxon of the same level; indeed they must have done. Therefore, the only Linnaean taxa that are not paraphyletic are those that have never given rise to daughter taxa of the same rank. For Linnaean families in the fossil record, between about the third and one half seem to be paraphyletic, the others becoming extinct without daughters (Valentine 1990). Paraphyly provides a nested aggregational hierarchy of taxa with similar ranks. Probably no system of classification has as much organizing power as a properly nested hierarchy (see Simon 1962).

An important feature of such a nested Linnaean hierarchy is that it provides a clear strategy to cope with the incompleteness of the fossil record. As taxa in each successive higher category have generally many individuals, a wider distribution and a longer geological range than lower taxa, their completeness of representation as fossils increases at each succeeding level (it may remain the same as a limit). If species records are too spotty and incomplete, we may simply employ generic records, which involve the same species, but clustered. Most Phanerozoic marine diversity studies have been prosecuted at the family level (see papers in Valentine 1985), but generic data are being compiled by Sepkoski, and some preliminary uses of this generic data set have appeared (Sepkoski 1986; 1988; Raup & Boyajian 1988).

The Linnaean hierarchy is particularly useful in considering the changing relative and absolute diversities of individual clades, whereby the taxonomic composition of the biosphere is altered. For such studies it is possible to consider a taxon of a given level as a unit (an 'individual' see Ghiselin 1974) in the macroevolution of a higher level; to consider genera as units within a class, for example. Such usage implies that taxa above the species level have certain properties: they must be able to reproduce, some of their features must be heritable and subject to sorting, and they must nevetheless suffer some changes that are themselves heritable and subject to sorting. As holophyletic taxa do not reproduce by definition, they cannot contribute to macroevolution among clades. In classifcations that admit only holophyletic taxa, families (for example) cannot derive from other families. If paraphyly is permitted, however, then there is no problem with clade evolution in principle (Valentine 1990). Families do emanate from other families. Of course a family arises from within a particular genus of its ancestral family, and moreover from a particular species within that genus. Indeed families arise, not from an entire species, but from some sample of a species population, a vicar or founder population, that may be as small as two genetic individuals (or even one in some cases), a sample that need have no effect whatsoever on the biology of the species (genus, family) from which it came. It seems that either species no more give rise to families than do other families, or that, units in both categories may be considered ancestral.

A Linnaean hierarchical system emphasizes the pattern of adaptive radiation that is so common at every evolutionary level, and in the fossil record it permits the employment of taxa above the species level to measure and to trace the tempo of evolutionary activity across the biosphere or of large realms within it, the clear strategy to best combat a spotty record. Furthermore, descendant Linnaean taxa do inherit features from ancestral ones, while still other features are derived; cladistic methods show this. The differential success of taxa with different derived features, then, constitutes macroevolution. A final objection posed to such use of taxa above the species level as evolutionary units is that they do not interact the way individuals do within populations, in particular that they do not interbreed (Wiley 1981). However, there are entire prokaryote Kingdoms that do not interbreed, yet such organisms certainly do evolve. I shall proceed to use Linnaean taxa here, for they are best suited to the present purpose.

4. MODEL LINNAEAN TAXA

Modelling of the behaviour of taxa in a Linnaean-style hierarchy has suggested how some of their interrelations and trends may be interpreted. Species lineages are permitted to grow within a model biosphere with finite resources represented as adaptive 'addresses', with attempted randomly chosen speciations or origins of higher taxa at stochastically constant rates. Each species occupies one adaptive address exclusively; attempted speciations onto occupied addresses fail. Random extinctions of species occur at a 'background' rate, so that species diversity reaches a stochastic equilibrium when successful speciations slow to the extinction rate. The success of originations is thus density-dependent (diversity-dependent). Progressively higher taxa requiring increasingly large, unoccupied blocs of adaptive addresses to originate. At speciation (S) and extinction (E) rates inferred from the fossil record, trees may be generated that resemble, in the frequency distributions of branches at all levels, many phylogenetic trees of living taxa (Valentine & Walker 1987). An interesting feature of this model is that with average E and S estimated from the fossil record, about one third of the adaptive addresses remain unoccupied (Walker & Valentine 1984). Three distinctive behaviours are exhibited by higher taxa during and subsequent to the rise of species diversities to a (stochastic) equilibrium at which S comes to equal E. The first behaviour occurs during the growth of species diversity towards an equilibrium; the diversities of higher taxa rise also. Secondly, when species reach a stochastic equilibrium with speciation and background extinctions approximately balanced, most higher taxa tend to decline in diversity; it is easier then to extinguish higher taxa than to originate new ones, for adaptive space is vacated by declining taxa piecemeal and it is usually taken over by surviving groups. Thirdly, those taxa that happen to become quite large relative to the capacity of the model biosphere become relatively immune to background extinction levels; it is taxa on the higher levels that first achieve such size, of course.

If two independent clades that have different ratios of E and S are both present in the same adaptive space, then the clade with the higher stochastic equilibrium will eventually exclude the other clade from that space (Walker 1984; Valentine 1990). If two clades have equal E/S ratios, but have different species turnover rates, then the clade with the lower turnover rate will eventually exclude the other (Valentine 1990).

5. MARINE INVERTEBRATE RECORD

Marine invertebrates furnish the longest and most complete fossil record. Marine invertebrate diversity displays three main phases, as identified faunally by Sepkoski (1981): a late Precambrian–Cambrian fauna I, during which species diversity rose from the earliest records of animals to a possible Late Cambrian 'plateau'; a fauna II that includes the remainder of the Paleozoic Era when species diversity rose again from the Late Cambrian levels to perhaps the end of the Ordovician, and thereafter fluctuated about a fairly constant or perhaps slightly rising mean until terminated at the severe Permian–Triassic extinction; and a Mesozoic–Cenozoic fauna III after the extinction, when diversity climbed to an all-time high during a rise that shows no evidence of slackening (mankind's depredations aside, that is). These Faunas were originally established at the family level, but are convenient for discussion of diversity trends across the entire taxonomic hierarchy (figure 1).

Phase I can be subdivided into two main portions. The late Precambrian ('Upper Vendian') fauna contains body fossils, most of which appear to be of cnidarian grade (see, for example, Jenkins 1984; Norris 1989). Bilaterians are chiefly represented by epifaunal creeping traces and horizontal shallow semi-infaunal or infaunal burrows that show a variety of behaviours but from which no very accurate diversity estimate may be drawn. Possible bilaterian body fossils include sprigginids and vendomiids that may represent segmented, haemocoelic and scleritized worms (Valentine 1989*a*) but that have also been considered to represent an extinct non-metazoan clade (Seilacher 1989). The Early Cambrian record contains evidence of an explosive radiation of higher invertebrate body plans that probably raised the number of phylum-level

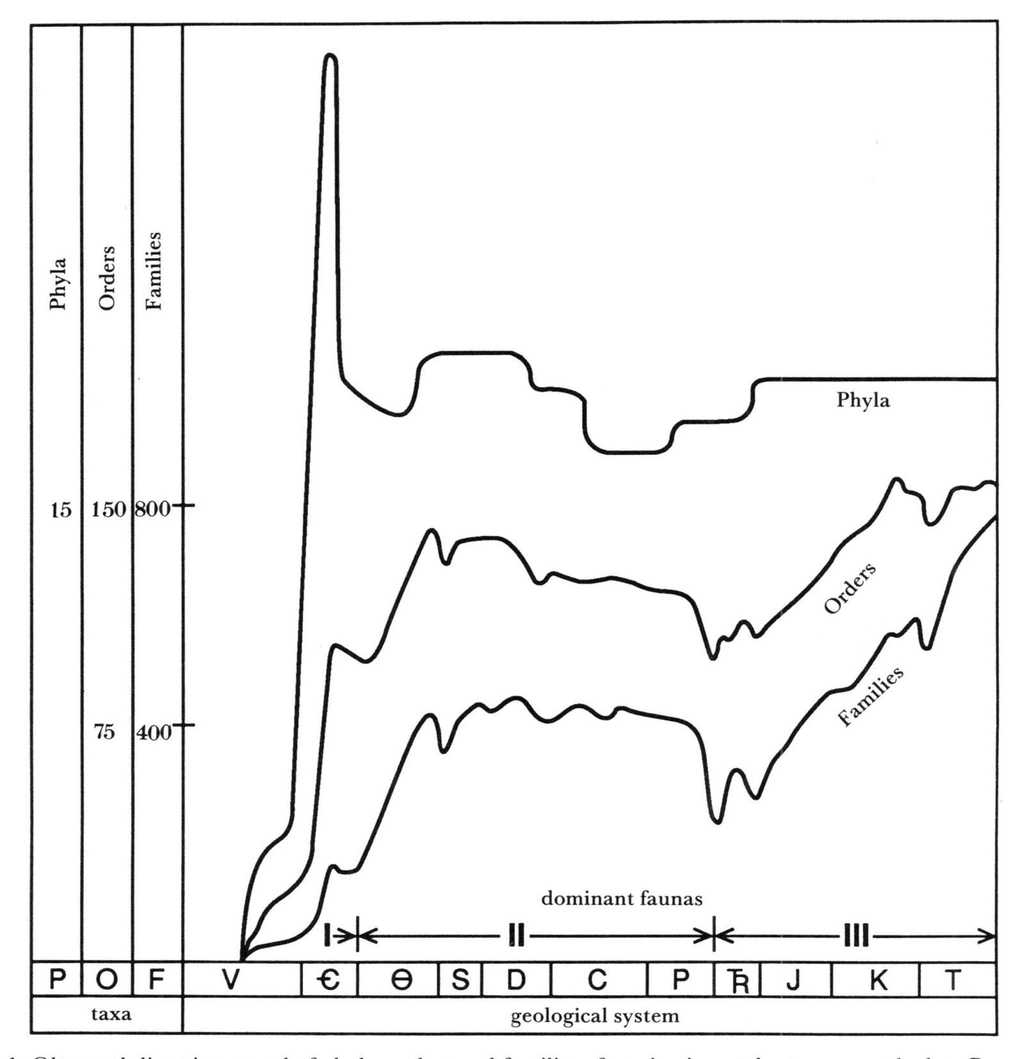

Figure 1. Observed diversity record of phyla, orders and families of marine invertebrates across the late Precambrian and the Phanerozoic. The Precambrian–Cambrian boundary is placed at 540 Ma ago; the remaining time divisions are as in Sepkoski (1981). Intervals of measurement are geological stages, which are not of equal lengths, and there are slight differences in the taxa used from level to level; neither of these features strongly biases the trends, however. The peaks in the Cambrian are from Lagerstätten, which include poorly skeletonized forms. Fauna I is dominated by trilobites, fauna II by articulate brachiopods and crinoids, and fauna III by molluscs. Families after Sepkoski (1981); Orders after Sepkoski (1978), amended by Valentine *et al.* (1990).

clades to the highest point in their history (Valentine 1969; Valentine *et al.* 1990), although species diversity must have been many times lower then than at present (Valentine 1969; Bambach 1977).The first two stages of the Lower Cambrian are particularly characterized by the appearance of novel body plans, and of 'skeletal plans' of extinct forms that have proved to represent unique body plans whenever their gross body morphology is discovered (e.g. wiwaxiids (Bengston & Conway Morris, 1984) and *Microdictyon* (Chen *et al.* 1989)). Perhaps between 35% and 50% of the phylum-level taxa and over 90% of the ordinal-level taxa have become extinct. The record of body fossil diversity at three taxonomic levels is shown in figure 1.

The Middle and Late Cambrian diversity decline that can be inferred for the class and ordinal levels can be interpreted as acompanying a stable or even somewhat increasing species diversity, a situation that occurs when species' S and E are approximately equal (Valentine & Walker 1987). The Vendian part of Fauna I, then, can be regarded as recording a radiation of diploblastic and acoelomate bilaterian grades, the establishment of segmentation, and a build-up of body-plan complexity that culminated in the establishment of several lineages of higher invertebrates near the Cambrian boundary. During Early Cambrian time the earliest coelomates appear (brachiopods), vertically penetrating burrows were established (the timing and extent of which are still in dispute) and higher invertebrates radiated explosively; some of the body plans involved durable skeletons to create the Precambrian–Cambrian boundary.

Fauna II began with a decided rise in diversity, nearly doubling at the Ordinal level (figure 1). In a study of this diversification, Sepkoski (1988) concluded that it resulted chiefly from a partitioning of resources on the community level, although accompanied also by a less important increase in species packing within communities. Provinciality was not believed to be a significant contributor. The reason that the shelf environment became the site of an important increase in habitat specialization at this particular time is not understood, although it may be associated with the faunal dynamics of the clades that rose to dominance at that time (see later). Phylum, class and ordinal diversity apparently declined through the remainder of the Paleozoic, probably punctuated at mass extinction events (Valentine 1969). There is some indication that familial and generic diversities, increasingly more volatile than higher taxa, fell slightly through that Era (Sepkoski 1981, 1988). Species diversity, even more volatile than generic diversity, may have increased somewhat overall during the Middle and Late Palaeozoic, with higher taxa becoming increasingly speciose (Valentine 1969), possibly one of the reasons that high-level diversities do not decrease markedly (Flessa & Jablonski 1985).

Following the Permian–Triassic extinctions, Fauna III began with few marine invertebrate species, perhaps fewer than at the beginning of Cambrian time (Valentine *et al.* 1978), but so far as we know with at least as many phyla and classes as exist now. A few new orders appear. At lower taxonomic levels, however, high rates of diversification continued right into the Quaternary, nearly doubling the number of families from their fauna II levels, and implying greater increases in genera and still greater ones in species. Species packing within communities is thought to have doubled from fauna II to fauna III (Bambach 1977), and this estimate may be conservative. There are as yet no quantitative estimates for increases (if any) in the numbers of communities within provinces. The number of provinces, however, began to increase following the Mesozoic breakup of Pangaea and this has probably been an important contributor to global marine diversity during fauna III, especially at the species level (Valentine 1967; Valentine *et al.* 1978).

Data from all Faunas may be synthesized into a generalized description of the dynamics of marine invertebrate diversity changes. The earliest (Vendian) faunas seem to consist of creeping vermiform and segmented haemocoeloic bilaterians, together with diploblastic sessile and pelagic forms. As the fauna is soft-bodied, taxonomic diversities cannot be estimated with confidence, although they appear to be relatively low at all levels and to have a far lower ratio of low- to high-level taxa than exists today. This fauna presumably records the rise in complexity of bilaterians

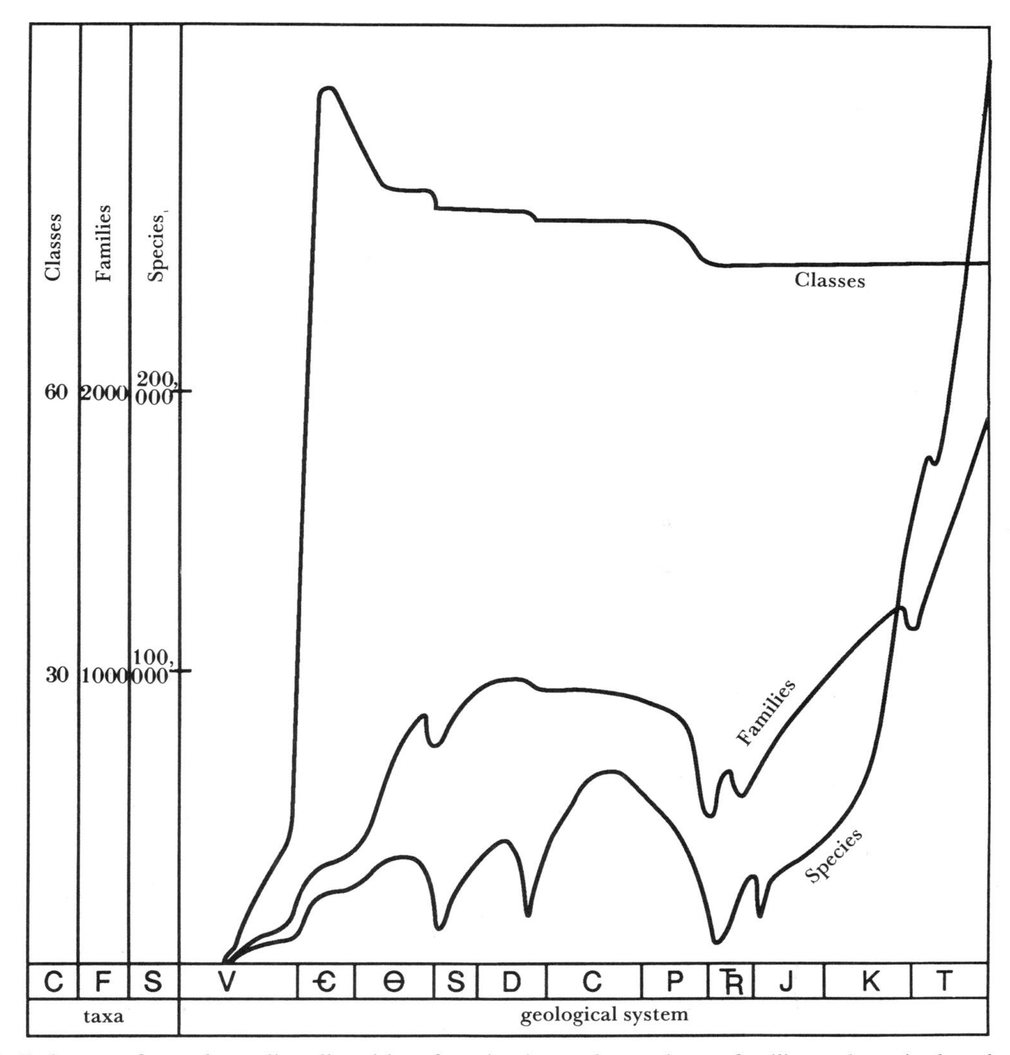

Figure 2. Estimates of actual standing diversities of marine invertebrate classes, families and species based on fossil diversities, trends in fossil diversities among taxonomic levels, and ratios of easily to poorly fossillized taxa at present and in Lagerstätten.

from simple acoelomate beginnings to rather complex segmented forms with hydrostatic skeletons. As the level of complexity appropriate to higher invertebrates is achieved, secondary body cavities appear in some lineages and both arthropod and annelid ancestors explode into a variety of body plans, chiefly segmented (Jacobs 1990). Probably the ratio between low-level and high-level taxa was never lower than during the middle to late Early Cambrian. Those lineages that happen to have a high speciation rate diversified rapidly in the relatively empty adaptive space of the times; the Cambrian fauna was eventually dominated by trilobites, which have high rates of S and E for invertebrates (Stanley 1979), especially the earlier lineages (Foote 1988), and which had relatively wide distributions across the Cambrian shelves. These broad associations were replaced by benthic communities dominated by brachiopods and crinoids, the members of which were more narrowly distributed and possessed slower turnover rates (see Stanley 1979; Walker 1984; Valentine 1990).

That a general increase in specialization should correlate with a general slowing of taxonomic turnover, requiring a concomitant lowering of extinction rates, seems contradictory at first glance: shouldn't more specialized organisms be at greater extinction risk? However, I argue here that in fact those lineages that are endowed for some reason with relatively high extinction resistance are more likely to produce successful specialized taxa (in macroevolutionary terms, those of longer durations). A clade with high extinction rates that produces specialists may practically commit suicide, and this has no doubt happened often. The specialists with the best chance to persist, then, may be precisely those that are the most extinction-resistant. So the high-turnover, generalized taxa dominating fauna I are replaced by lower-turnover, more specialized taxa of fauna II.

The reign of fauna II was brought to a spectacular close by the end-Permian extinctions. From the Cambrian example, one would expect the rediversification to be dominated by faster-turnover clades, and indeed this is observed (Van Valen 1984). The early Mesozoic success of higher-turnover ammonoid lineages is a case in point. More successful in the long run, however, are the gastropods, bivalves, and other important members of fauna III that have even lower turnover rates than the fauna II dominants (see rates in Stanley (1979); Raup & Sepkoski (1982); Raup & Boyajian (1988)). Furthermore, the fauna III organisms appear to be still more specialized than those of fauna II, since species packing within communities rises significantly. The same generalities that apply to the fauna I–II shift apply to the fauna II–III shift. In addition, the effects of high species diversities achieved within communities have been multiplied by provincialization to reach a great Quaternary diversity high (Valentine 1967). Figure 2 shows estimates of actual standing diversities of classes, families and species of the marine realm during the Vendian and the Phanerozoic.

6. TERRESTRIAL TETRAPOD RECORD

Perhaps because the terrestrial record seems less complete, fewer diversity studies involve terrestrial than marine organisms. Some of the problems involved in such analyses are reviewed in Padian & Clemens (1985), who present Phanerozoic patterns at the generic and ordinal levels. Tetrapod diversities at the family level have been considered by Benton (1985), who has tabulated family geologic ranges (Benton 1987). These sources provide the chief basis of the following discussion.

Three assemblages of tetrapod families are recognized: a primitive assemblage I of chiefly late Palaeozoic dominance; a Mesozoic assemblage II (early diapsids, dinosaurs, pterosaurs); and assemblage III, including mammals and birds, that began in the early Mesozoic but assumed dominance only in the Late Cretaceous (figure 3). A metric that estimates completeness (the ratio of all stages in which a taxon is found to all stages that it spans) suggests that faunal completeness is not particularly time-dependent for the tetrapod groups; for example, the record of the small lepidosaurs and lissamphibians of assemblage III is poor, while the record of mammal-like reptiles is judged to be rather complete (Benton 1987). The strong Tertiary diversity rise, then, is presumably real. Per-taxon extinction rates have been calculated for families from within various tetrapod groups from Benton's data (Valentine *et al.* in preparation). Early amphibians display faster turnover rates than late amphibians, early reptiles than later reptiles and dinosaurs, and mammal-like reptiles (therapsids) than mammals, even though the mammalian record includes short-lived families associated with the early Cenozoic radiation. However, dinosaur turnover rates are only a little faster than those for mammals.

The contributions of various forms of diversity accommodation to the increased Cenozoic tetrapod diversity has not been studied, except that biogeographic endemism must play a role. The work of

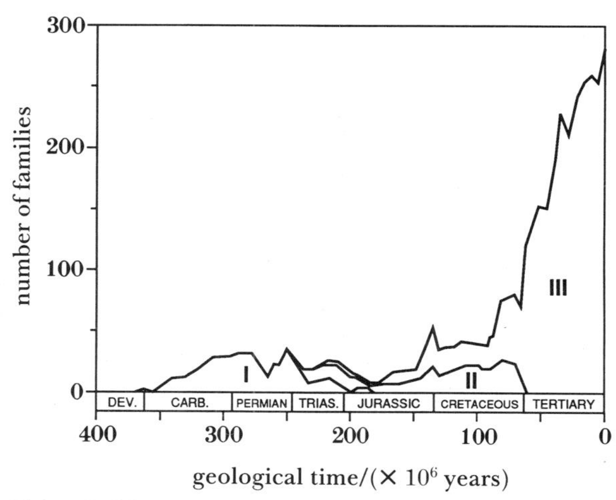

Figure 3. Observed Phanerozoic diversity record of terrestrial tetrapod families. Fauna I includes early amphibians, anaspids and mammal-like reptiles; fauna II, early diapsids, dinosaurs and pterosaurs; fauna III, 'modern' amphibians, reptiles, birds, mammals. After Benton (1987).

Flessa (1975) and Flynn (1986) make it clear that the distinctiveness of regional faunas contributes significantly to mammalian diversity at both generic and familial levels. These studies conclude that the arrangement and number of land regions is a primary control on diversity, while latitudinal environmental variation has also been important. As continental separations have increased and climatic zonation has been enhanced during the Cenozoic, geographic factors must have become increasingly important as diversity controls both intercontinentally and in terms of habitat patchiness intracontinentally. Estimates of the extent to which tetrapod species packing within environments has increased or varied seems not to have been attempted. Whether increasing specialization has accompanied slowing in tetrapod taxonomic turnover rates cannot be addressed as yet. The data thus suggest possible similarities with marine faunas but are hardly conclusive at this time.

7. VASCULAR PLANT RECORD

Data from land plant (tracheophyte) diversities have been summarized by Niklas *et al.* (1980, 1983, 1985) for species within major plant groups. Four floras were recognized (figure 4), which tend to be composed chiefly of groups that are successively freed from dependence on moisture for reproduction. It is the experience of palaeobotanists that the later local floras are more speciose than the earlier except during the Permian and Triassic, so that increased species packing within plant associations is probably important (Niklas *et al.* 1980). That habitat patchiness has increased as Cenozoic climates have developed seems likely also (Tiffney & Niklas 1990) although habitat evolution has not been thoroughly studied in this context. On the larger geographic scale, the increased climatic zoning of Cenozoic time was probably quite important, but there seems to be no clear signal that plant diversity is tied to the number of land regions (Tiffney & Niklas 1990).

A striking feature of vascular plant evolution, at least to one accustomed to marine invertebrate history, is that the pace of evolutionary turnover has increased from flora to flora (Niklas *et al.* 1983, 1985), precisely opposite to the turnover trajectory of marine forms and certainly different from tetrapods also. Elements of the later floras display faster rather than slower turnover rates; the dependence on extinction resistance (low E) for long-term success in invertebrate faunas is replaced by a dependence on high speciation rates (high S).

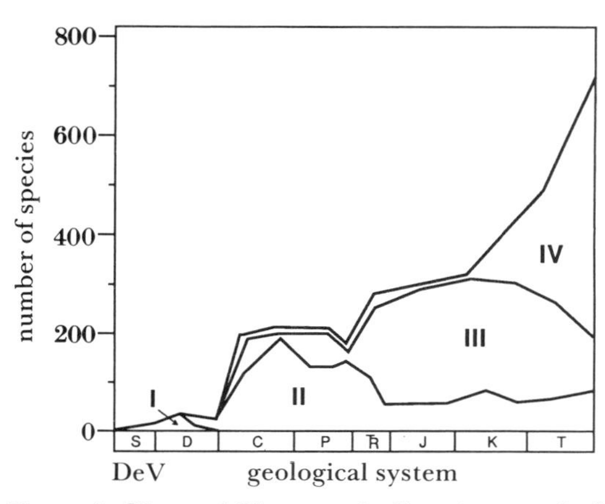

Figure 4. Observed Phanerozoic diversity record of vascular land plants at the species level (excluding spore and pollen floras). Flora I embraces early vascular plants; flora II, pteridophytes; flora III, gymnosperms; flora IV, angiosperms. After Niklas *et al.* (1983).

It would seem that, among the best-known groups of organisms, trachaeophytes are contrarians. If their increasing diversity and presumably increasing partitioning of adaptive space indicates increasing specialization, the response to any accompanying increase in extinction rates has been to favour lineages with higher S. This is a possible strategy, but why animals seem to have adopted low E and plants high S strategies is not clear. One possibility is that animal radiations have tended to produce the major clades early, and then those taxa are sorted out in favour of slow turnover groups over tens to hundreds of millions of years. In plants, by contrast, the major groups have appeared sequentially, invading new environments over hundreds of millions of years. The incoming floras may have favoured high speciation rates so as to cope with colonization of increasingly hostile environments (Valentine *et al.* in preparation). Why plants have not been able to increase their extinction resistance, and thus allow a lowering in turnover rates, is not at all clear, but at least these studies permit the question to be posed.

It is a pleasure to acknowledge the help of Bruce Tiffney with tetrapod and especially plant data, and of Andy Wyss for information on the tetrapod record. Typing was by Ellie Dzuro. This research was supported by NSF Grants EAR 84-17011 and EAR 90-15444.

REFERENCES

Bambach, R. K. 1977 Species richness in marine benthic habitats through the Phanerozoic. *Paleobiology* **3**, 152–167.
Bengston, S. & Conway Morris, S. 1984 A comparative study of Lower Cambrian *Halkieria* and Middle Cambrian *Wiwaxia*. *Lethaia* **17**, 307–329.
Benton, M. J. 1985 Mass extinction among non-marine tetrapods. *Nature, Lond.* **316**, 811–814.
Benton, M. J. 1987 Mass extinction among families of non-marine tetrapods: the data. *Mèm. Soc. geol. France* **150**, 21–32.
Chen, J.-Y., Hou, X.-G., & Lu, H.-Z. 1989 Early Cambrian netted scale-bearing worm-like sea animal. *Acta Palaeont. Sinica* **28**, 1–16.
Eldredge, N. & Salthe, S. N. 1984 Hierarchy and evolution. *Oxford Surv. Biol.* **1**, 184–208.
Flessa, K. W. 1975 Area, continental drift and mammalian diversity. *Paleobiology* **1**, 189–194.
Flessa, K. W. & Jablonski, D. 1985 Declining Phanerozoic background extinction rates: effect of taxonomic structure? *Nature, Lond.* **313**, 216–218.
Flynn, J. J. 1986 Faunal provinces and the Simpson coefficient. *Univ. Wyoming Contr. Geol.* Spec. Pap. no. 3 317–338.
Foote, M. 1988 Survivorship analysis of Cambrian and Ordovician trilobites. *Paleobiology* **14**, 258–271.
Ghiselin, M. T. 1974 A radical solution to the species problem. *Syst. Zool.* **23**, 536–544.
Jacobs, D. K. 1990 Selector genes and the Cambrian

radiation of Bilateria. *Proc. natn. Acad. Sci. U.S.A.* **87**, 4406–4410.

Jenkins, R. J. F. 1984 Interpreting the oldest fossil cnidarians. In *Recent advances in the paleobiology and geology of the cnidaria* (ed. W. A. Oliver), (*Palaeontograph. Americ.* **54**) pp. 95–104.

Niklas, K. J., Tiffney, B. H. & Knoll, A. H. 1980 Apparent changes in the diversity of fossil plants. *Evol. Biol.* **12**, 1–89.

Niklas, K. J., Tiffney, B. H. & Knoll, A. H. 1983 Patterns in vascular land plant diversification. *Nature, Lond.* **303**, 614–616.

Niklas, K. J., Tiffney, B. H. & Knoll, A. H. 1985 Patterns in vascular land plant diversification: an analysis at the species level. In *Phanerozoic diversity patterns: profiles in macroevolution* (ed. J. W. Valentine), pp. 97–128. Princeton University Press.

Norris, R. D. 1989 Cnidarian taphonomy and affinities of the Ediacaran biota. *Lethaia* **22**, 381–393.

Padian, K. & Clemens, W. A. 1985 Terrestrial vertebrate diversity: episodes and insights. In *Phanerozoic diversity patterns: profiles in macroevolution* (ed. J. W. Valentine), pp. 41–96. Princeton University Press.

Patterson, C. & Smith, A. B. 1987 Is the periodicity of extinctions a taxonomic artefact? *Nature, Lond.* **330**, 248–251.

Raup, D. M. & Sepkoski, J. J., Jr. 1982 Mass extinctions in the marine fossil record. *Science, Wash.* **215**, 1501–1503.

Raup, D. M. & Boyajian, G. E. 1988 Patterns of generic extinction in the fossil record. *Paleobiology* **14**, 109–125.

Sadler, P. M. 1981 Sediment accumulation rates and the completeness of stratigraphic sections. *J. Geol.* **89**, 569–584.

Schindel, D. E. 1980 Microstratigraphic sampling and the limits of paleontologic resolution. *Paleobiology* **6**, 408–426.

Seilacher, A. 1989 Vendozoa: organismic construction in the Proterozoic biosphere. *Lethaia* **22**, 229–239.

Sepkoski, J. J., Jr 1978 A kinetic model of Phanerozoic taxonomic diversity. I. Analysis of marine orders. *Paleobiology* **4**, 223–251.

Sepkoski, J. J., Jr 1981 A factor analytic description of the Phanerozoic marine fossil record. *Paleobiology* **7**, 36–53.

Sepkoski, J. J., Jr 1986 Phanerozoic overview of mass extinction. In *Patterns and processes in the history of life* (ed. D. M. Raup & D. Jablonski), pp. 277–295. Berlin: Springer–Verlag.

Sepkoski, J. J., Jr 1988 Alpha, beta, or gamma: where does all the diversity go? *Paleobiology* **14**, 221–234.

Simon, H. A. 1962 The architecture of complexity. *Proc. Am phil. Soc.* **106**, 467–482.

Stanley, S. M. 1979 *Macroevolution, pattern and process.* San Francisco: Freeman.

Tiffney, B. H. & Niklas, K. J. 1990 Continental area, latitudinal distribution and topographic variety: a test of correlation with terrestrial plant diversity. In *Causes of evolution: a paleontological perspective* (ed. R. Ross & W. Allmon). University of Chicago Press. (In the press.)

Valentine, J. W. 1967 The influence of climatic fluctuations on species diversity within the Tethyan provincial system. In *Aspects of tethyan biogeography* (ed. C. G. Adams & D. V. Ager), pp. 153–166. London: Systematics Association.

Valentine, J. W. 1969 Patterns of taxonomic and ecological structure of the shelf benthos during Phanerozoic time. *Palaeontology* **12**, 684–709.

Valentine, J. W. (ed.) 1985 *Phanerozoic diversity patterns: profiles in macroevolution.* Princeton University Press.

Valentine, J. W. 1989*a* Bilaterians of the Precambrian–Cambrian transition and the annelid-arthropod relationship. *Proc. natn. Acad. Sci. U.S.A.* **86**, 2272–2275.

Valentine, J. W. 1989*b* How good was the fossil record? Clues from the Californian Pleistocene. *Paleobiology* **15**, 83–94.

Valentine, J. W. 1990 The macroevolution of clade shape. In *Causes of evolution: a paleontological perspective* (ed. R. Ross & W. Allmon). University of Chicago Press. (In the press.)

Valentine, J. W. & Walker, T. D. 1987 Extinctions in a model taxonomic hierarchy. *Paleobiology* **13**, 193–207.

Valentine, J. W., Foin, T. C. & Peart, D. 1978 A provincial model of Phanerozoic marine diversity. *Paleobiology* **4**, 55–66.

Valentine, J. W., Awramik, S. M., Signor, P. W. & Sadler, P. M. 1990 The biological explosion at the Precambrian–Cambrian boundary. *Evol. Biol.* **25** (In the press.)

Van Valen, L. M. 1984 A resetting of Phanerozoic community evolution. *Nature, Lond.* **307**, 50–52.

Vrba, E. S. & Eldredge, N. 1984 Individuals, hierarchies and processes: towards a more complete evolutionary theory. *Paleobiology* **10**, 146–171.

Walker, T. D. 1984 The evolution of diversity in an adaptive mosaic. Ph.D. thesis, University of California, Santa Barbara.

Walker, T. D. & Valentine, J. W. 1984 Equilibrium models of evolutionary species diversity and the number of empty niches. *Am. Nat.* **124**, 887–889.

Wiley, E. O. 1981 *Phylogenetics: the theory and practice of phylogenetic systematics.* New York: John Wiley.

Discussion

P. J. Grubb (*Botany School, University of Cambridge, U.K.*) What is the order of magnitude of the differences in longevity of particular genetic lines of animals or plants early and late in the record? Is there any evidence on whether the shorter-lived lines of animals found early on were made up of shorter-lived individuals?

J. W. Valentine. Average extinction rates for marine invertberate species appear to fall by nearly an order of magnitude between fauna I and fauna III, though data are spotty and there is considerable variation among clades. Land plant species increase extinction rates by about the same amount (a factor of 8 or so) between flora I and flora IV.

Unfortunately, data on individual lifespans are so spotty for early animals that they can't be meaningfully compared with the spans of living forms. Incidentally, Bruce Tiffney tells me that with land plants, much of the increase in turnover in flora IV is because of angiosperms of herbaceous habit and that as herbs have shorter lives than plants with secondary growth, the average generation and species turnovers are positively correlated in this case.

Presence and absence of density dependence in a neotropical tree community

STEPHEN P. HUBBELL,[1,2] RICHARD CONDIT[1] AND ROBIN B. FOSTER[2,3]

[1] *Department of Ecology and Evolutionary Biology, Princeton University, Princeton NJ 08544 U.S.A.*
[2] *Smithsonian Tropical Research Institute, Box 2072, Balboa, Panama*
[3] *Department of Botany, Field Museum, Chicago IL 60607 U.S.A.*

SUMMARY

We report dynamic data on the spatial pattern of sapling recruitment over a three-year interval in a 50 ha† mapped plot Barro Colorado Island (BCI), Panama. We analysed sapling recruitment of a given tree species against recruitment of all other competing tree species, as a function of distance to nearest conspecific adult tree. Strong negative conspecific effects of large trees on saplings were detectable in a few very common species, but not in many others. The power to detect conspecific effects was evaluated in model tree populations in which the strength of these effects was known *a priori*. The measured conspecific effects appear strong enough in the densest species to prevent them from assuming complete dominance. However, many species do not show these effects, and we conclude that these effects play only a limited contributing role to the maintenance of tree species diversity in the BCI forest.

INTRODUCTION

The regulation of plant populations, particularly of long-lived plants such as trees, is very poorly understood. By regulation, we refer to the idea that there are clear limits to population size, and that the processes that generate these limits operate in a density-dependent manner (Silvertown 1982). In principle, density-dependent processes in plant populations can affect per head rates of birth, death and migration, or any combination of these rates. In practice density dependence is a more problematic concept to apply to sessile plants than to mobile animals.

As rooted organisms, sessile plants probably rarely if ever experience competition for resources from the population as a whole, but the spatial scales over which competition actually operates are poorly known. Most populations, even those of highly agile animals, are subdivided with respect to competition for resources, but competition is often an especially localized phenomenon in plants, which compete mainly with their immediate neighbours for limited light, water, and soil nutrients. For this reason, density is probably not the variable of choice in constructing neighbourhood models of plant competition. Rather, the number, size, and proximity of competing plants may be more appropriate variables (Pacala & Silander 1985; Silander & Pacala 1985). Most of the conceptual advances in the field of density dependence in plant populations have come through studies of monocultures, leading to the discovery of the limiting relation between stem density and mean weight per plant in self-thinning populations (White 1981). We know considerably less about the prevalence and mechanics of density dependence in species-rich plant communities, such as tropical forests.

Nevertheless, it would be incorrect to argue that plants never compete at a distance. Non-neighbouring plants have been shown, for example, to compete for the attention of pollinators and seed dispersal agents in several species (see, for example, Howe & Vande Kerchove 1979). The question is whether such effects are ever strong enough to limit plant populations. We still lack a sufficient number of empirical studies to generalize about the spatial scales over which neighbourhood competitive effects operate on different plant life history stages and processes. Whatever the appropriate spatial scales turn out to be, it is clear that we must adopt an explicitly spatial approach to understanding density dependence in mixed-species plant communities, with mapped populations, at least until we better understand the connection between local neighbourhood effects and the regulation of plant populations as a whole.

This paper addresses the question of the prevalence and importance of density dependence in limiting the abundance of tree populations in a species-rich neotropical forest in Panama. The evidence on density dependence is phenomenological and non-experimental, and therefore our case is circumstantial. Nevertheless, most hypotheses for density dependence in plant populations have explicit spatial consequences, so the signature of density-dependence should be readily apparent in the unfolding fate of individual plants in precisely mapped plant communities.

† 1 hectare = 10^4 min^2.

THE BARRO COLORADO ISLAND FOREST: A CASE STUDY

Over the past decade we have been studying the population and community dynamics of a mapped, 50 ha plot of old growth forest on Barro Colorado Island (BCI), Panama (Hubbell & Foster 1983, 1986*a*, *b*). The organizing objective of this research has been to understand the maintenance of tree diversity in the BCI forest and the factors that determine patterns of relative tree species abundance. As part of this objective, we have assessed the phenomenological evidence for density dependence in tree populations in the plot, first from the static data on tree population density and dispersion obtained from the primary census and map completed in 1982 (Hubbell & Foster 1986*a*), and then from the short-term dynamic data on survival, growth, and recruitment obtained from the 1985 recensus (Condit *et al.* 1990). The plot is now being completely censused for the third time (1990), but the results of this recensus will not be known until next year.

BCI is an island of about 15 km² in artificial Gatun Lake in the zone of the Panama Canal. When the canal was finished, BCI was declared a nature reserve and it is now managed as a biological research station by the Smithsonian Tropical Research Institute. In 1980 we laid out the 50 ha plot in old-growth forest on the summit of the island (Hubbell & Foster 1983). The plot is on relatively uniform terrain, with gentle slopes on the eastern and southern margins, and with a seasonal swamp in the centre of about 2 ha in extent (figure 1). With the exception of a small piece of secondary forest about 75 years old at the north edge of the plot, the remaining *ca.* 49 ha is covered with old-growth forest that dates from precolombian times. Thorough palaeoecological surveys have revealed that the forest was never cleared for slash-and-burn agriculture, but occasional small clearings, probably seasonal hunting camps, were made in the forest prior to 550 years BP (Piperno, personal communication). The BCI climate is seasonal, with a pronounced dry season from late December until April (Leigh *et al.* 1982).

Because the BCI flora is so well known (Croat 1978), BCI was an ideal site for a large-scale study of tropical forest dynamics. The 50 ha plot contains approximately 242000 free-standing woody plants with stem diameters of 1 cm diameter at breast height (dbh) or larger, each of which has been tagged, mapped and identified to one of 306 species in the plot (Foster & Hubbell 1990). Species abundances range over 4.6 orders of magnitude, from 21 extremely rare species represented by a single individual apiece, to a very common shrub, *Hybanthus prunifolius* (Violaceae), represented by over 40000 individuals in 1985 (figure 1) (Hubbell & Foster 1986*b*).

In this paper, the phenomenological evidence for density dependence in BCI tree species from the first and second censuses is reviewed. We exploit natural variation in local tree density to test hypotheses of density dependence. Tree densities both within and among species vary widely, as revealed when the 50 ha plot is tessellated into nearest neighbour polygons. For example, figure 2 is the Dirichlet tessellation for all canopy trees over 30 cm dbh, and is shown to convey a sense of the great range in local stand densities found within the plot. We then analyse density dependence in greater detail in two very common species, one exhibiting strong density dependence (*Trichilia tuberculata*, Meliaceae), and the other no detectable density dependence (*Quararibea asterolepis*, Bombacaceae). We conclude by modelling the dynamics of *Trichilia tuberculata*, and argue that the density dependence observed in this species is sufficient to limit the species at or near its present abundance. However, we also conclude that most of the remaining species in the BCI forest are probably nowhere near the densities at which

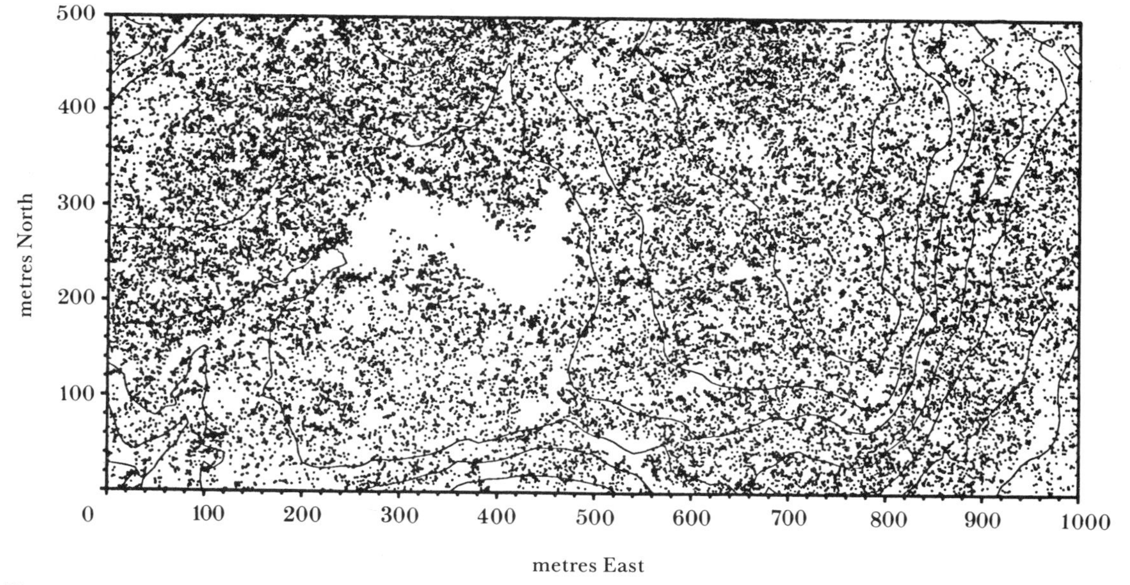

Figure 1. Map of the commonnest species in the 50 ha plot on Barro Colorado Island, the shrub, *Hybanthus prunifolius* (Violaceae), with 41106 individuals in the 1985 census. This species exhibited no detectable negative density dependence in recruitment.

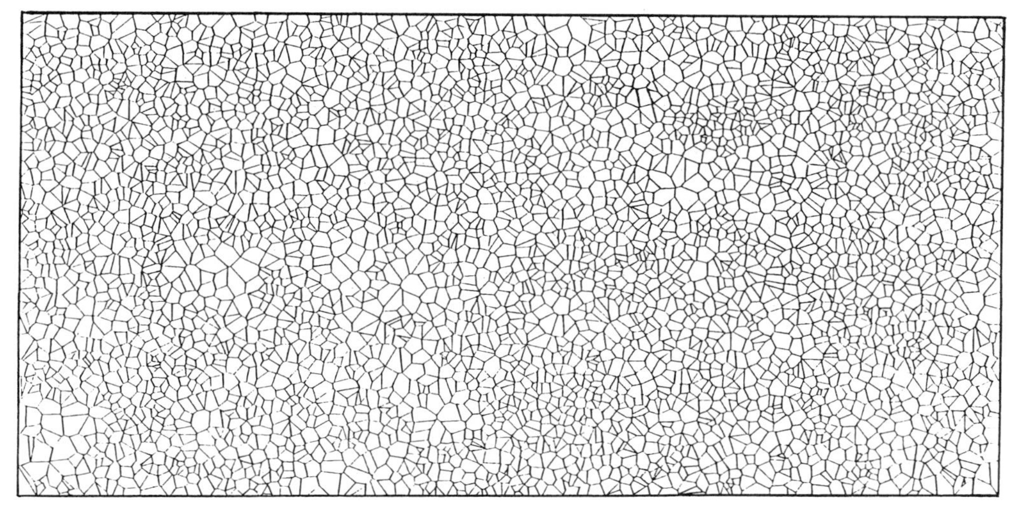

Figure 2. Dirichlet tessellation of the 50 ha plot for canopy trees greater than 30 cm dbh, showing the great variation in local tree densities in the plot. Each polygon defines the boundaries of the nearest neighbour region surrounding each canopy tree.

density dependence would begin to strongly influence their population dynamics.

THE SEARCH FOR DENSITY DEPENDENCE IN BCI TREE SPECIES

Are there density-dependent limits to the abundance of tree species in the BCI forest? The short answer is of course yes. An ultimate limit is set by the self-thinning law for each species growing in monoculture. However, the more interesting and difficult question is: are there non-trivial limits to tree abundance short of reaching a state of single-species dominance? If so, at what population densities of a given species in a mixed-species stand are these limits approached? What are the life history stages and processes at which density-dependence regulates the population? We cannot at present answer all these questions satisfactorily, but analyses of both the static, first-census data and the recensus data on short-term dynamics led us to the conclusion that only two or three of the most abundant tree species in the BCI forest are prevented by density dependence from increasing above present abundance levels in the plot (Hubbell & Foster 1986*a*, 1990*a*).

(*a*) *Evidence from Population Dispersion*

With the primary census data, we began looking for the spatial signature of density dependence at several spatial scales within the limits set by the 50 ha plot. The first study was a quadrat-based analysis of the relative dispersion of saplings and adults. We analysed how the adult (per head) production of saplings varied with local adult tree density, in quadrats ranging in size from 4×10^2 m^2 to 10^5 m^2 (10 ha) (Hubbell & Foster 1986*a*). We found a strong local depression in the per head number of juveniles in the densest patches of adults in the two most common mature-phase, canopy tree species, *Trichilia tuberculata* (Meliaceae) and *Alseis blackiana* (Rubiaceae), particularly in the range of quadrat sizes of 1 ha and larger. Of the remaining 46 canopy species examined, roughly half also showed significant but weak negative second-order terms in a quadratic regression of the number of small saplings per hectare on number of adults per hectare (Hubbell & Foster 1986*a*). However, these second-order effects, although statistically significant, were judged unlikely to be important to population regulation because they are grossly too small to limit populations to observed abundances, sometimes by several orders of magnitude. The density effects detected in the static data analysis were not strong enough to prevent any species, with the possible exception of *Trichilia* and *Alseis*, from assuming complete monospecific dominance or increasing to impossibly supersaturated tree densities.

We re-analysed the primary census data from a focal tree perspective, asking which species were likely to replace the current adult trees in the canopy, by using a tree replacement analysis modified from Horn (1975). Based on the distribution of saplings growing beneath the adults of each canopy tree species, we estimated the probabilities of self-replacement or replacement by other species in the next generation (Hubbell & Foster 1986*c*). The major conclusions from these analyses were that: (i) all species were capable of in-situ self-replacement; (ii) nevertheless, the probability of self-replacement was very low, as was the probability of replacement by any other specified species, and (iii) this low replacement rate was not because of any inability to self-replace but simply to the low frequency of any given species and the high species richness in the forest. We were unable to reject the null hypothesis that the collective set of probabilities of self-replacement were a random draw of the probabilities of any arbitrary pairwise species replacement. However, when the 10 commonest species were analysed individually, strong sapling avoidance of adults was found in two species, *Trichilia* and *Alseis*, and is visibly evident in *Trichilia* (figure 3).

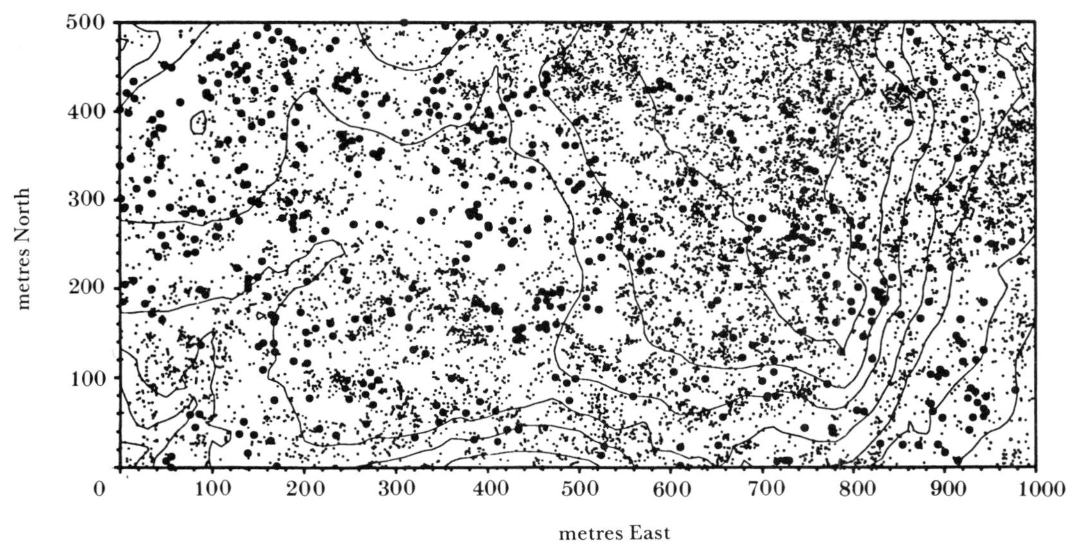

Figure 3. Map of the commonnest canopy tree in the 50 ha plot, *Trichilia tuberculata* (Meliaceae). Large dots are adult trees greater than 30 cm dbh; small dots are all smaller *Trichilia* plants.

(*b*) *Evidence from growth and survival patterns*

When the plot was recensused in 1985, we had our first opportunity to analyse short-term rates of survival, growth, and recruitment of new saplings. In analysing the dynamic data, we again took a focal tree approach and analysed sapling performance as a function of adult tree proximity. First, we compared the growth and survival rates of 1–4 cm dbh saplings located beneath large trees of either the same or of different species, for canopy tree species (Hubbell & Foster 1990*a*). When all species were pooled and tested together, there was a 5.3 % reduction in survival rate ($p < 0.0001$), and 23.5 % reduction in annual growth rate ($p < 0.0003$), in saplings growing beneath conspecific vs. beneath heterospecific large trees.

When separate analyses of individual species were conducted, however, the results were less clear cut. It now appears that the pooled species results were driven in part by the behaviour of the two commonnest species, which show strong conspecific effects. Table 1 compares the growth and survival performance of saplings beneath conspecific and heterospecific large trees for the 11 most common canopy species. Sample size were adequate to compare growth rate differences in three juvenile size classes, 1–2, 2–4 and 4–8 cm dbh, but sample sizes limited survival comparisons to one size class, 1–4 cm dbh.

Several conclusions can be drawn from these results: first, all significant conspecific effects on sapling growth and survival were negative in sign. Secondly, the only two species to show significant negative conspecific effects on both survival and growth were *Trichilia* and *Alseis*, the two most abundant canopy tree species, and the same two that showed local density dependence in the static first-census data. Thirdly, most common species did not exhibit detectable conspecific effects, and most were a long way from significance ($p > 0.5$), not for lack of sample sizes (all species tested had more than 500 total saplings). Even species as abundant as the fifth most common canopy tree, *Quararibea asterolepis* (Bombacaceae), showed no effect of neighbouring conspecific adult on sapling growth or survival (table 1). Fourth, being beneath a conspecific adult was more likely to affect sapling growth rate than survival rate. Finally, significant negative conspecific effects on growth occurred more often in the smaller size classes, which was not unexpected. More surprising was the suggestion that these negative effects may persist into quite large juvenile size classes (2–4 cm dbh), at least in some species. Overall, negative conspecific effects on growth were much more common than positive effects ($p < 0.001$), although there was no difference in the case of survival (table 1).

These conspecific effects, when detected, appeared to be real and not due to spurious correlations. We tested all reasonable alternative hypotheses. For example, if conspecific adult trees happened to be closer on average to the test saplings than the heterospecific adult trees, or happened to be larger. In these cases, apparent conspecific effects could be due to competition from a closer or larger tree instead. We tested four alternative hypotheses for these effects: (i) conspecific adult trees are nearer; (ii) conspecific adult trees are larger; (iii) conspecific sites are shadier (higher canopy), and (iv) conspecific sites have higher overall competition because of higher stand densities. Although there were some significant differences in these variables, without exception the significant differences were not in the right direction to explain significant conspecific effects (table 2). Indeed, when negative conspecific effects were detected, they were found in spite of incidental effects in the opposite direction. For example, negative growth and survival effects were found in *Alseis* and *Trichilia* in spite of the fact that conspecific adult trees were on average farther from the test saplings and smaller in diameter, than were heterospecific large trees (table 2). Partial regression analyses confirmed the results of these univariate tests.

Table 1. *Summary of juvenile growth and survival beneath conspecific against heterospecific large trees, over the interval 1982–1985 in the BCI 50 ha plot*

(Growth rate differences are reported for three sapling size classes, 1–2 cm dbh, 2–4 cm dbh, and 4–8 cm dbh. Survival rate differences are reported for a single size class, 1–4 cm dbh. The entry of a 'less-than' symbol means that juvenile performance was worse in saplings beneath trees of the same species than beneath trees of other species. A 'greater-than' symbol means that performance was better beneath trees of the same species. Unbracketed symbols indicate significant differences in performance at $p < 0.05$; symbols in brackets show non-significant differences. Note that all individually significant growth and survival rate differences represent poorer performance beneath conspecifics. Also, for growth rates, there are significantly more 'less-than' symbols (24) then 'greater-than' symbols (9) ($p < 0.01$, χ^2). However, there are roughly equal numbers of species that show better survival beneath conspecifics (5 species) than beneath heterospecifics (6 species).

		performance of different diameter classes (cm dbh)			
		growth rate			survival rate
tree species	family	1–2	2–4	4–8	1–4
Alseis blackiani	Rubiaceae	<	(<)	(<)	<
Beilschmiedia pendula	Lauraceae	<	<	(>)	(<)
Drypetes standleyi	Euphorbiaceae	(<)	(<)	(<)	(>)
Guaterria dumetorum	Annonaceae	(>)	(>)	(>)	(>)
Ocotea skutchii	Lauraceae	<	(<)	(<)	(<)
Poulsenia armata	Moraceae	(<)	<	(>)	(<)
Prioria copaifera	Leguminosae	(<)	(<)	(>)	(>)
Quararibea asterolepis	Bombacaceae	(>)	(<)	(>)	(>)
Tabernaemontana arborea	Apocynaceae	<	(>)	(<)	(<)
Tetragastris panamensis	Burseraceae	(<)	(<)	(<)	(>)
Trichilia tuberculata	Meliaceae	<	<	(<)	<

Table 2. *Tests of four alternative hypotheses to explain apparent conspecific effects, including (i) distance to neighbouring adult tree; (ii) size (diameter) of neighbouring adult tree; (iii) canopy height over conspecific against heterospecific site (shade), and (iv) stand density at conspecific against heterospecific site.*

(In (i), a 'less-than' symbol means that the conspecific tree is on average closer than the heterospecific tree to the test sapling. In (ii), a 'less-than' symbol means that the conspecific tree is on average smaller in diameter than the heterospecific tree. In (iii), a 'less-than' symbol means that the conspecific site on average has lower canopy (younger forest) than does the heterospecific site. In (iv), a 'less-than' symbol means that the conspecific site on average has lower stand density than does the heterospecific site. Only differences significant at $p < 0.05$ are shown; dashes show non-significance. The column headings 1, 2, and 4 refer to the juvenile diameter classes, 1–2, 2–4 and 4–8 cm dbh. See table 1 for the full species names.)

	four alternative hypotheses											
	adult distance (cm dbh)			adult diameter (cm dbh)			Canopy height (cm dbh)			stand density (cm dbh)		
species	1	2	4	1	2	4	1	2	4	1	2	4
Alseis	>	>	>	<	<	<	<	<	–	–	>	–
Beilschmiedia	–	–	–	>	–	–	<	–	–	–	–	–
Drypetes	–	–	–	<	–	–	–	–	–	–	–	–
Guatteria	–	–	–	–	–	–	–	–	–	–	–	–
Ocotea	–	–	–	>	>	–	–	–	–	–	–	–
Poulsenia	<	–	–	–	–	–	–	–	–	–	–	–
Prioria	–	–	–	–	–	–	–	–	–	–	–	–
Quararibea	–	–	–	>	>	>	–	–	–	–	–	–
Tabernaemontana	–	–	–	<	–	–	<	–	–	–	–	–
Tetragastris	–	–	–	–	–	–	–	–	–	–	–	–
Trichilia	>	>	>	<	<	<	<	–	–	–	–	–

(c) *Evidence from recruitment patterns*

We examined sapling recruitment patterns for density dependence, again taking a focal adult approach. We tested whether sapling recruitment was inhibited near conspecific adult trees (Condit *et al.* 1990). We defined a recruit as any plant < 4 cm dbh in 1985 that was not present in the first census, and therefore was < 1 cm dbh in 1982. To set this analysis in a community context, we compared new recruitment in a given species relative to the collective recruitment of all its competitors. We calculated the ratio of recruits of focal species i to all recruits, including the same and competing species, at each location as a function of distance from the nearest adult tree of the focal species. This relative recruitment analysis represented an improvement over most previous anslyses, including ours, which had treated each species in isolation from

Table 3. *Recruitment pattern of new saplings relative to nearest conspecific adult in 15 shrub species, 19 understory treelet species, 25 midstory tree species, and 22 canopy tree species (modified from Condit et al. 1990)*

('Attracted pattern' means that recruits were more abundant than expected near conspecific adults. 'Repelled pattern' means that recruits were less abundant than expected near conspecific adults. 'Near' is defined as one adult canopy radius. The proportion of species repelled within 1–5 crown radii is also reported).

recruitment pattern	shrubs	understory	trees midstory	canopy	total
attracted pattern	13	8	3	3	27
repelled pattern	0	3	4	6	13
partially repelled pattern	1	2	4	7	14
no pattern (indifferent)	1	6	14	6	27
total number of species	15	19	25	22	81
proportion of species repelled or partially repelled	0.07	0.26	0.32	0.59	0.33
proportion of species repelled within 1–5 canopy radii	0.07	0.16	0.28	0.36	0.23

its competitors. We also, for the first time, analysed recruitment in species of all growth forms, from shrubs to canopy trees. We recognized four recruitment patterns: repelled, partially repelled, attracted and indifferent (Condit *et al.* 1990). Species exhibiting repelled patterns showed significant deficits in the number of recruits close to conspecific adults, and consistent repulsion to some distance from the adult tree. Partially repelled species showed inconsistent repulsion, with significant deficits of recruits at some nearby distance intervals, but not others. Attracted species showed significant over-representation of recruits close to adults. Indifferent species exhibited patterns that were indistinguishable from the random patterns of recruitment exhibited by all species collectively.

We also compared relative recruitment rates within one crown radius of the adult trees, and then within 1–5 crown radii, in the immediate neighbourhood. The results showed that several canopy tree species inhibited their own recruitment rate relative to that of other species immediately beneath their own canopy (table 3). However, this negative effect was much less prevalent among shrubs, understory treelets, and midstory trees. There was no effect even in the most common species in the plot, the shrub, *Hybanthus prunifolius* (figure 1), which accounts for almost one out of every six plants > 1 cm dbh in the plot. When the effect was present, it generally disappeared at the edge of the crown in most species. Only a few tree species inhibited recruitment beyond their crown in the near neighbourhood from 1–5 crown radii away, including *Trichilia*. Most species in fact showed the opposite, with significant excesses of recruitment in the immediate neighbourhood adjacent to the focal tree (Condit *et al.* 1990). We concluded that these negative conspecific effects, when they occurred, were too local to prevent the given species from becoming very common. Indeed, even strong repulsion effects in *Trichilia*, extending out several canopy radii, did not prevent this species from becoming the most abundant species in the BCI forest.

In theoretical models, confining conspecific inhibition of recruitment just to prevention of self-replacement results in an equilibrium tree community containing only three tree species (Hubbell 1980).

ANALYSING DENSITY DEPENDENCE IN *TRICHILIA* AND *QUARARIBEA*

The preceding analyses of static pattern and short-term dynamics all led to a similar conclusion, namely that detectable negative conspecific effects in BCI tree species are limited mainly to a few canopy trees, particularly those that are extremely abundant, and are less often found in shrubs and subcanopy tree species, regardless of abundance. However, they fall short of proof that these conspecific effects generate sufficiently strong density dependence to limit populations even of the most common canopy trees, let alone populations of the less common species. In the absence of experimental tests, the best remaining avenue for analysis is to model the observed local conspecific effects and evaluate their consequences for population-level regulation of total abundance.

We chose to examine two species in detail, *Trichilia tuberculata*, the most abundant canopy tree species in the BCI plot with 13164 individuals > 1 cm dbh in 1985, and *Quararibea asterolepis*, the fifth most abundant canopy tree, with 2382 individuals greater than 1 cm dbh. The full details of the analyses and results will be presented in a separate paper (Condit *et al.* in preparation), but the main results are outlined here.

(*a*) *Density effects on recruitment*

To estimate density-dependence in recruitment rates, for each newly recruited sapling in 1985, we calculated the number of large conspecific trees (> 16 cm dbh) that were present in 1982 within 5, 10, 30, 50 and 75 m of the given sapling. We used the 1982 data for large trees because virtually all of the 1985 recruits entering the census at a diameter of 1 cm were alive in

1982. Because the number of recruits was large, it was possible to isolate and directly measure the effects of large conspecifics at different distances. We also considered whether the presence of large subadult trees also influenced recruitment rate.

(*b*) *Density effects on survival and growth*

We estimated density-dependence in survival and growth in plants in the following six size classes: 1–2, 2–4, 4–8, 8–16, 16–32 and 32 cm dbh and above. The effect of conspecific density of three size classes on the focal plant were considered; these size classes were 1–4, 4–16 and 16 cm dbh and above. Each focal plant was categorized by the number of conspecifics present within various distance intervals. In *Trichilia*, the distances began at 5 m, then increased to larger values (10, 15 and 50 m), while holding density within 5, 10 or 15 m at zero. In *Quararibea*, the same analyses were done, but the distance intervals had to be adjusted because of smaller sample sizes, to 7, 14, 21 and 50 m, respectively. With these adjustments, we were able to directly isolate the effect of conspecific density on one of the three size classes while holding density of the other two constant. For example, we allowed density of 1–4 cm dbh neighbours within 5 m to vary while holding the density of plants greater than 4 cm dbh within 5 m at zero. Statistical significance in growth or mortality was assessed both by chi-square comparisons and by multiple regression, to minimize errors due to correlated effects. There were 12 independent variables: the number of neighbours in the three size classes (1–4, 4–16, and 16+ cm dbh) and four distance intervals. Results of the multiple regression analysis bore out all conclusions from the chi-square tests, so we report only the χ^2 tests here.

(*c*) *Density dependent life tables*

The BCI plot was divided into 50, 100 × 100 m (1 ha) subplots. These hectares were grouped into six categories of adult density, based on the density of *Trichilia* trees greater than 30 cm dbh present in 1982, and five categories for *Quararibea*. For each of these density regions, a life table was constructed, based on approximately doubling size classes. Transition probabilities for each pair of size classes, and mortality data for each, were estimated from the 1982–85 census data. Transitions to smaller size classes were disallowed in this analysis, although this is an oversimplification since a small fraction of trees that lose their trunk manage to resprout and survive. All recruitment was attributed to trees > 30 cm dbh. An estimate of recruitment rate was determined for each of the six density regions by dividing the number of 1985 recruits by the number of 1982 trees greater than 30 cm dbh. The population rate of increase, λ, was calculated for each life table following standard procedures. To test whether differences in population growth rate were significant, we used a simple bootstrap procedure. We calculated λ values 100 times for life tables created from random subsets of hectares, and set the 5% α level by the fifth largest λ value.

(*d*) *Population simulations*

A life table model was created that took into account density dependent effects on recruitment and size-specific survival and growth. The model kept track of individual plants in an explicitly spatial fashion within a 16 ha (400 × 400 m) model plot. Six doubling size classes were used: 1–2, 2–4, 4–8, 8–16, 16–32, and 32 cm dbh and above. Initial conditions assumed random dispersion of 25 plants divided randomly among the six size classes. We then calculated survival and growth probabilities for each individual as a function of the number of its conspecific neighbours. Density dependence in survival and growth was modelled by first defining baseline (maximal) growth and survival rates for individuals unaffected by conspecific neighbours. Growth and survival were then adjusted downward as the number of neighbours increased, assuming a linear effect of number of neighbours in each distance interval. The slope of the effect was estimated by regressing growth or mortality rate on number of neighbours.

Density dependence in recruitment was more problematic to model. We defined a baseline recruitment rate by calculating the density of recruits in regions of the plot with the lowest density of large conspecific neighbours. This assumes that regions that do not currently have many adult trees are suitable regeneration sites for saplings. This seems reasonable at least in *Trichilia* because these regions showed the highest recruit density recorded, which we call r_{max}. Let a be the number of 32+cm dbh trees in the plot. Then $50 \cdot r_{max}/a$ estimates the recruitment rate per head in the absence of density-dependent effects, such as when the population density is very low. Next we calculated the proportion of the 16 ha that fell within given distances and densities of large trees. The number of recruits in this area was reduced from the maximum recruit density by the appropriate density-dependent factor based on the regression analyses.

Each cycle of the model corresponded to a time step of three years as all parameter values were based on the three year intercensus period, 1982–85. Simulations were continued for up to 4000 cycles, and total population and numbers in each size class were recorded every 100 cycles.

DENSITY DEPENDENCE IN *TRICHILIA*, AND ITS ABSENCE IN *QUARARIBEA*

(*a*) *Density dependence in recruitment*

The density of recruits in *Trichilia* was much lower in regions close to large conspecifics (figure 4*a*). In areas within 5 m of a large *Trichilia* tree, there were fewer than 15 recruits per hectare, whereas the mean for the whole plot was 32.6 ha^{-1}. We were surprised to discover that the density of large conspecifics from 10–30 m away, 30–50 m away, and 50–75 m away all had significant, independent negative effects on recruit density in *Trichilia*. We did not test beyond 75 m because of small sample sizes. Figure 4*b* summarizes the effects from 10–75 m. In regions of the plot with fewer than 25 large plants within 75 m (and none

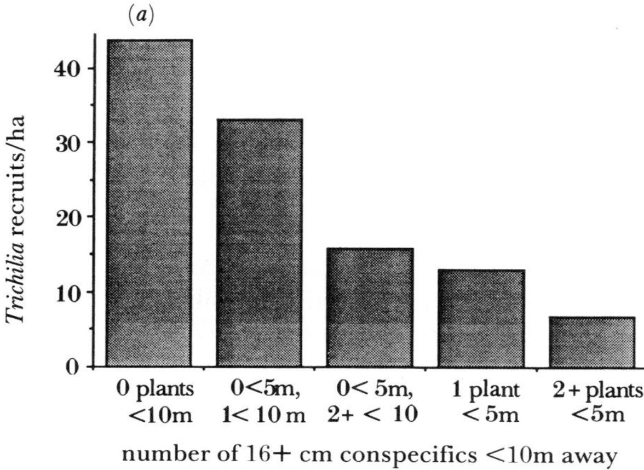

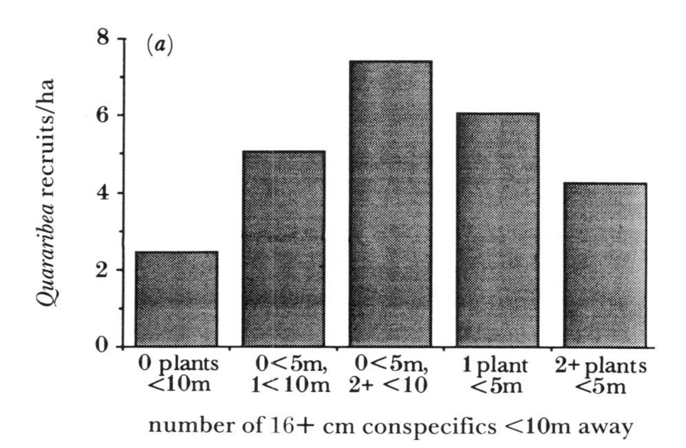

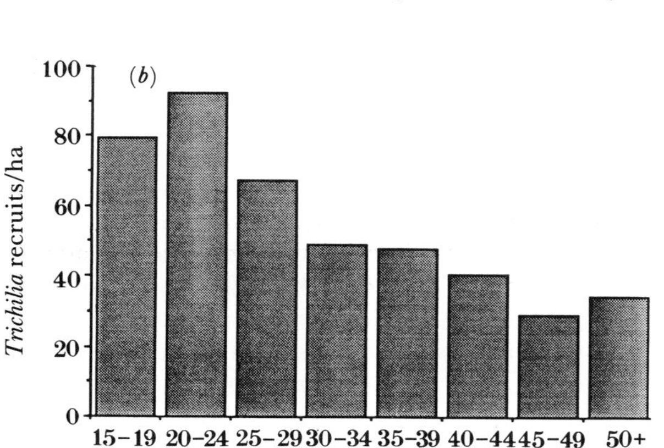

Figure 4. Effect of conspecific neighbours on recruitment rate per hectare in *Trichilia tuberculata*. (*a*) Effect of the number of large (16+cm dbh) conspecifics less than 10 m away. (*b*) Effect of the number of large (16+cm dbh) conspecifics at distances between 10 and 75 m away. There were none within 10 m for these calculations.

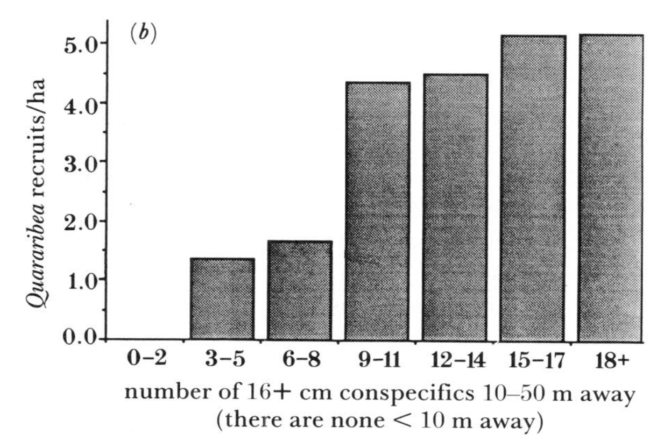

Figure 5. Effect of conspecific neighbours on recruitment rate per hectare in *Quararibea asterolepis*. (*a*) Effect of the number of large (16+cm dbh) conspecifics less than 10 m away. (*b*) Effect of the number of large (16+cm dbh) conspecifics at distances between 10 and 75 m away. There were none within 10 m for these calculations.

within 10 m), recruit density exceeded 80 ha^{-1}, more than tenfold greater than in regions very close to large neighbours (figure 4*a* the lowest point). The effect of local density of neighbouring 4–16 cm dbh plants on recruit density in *Trichilia* was just the opposite: in regions with more mid-sized neighbours, there were also more recruits. For example, in regions with no mid-sized neighbour within 5 m, recruit density was 32.3 ha^{-1}, whereas in regions with more than three mid-sized neighbours, the figure was 48.6. This difference was statistically significant. In *Quararibea*, there was no suggestion that large conspecifics were negatively associated with recruitment. Both large and mid-sized conspecifics showed positive associations with recruits in this species (figure 5*a*, *b*).

(*b*) *Density dependence in survival and growth*

There were extensive local density effects on survival in *Trichilia*. The most pronounced effects were negative, large trees within 5 m greatly reduced survival. Within 5 m, this effect persisted until young trees were 8 cm dbh (figure 6*a*, table 4). In addition, the effect of large neighbours on the smallest size class persisted as far as 15 m (table 4). Moreover, local density of mid-sized trees was also associated with reduced survival of the two smallest size classes (figure 6*b*). In contrast to the effects of large and mid-sized conspecifics, local density of small plants tended to be positively associated with survival in *Trichilia* (table 5). Growth effects were rather inconsistent, however, with some positive and some negative effects; and none were as great in magnitude as the strongest survival effects (table 5).

In *Quararibea*, we concluded that there is no credible evidence of systematic effects of local density on survival (figure 7, table 6). There were six tests showing a significant effect, but these were scattered, and as many were positive as were negative (table 6). The effects of local density on growth in *Quararibea* were similar, with only four tests showing significance, one positive and three negative.

In both *Trichilia* and *Quararibea*, density of large conspecifics between 10 and 50 m had no effect on survival or growth of any size class. Because there was no effect of the largest size class, we did not repeat the test with mid-sized trees as the independent variable.

(*c*) *Density dependence in life tables*

Figure 8 shows the hectares with highest and lowest density of adult trees in *Trichilia*. In the 16 lowest

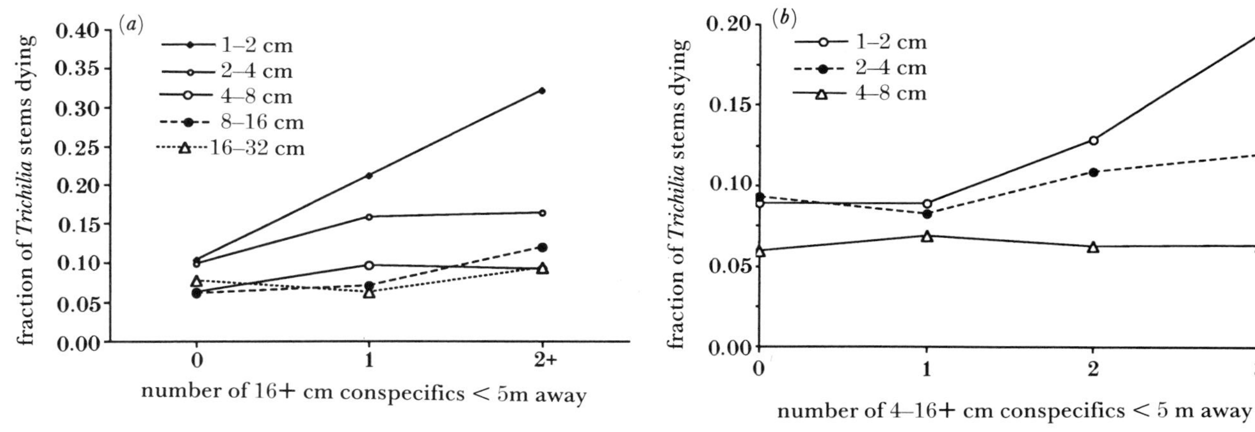

Figure 6. Effect of conspecific neighbours on mortality rate of different size classes of *Trichilia tuberculata*. (*a*) Fraction of plants dying between 1982 and 1985 as a function of the number of large (16+cm dbh) conspecifics less than 5 m away. (*b*) Fraction of plants dying between 1982 and 1985 as a function of the number of medium-sized (4–16 cm dbh) conspecifics less than 5 m away.

Table 4. *Conspecific density effects on survival in Trichilia*

(The right-most columns are slopes of regressions of survival on density in a particular distance range. In doing the regressions, terminal size classes were combined so that sample size in any class was never below 20. A positive sign indicates survival increased with distance, a negative sign indicates a decline. Blanks indicate no significant effect, based on a χ^2 test. Boldface entries were used in the simulation model; others were not.)

		distance/m			
effect of 1–4 cm dbh size class					
baseline survival	size class affected	0–5	5–10	10–15	15–20
0.903	1–2 cm	—	+0.022	—	—
0.911	2–4 cm	+0.014	—	—	—
0.949	4–8 cm	—	—	—	—
0.950	8–16 cm	—	—	—	—
0.938	16–32 cm	—	—	—	—
0.843	32+cm	+0.087	+0.013	—	—
effect of 4–16 cm dbh size class					
	size class affected	0–5	5–10	10–15	15–20
	1–2 cm	**−0.036**	—	−0.024	—
	2–4 cm	−0.011	—	—	—
	4–8 cm	—	—	—	—
	8–16 cm	—	—	—	—
	16–32 cm	—	—	—	—
	32+cm	—	—	—	—
effect of 16+cm dbh size class					
	size class affected	0–5	5–10	10–15	15–20
	1–2 cm	**−0.109**	**−0.022**	**−0.020**	—
	2–4 cm	**−0.033**	**−0.013**	—	—
	4–8 cm	**−0.015**	**−0.006**	—	—
	8–16 cm	—	—	—	—
	16–32 cm	—	−0.029	—	—
	32+cm	—	—	—	—

ranking hectares, the population growth rate calculated from the corresponding life table was $\lambda = 1.026$, whereas in the four highest ranking hectares, $\lambda = 0.917$. Our bootstrap analyses indicated that a difference between two λ values of 0.056 would be statistically significant at the 5% level, so the difference between high and low density regions is very unlikely to be because of chance. There was a nearly monotonic decrease in λ from low to high density regions (figure 8). Population growth rate was uncorrelated with density in *Quararibea*. Although the lowest λ was found at the lowest density, the greatest difference between λs (0.04) was less than the critical value for statistical significance ($p < 0.05$).

(*d*) ***Simulation results***

Because of the fact that *Quararibea* did not exhibit density dependence, we simulated only the density-dependent growth of *Trichilia*. The parameters used to model density dependence in *Trichilia* survival and growth are given in tables 5 and 6. In the model we

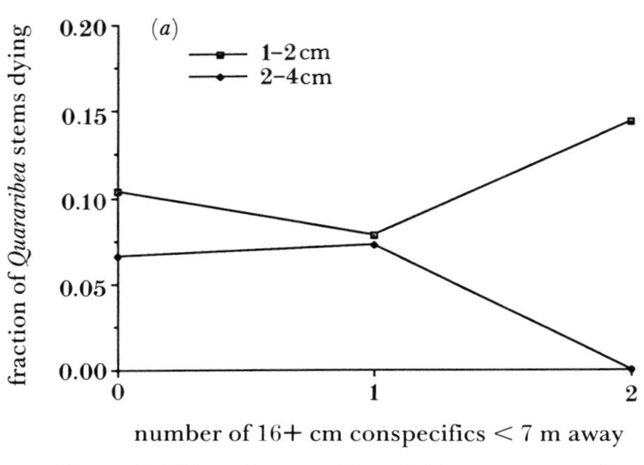

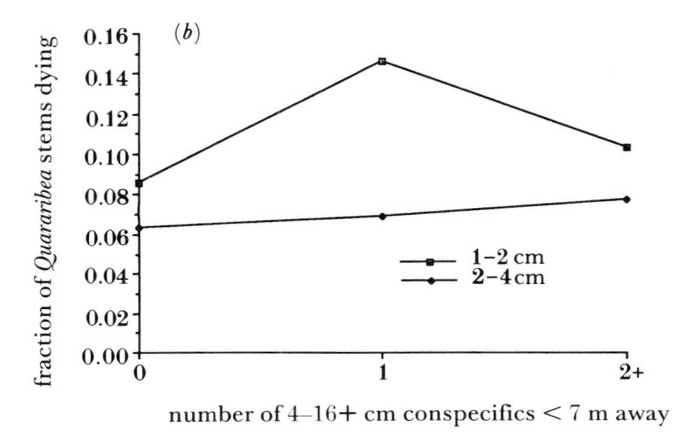

Figure 7. Effect of conspecific neighbours on mortality rate of different size classes of *Quararibea asterolepis*. (*a*) Fraction of plants dying between 1982 and 1985 as a function of the number of large (16+cm dbh) conspecifics less than 7 m away. (*b*) Fraction of plants dying between 1982 and 1985 as a function of the number of medium-sized (4–16 cm dbh) conspecifics less than 7 m away.

Table 5. *Conspecific density effects on growth in Trichilia*

(See explanation for table 4. Growth rate of the largest size class is zero by definition)

		distance/m			
effect of 1–4 cm dbh size class					
baseline growth	size class affected	0–5	5–10	10–15	15–20
0.123	1–2 cm	—	—	—	—
0.104	2–4 cm	—	—	—	—
0.059	4–8 cm	−0.017	—	—	—
0.040	8–16 cm	—	—	—	—
0.062	16–32 cm	−0.013	—	—	—
0.00	32+cm	—	—	—	—
effect of 4–16 cm dbh size class					
	size class affected	0–5	5–10	10–15	15–20
	1–2 cm	—	+0.011	—	−0.024
	2–4 cm	—	—	−0.032	—
	4–8 cm	−0.011	—	—	—
	8–16 cm	−0.019	—	—	—
	16–32 cm	—	—	—	—
	32+cm	—	—	—	—
effect of 16+cm dbh size class					
	size class affected	0–5	5–10	10–15	15–20
	1–2 cm	**−0.026**	—	—	—
	2–4 cm	−0.030	—	—	—
	4–8 cm	—	+0.022	—	—
	8–16 cm	+0.041	—	—	—
	16–32 cm	—	—	—	—
	32+cm	—	—	—	—

used the clearest density effects (boldface in tables 5 and 6), those which were consistent with a simple, *a priori* hypothesis, namely that the largest size class should have its largest effects on the smallest size classes and at the shortest distances. The baseline survival and growth figures, those in the absence of local neighbours, are also given in tables 5 and 6.

The maximum recruit density in *Trichilia* was about 80 recruits per hectare, found in regions with the lowest density of large *Trichilia* trees (figure 4*b*). If this density of recruits were found throughout 50 ha, there would be 4000 recruits, or 8.5 recruits per adult. Depression factors were: 0.85 for regions which had one or more large conspecific within 5 m, 0.66 if one was 5–10 m away (but none within 5 m), and 0.52 for regions that had more than 30 large conspecifics within 75 m (but none within 10 m). Thus, recruitment was almost completely eliminated within 5 m of large trees (85% eliminated), and reached its maximum density only in regions with less than 30 large trees within 75 m. These figures come directly from the data in figure 4, with some regions combined.

The model *Trichilia* population grew from 25 trees at the beginning to a maximum density of about 4200 trees in 16 ha in 500 cycles (1500 years). It remained at about this level for the remaining 1000 cycles of the model (figure 9*a*). Mortality rate of the smallest size class increased from 0.097 during the first 100 cycles to

Table 6. *Conspecific density effects on survival in Quararibea*

(See explanation for table 4. In this species, the baseline survival rate was simply the mean rate for the entire plot, since there was very little density dependence apparent.)

		distance/m		
effect of 1–4 cm dbh size class				
baseline survival	size class affected	0–7	7–14	14–21
0.895	1–2 cm	—	—	—
0.935	2–4 cm	—	—	—
0.940	4–8 cm	—	—	+0.048
0.959	8–16 cm	—	—	—
0.947	16–32 cm	—	—	—
0.928	32+cm	—	—	+0.047
effect of 4–16 cm dbh size class				
	size class affected	0–7	7–14	14–21
	1–2 cm	—	—	—
	2–4 cm	—	−0.083	—
	4–8 cm	—	—	—
	8–16 cm	+0.029	—	—
	16–32 cm	−0.044	—	—
	32+cm	—	—	—
effect of 16+cm dbh size class				
	size class affected	0–7	7–14	14–21
	1–2 cm	—	—	—
	2–4 cm	—	—	—
	4–8 cm	—	—	—
	8–16 cm	−0.039	—	—
	16–32 cm	—	—	—
	32+cm	—	—	—

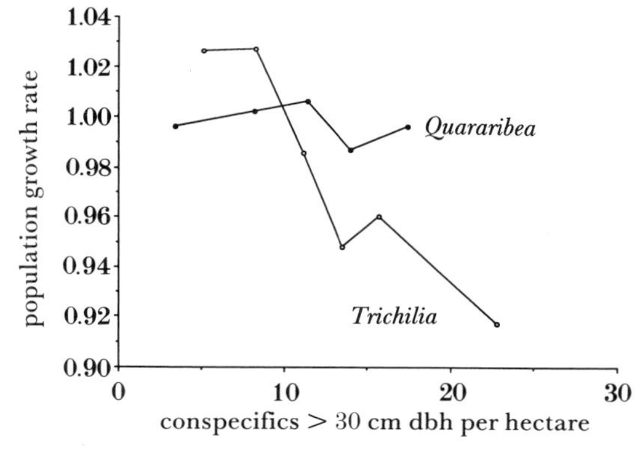

Figure 8. Comparative population growth rate (λ), as a function of the density of large (> 30 cm dbh) conspecific trees per ha in *Trichilia tuberculata* and *Quararibea asterolepis*.

0.14 when the population peaked, and recruitment rate declined from 8.5 to 5.3. Figure 10*b* shows the relation between these two parameters and the total density of large trees. If all density dependence was removed from the model, then the population grew exponentially at an observed rate $\lambda = 1.017$ (figure 9*a*). When we calculated λ from the set of life table parameters derived directly from field measurements, we obtained $\lambda = 1.016$. Moreover, the population levelled off at a realistic population density and with similar size class distributions to those observed in the 50 ha BCI plot (figure 10). Thus we believe we have accurately captured the essential features of density-dependent population growth in *Trichilia*.

We do not yet know the causes of the strong density-dependence discovered in *Trichilia*, but we suspect that heart-rot fungi may be involved. In mahogany, also a member of the Meliaceae, heart-rot fungi are a major cause of tree death, especially in saplings near infected adults. Moreover, the fungus continues to live for a long time as a saprophyte on the wood of trees that it has killed, so it may contaminate areas previously occupied by the host species and prevent recruitment for several years or decades. *Trichilia tuberculata* does harbour a heart-rot fungus, but we know nothing about its life history or host relations. This interesting possibility merits further research.

CONCLUSIONS

In this paper we have endeavoured to summarize the evidence for density dependence in regulating tree populations in a neotropical forest. Although the study is now a decade old, our conclusions must all be regarded as tentative because although we have monitored the forest for a decade, this only a small fraction of the lifespan of many of its member tree species. Nevertheless, there is a remarkable consistency among the answers we are getting, whether they are derived from analyses of the static, first-census data, from recensus data on short-term dynamics of survival, growth, and recruitment, or from detailed studies of density effects in individual BCI tree species.

The picture emerging from these studies is that, with the exception of one or two extremely abundant canopy and possibly midstory tree species, the vast majority of the species in the BCI forest are nowhere

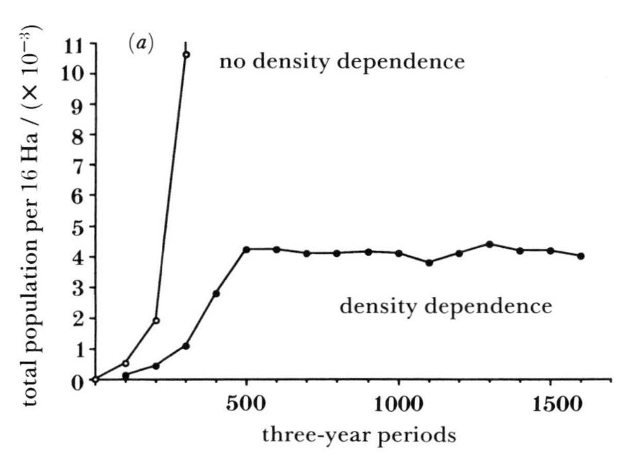

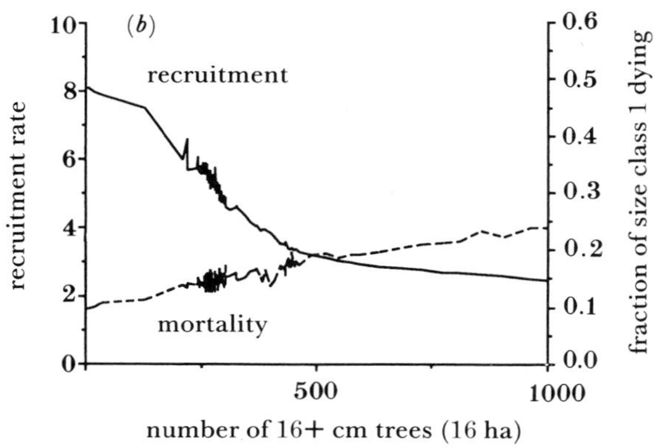

Figure 9. Simulation results for the growth of *Trichilia tuberculata* model populations in a 16 ha area (400 × 400 m). (*a*) Population growth curves with and without density dependence on survival, growth, and recruitment, as measured in the natural population. (*b*) Relation between recruitment rate per 16 ha and mortality rate per three-year cycle, as a function of the number of large (16 + cm dbh trees) per 16 ha.

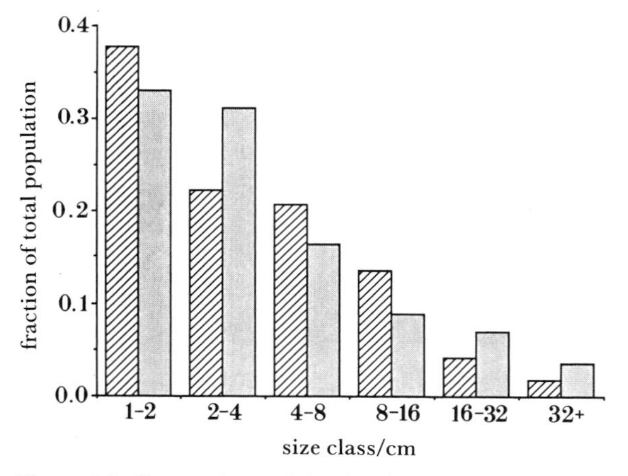

Figure 10. Comparison of the simulated (hatched bars) and observed (solid bars) size structure of the model and natural *Trichilia tuberculata* populations, after the model population reached more or less stable size structure (1000 three-year cycles).

near the densities at which density-dependent feedbacks can limit further population growth. One might argue that any density dependence at all, even density dependence too weak for us to detect, will ultimately limit a population. While certainly true in principle, these theoretical limits may be unrealistically high for a space-limited plant population.

A plant population cannot increase beyond the ultimate limit set by the self-thinning law in a monospecific stand, the density at which local and global competitive effects become one and the same. The limits we seek cannot be the same as those setting the self-thining limits for monocultures; otherwise the population should grow to these limits. But higher limits are simply not credible for any real plant population. Therefore the limits, if they exist, must be lower. However, our current evidence is that the measured strength of density dependence in most BCI trees fails to meet this criterion. The exceptions are a few species such as *Trichilia* in which lower density limits do exist, set by as yet unknown density dependent factors that are demonstrably sufficient to regulate the population at observed densities.

Our current working hypothesis is that density dependence sets upper, often unattainably high, limits to the abundance of trees in the BCI forest, but that most species are far below these limits, too rare to encounter each other often or experience significant conspecific density effects. This is not to argue against effects at larger spatial scales than we have examined. For example, there might well be inverse-density effects when tree populations become so sparse that they suffer from Allee effects because of limited attention from pollinators and seed dispersal agents.

We thank the large numbers of students who have helped with the field work on the project over the past decade. We thank the National Science Foundation, the Smithsonian Institution (Scholarly Studies Program), the World Wildlife Fund (U.S), the Geraldine R. Dodge Foundation, The W. Alton Jones Foundation, and Earthwatch, for their support.

REFERENCES

Condit, R., Hubbell, S. P. & Foster, R. B. 1990 The recruitment of conspecific and heterospecific saplings near adults and the maintenance of tree diversity in a neotropical forest. *Am. Nat.* (In the press.)

Croat, T. B. 1978 *The Flora of Barro Colorado Island* (943 pages.) Stanford University Press.

Foster, R. B. & Hubbell, S. P. 1990 The floristic composition of a 50 ha plot on Barro Colorado Island. In *Four neotropical forests* (ed. A. Gentry). New Haven, Connecticut: Yale University Press.

Horn, H. S. 1975 Markovian processes of forest succession. In *Ecology and evolution of communities* (ed. M. L. Cody & J. Diamond), pp. 196–211. Cambridge, Massachusetts: Belnap Press.

Howe, H. F. & Vande Kerchhove, G. A. 1979 Fecundity and seed dispersal of a tropical tree. *Ecology* **60**, 180–189.

Hubbell, S. P. 1980 Seed predation and the coexistence of tree species in tropical forests. *Oikos* **35**, 214–299.

Hubbell, S. P. & Foster, R. B. 1983 Diversity of canopy trees in a neotropical forest and implications for conservation. In *Tropical rain forest: ecology and management* (ed. S. L. Sutton, T. C. Whitmore & A. C. Chadwick), pp. 25–41. Oxford: Blackwell Scientific.

Hubbell, S. P. & Foster, R. B. 1986*a* Biology, chance, and history and the structure of tropical rain forest tree communities. In *Community ecology* (ed. J. Diamond & T. J. Case), pp. 314–329. New York: Harper and Row.

Hubbell, S. P. & Foster, R. B. 1986*b* Commonness and rarity in a neotropical forest: implications for tropical tree conservation. In *Conservation Biology: science of scarcity and diversity* (ed. M. Sonlé), pp. 205–231. Sunderland, Massachusetts: Sinauer Associates.

Hubbell, S. P. & Foster, R. B. 1986*c* The spatial context of regeneration in a neotropical forest. In *Colonization, succession and stability* (ed. A. Gray, M. J. Crawley & P. J. Edwards), pp. 395–412. Oxford: Blackwell Scientific.

Hubbell, S. P. & Foster, R. B. 1990*a* Structure, dynamics, and equilibrium status of a neotropical forest. In *Four neotropical forests* (ed. A. Gentry), pp. 522–541. New Haven, Connecticut: Yale University Press.

Hubbell, S. P. & Foster, R. B. 1990*b* Short-term population dynamics of trees and shrubs in a neotropical forest: El Niño effects and successional change. *Ecology* (In the press.)

Hubbell, S. P. & Foster, R. B. 1990*c* The fate of juvenile trees in a neotropical forest: implications for the natural maintenance of tropical tree diversity. In *Reproductive Biology of tropical forest plants* (ed. M. Hadley & K. S. Bawa) (Man in Biosphere series, vol. 7), pp. 325–349. Carnforth: UNESCO/IUBS, Paris and Parthenon Publishers.

Leigh, E. G., Jr., Standley Rand, A. & Windsor, D. M. 1982 *The ecology of a tropical forest: seasonal rhythms and long-term changes.* (468 pages.) Washington, D.C.: Smithsonian Institution Press.

Pacala, S. W. & Silander, J. A. 1985 Neighborhood models of plant population dynamics. I. Single-species models of annuals. *Am. Nat.* **125**, 385–411.

Silander, J. A. & Pacala, S. W. 1985 Neighborhood predictors of plant performance. *Oecologia* **66**, 256–263.

Silvertown, J. W. 1982 *Introduction to plant population ecology* (209 pages.) New York: Longman.

Welden, C. W., Hewett, S. W., Hubbell, S. P. & Foster, R. B. 1990 Survival, growth, and recruitment of sapling in relation to forest canopy height on Barro Colorado Island, Panama. *Ecology* (In the press.)

White, J. 1981 The allometric interpretation of the self-thinning rule. *J. Theor. Biol.* **89**, 475–500.

Discussion

P. J. Grubb (*Botany School, Cambridge University*). The point I wished to make in conversation earlier was that in some genera of wody plants represented by many species in a single forest stand the various species are strikingly various in leaf form, e.g. in *Diospyros* and *Eugenia* in various parts of the world, and they might plausibly be expected to respond to environmental conditions differently; and that in genera where this is not so, e.g. in *Strychnos* in Ghana, there is commonly wide variation between species in the size and form of the fruit and/or seeds, which seems likely to result in different conditions favouring establishment.

S. P. Hubbell. My challenge to you is to show that these differences are of any material significance.

P. J. Grubb. I accept the challenge!

C. D. Thomas. (*Centre for Population Biology, Imperial College at Silwood Park, U.K.*). Many insects feed on more than one species of plant. Potentially, such insects could regulate the combined densities of all host plants, perhaps limiting the total density of individuals in plant genera or families. Has Professor Hubbell looked for density- or distance dependence at generic or family level?

S. P. Hubbell. We have done some analyses pertinent to this issue. What we have found is that species in the same genus (*Inga* and *Ocotea*) and family (Lauraceae) do not interact negatively with one another, based on the same sorts of analyses presented here. For example, *Beilschmiedia pendula* and *Ocotea skutchii* in the Lauraceae have recruits closer to adults of the other species than one would expect by chance. This is probably because they fruit at the same time and share dispersers (toucans). In addition, nearby adults of one do not reduce survival of saplings of the other. We have not, however, combined all species within a family and looked for density dependence, and that is an interesting suggestion.

Species richness and population dynamics of animal assemblages. Patterns in body size: abundance space

J. H. LAWTON

Centre for Population Biology, Imperial College, Silwood Park, Ascot SL5 7PY, U.K.

SUMMARY

Links between population dynamics, the relative abundance of species and the richness of animal communities are reviewed within the framework of a simple conceptual model, based on body size and abundance. Populations of individual species occupy positions in this body size: abundance space. Problems of relative abundance and absolute species-richness revolve around a number of simple questions, including: what determines the upper and lower bounds (maximum and minimum population densities) of species in the assemblage; what determines the overall density of points (number of species) within these bounds; and how are the vertical and horizontal partitioning rules between species decided? The answers to these, and related questions, are briefly reviewed.

1. INTRODUCTION

A simple conceptual model based on body size and abundance links population dynamics with the richness and relative abundance of animal species within a local community (figure 1). Although it is simple, it allows us to bring together and explore a wide variety of processes hitherto considered largely in isolation. Each point in figure 1 represents the average abundance and average body size of a species in the assemblage. Fluctuations in density and changes in body size are a complication dealt with briefly towards the end of this paper. The model is simplest to interpret when applied to species from one trophic level, although data are too restricted for this ideal to be fully realized at the moment. It seems unlikely that the model can be usefully applied to animals with modular growth, such as corals or bryozoans.

If data are collated for animal populations from a wide variety of published studies across many communities and ecosystems, average population densities tend to be inversely correlated with body size (Peters & Wassenberg 1983; Damuth 1987). In contrast, collections of organisms from within local assemblages reveal the more complex shape shown in figure 2 and idealized in figure 1 (Lawton (1989)) and references therein; Blackburn *et al.* 1990), as do data for primates in different habitats (Clutton-Brock & Harvey 1977) and birds throughout the North American continent (Brown & Maurer 1987). This pattern is remarkably consistent across a wide range of taxa, habitats and geographic areas. Possible reasons for the differences between data assembled by Damuth (1987) and by Peters & Wassenberg (1983), and these more taxonomically and geographically restricted studies are discussed in Lawton (1989) and Blackburn *et al.* (1990).

This paper takes the pattern summarized in figure 1 as the norm and asks three questions.

(i) What determines maximum average abundances for species in the assemblage, i.e. what determines the slope and amplitude of the upper-bound *A–B*?

(ii) Below some critical density, *C–D*, species become so rare that they disappear from the community. What determines minimum viable population densities?

(iii) How is the density of points within body size: abundance space determined? This problem embraces at least two, interrelated subquestions, namely what are the horizontal and vertical partitioning rules that determine the positions and numbers of data points within the defined bounds?

Two details about figure 1 are not considered here. The first is the range of body sizes (*E–F*) from the smallest to the largest species in the assemblage. In its more interesting form, this range will be set by evolutionary design constraints; less interestingly, it will have been set arbitrarily by the investigator. Neither is a question of population regulation or

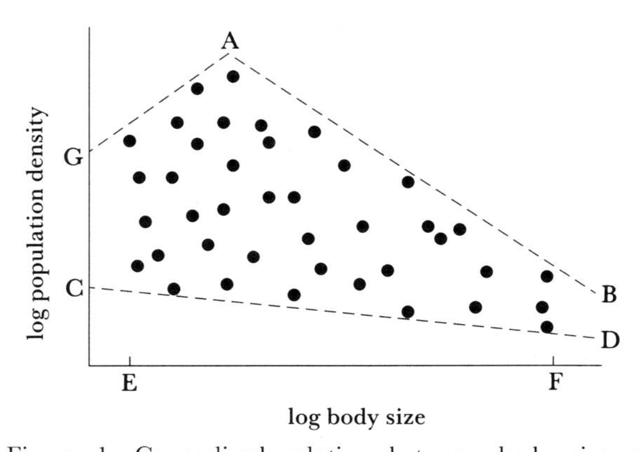

Figure 1. Generalized relation between body size and population density for animals in a local assemblage. Each dot represents the average body size and abundance of a single species in the assemblage. The limits (*A–B*, *E–F* etc.) are discussed in the text.

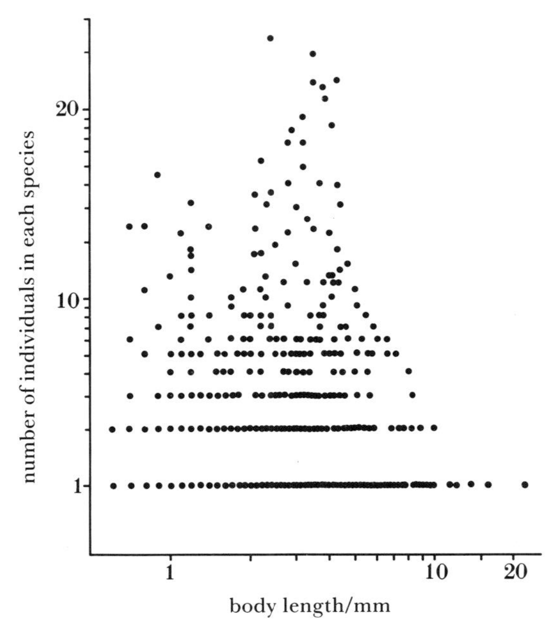

Figure 2. Data on 859 species of adult beetles plotted in body size: abundance space. The beetles were collected in single insecticide foggings from the canopies of ten Bornean lowland rainforest trees (Morse *et al.* 1988). Unlike the hypothetical data in the preceding figure, the lower bound (*C–D* in figure 1) is horizontal, because large numbers of species are represented by single individuals in the collection. It is not clear how many of these single individuals are 'tourists' without established local populations, or how the shape of the lower bound might be altered by making much larger, or sequential collections in an attempt to distinguish statistical sampling effects from biological ones.

dynamics. Second, there is a small region *G–A* where the upper bound declines as average body sizes decline. This small region is not well understood, although it is present in all the body size: abundance plots for local assemblages that I have seen. It may simply be a sampling phenomenon (Blackburn *et al.* 1990), or it may be something more interesting (see Brown & Maurer (1987, 1989) for further discussion).

2. MAXIMUM POPULATION DENSITIES AND BODY SIZE

Species whose populations lie along, or close to the upper bound (*A–B*) of figure 1 are, by definition, the most abundant in the assemblage. Being the most abundant, it seems reasonable to suppose that they are resource limited. If so, the predicted slope of the line *A–B* depends on average resource use. For many animals, individual metabolic rates, and hence food requirements scale as $W^{0.75}$ (Peters 1983), where W is body mass. Other things being equal (not least the total availability of food for animals of different body sizes), maximum population densities will therefore scale as $W^{-0.75}$. Similar arguments can be developed for species that are limited by the space or volume of the habitat (see, for example, Duarte *et al.* (1987)), or by territoriality (Schoener 1968; Clutton-Brock & Harvey 1977).

Discovering whether the upper bound in a diagram such as figure 1 conforms to or departs from theoretical expectations presents statistical difficulties, and I know of no formal attempts to carry out such a test; less formal inspection reveals an upper bound for birds in Brown & Maurer (1987) that is close to -0.75 (J. H. Brown, personal communication). Slopes of log body size: log abundance plots compiled from the literature, for species from a wide variety of different communities sometimes also lie close to -0.75 (Damuth 1987), which is to be expected if such data are biased towards the commonest species. But other studies report slopes that are clearly not -0.75 (Peters & Wassenberg 1983; Peters & Raelson 1984; Brown & Maurer 1986), perhaps reflecting the efficiency with which rarer species have been included in the surveys (Brown & Maurer 1986; Lawton 1989), as well as different ways of grouping species before analysis.

Surprisingly, even less is known about the magnitude of the upper bound. It seems reasonable to suppose that *A–B* will be higher in more productive ecosystems, but I know of no explicit test of this hypothesis. I return to the question of the magnitude of the upper bound in a later section.

3. MINIMUM VIABLE POPULATION DENSITIES

Defining the lower bound, *C–D*, in figure 1 is difficult for two reasons, one practical, the other conceptual. Practically, rare populations are by their nature difficult to study. Conceptually, the lower bound is elusive because many of the rare species taken in samples of a community may be transients; there is a continuum from vagrants, through species that occasionally breed in the community, to rare but permanent members with self-sustaining populations (see legend to figure 2); and the position of species along this continuum will depend on the size of the area being investigated. Nevertheless, it is worth considering what we know about the determinants of minimum viable population densities, at least in principle. The tentative suggestion in figure 1 is that minimum viable densities are higher for small-bodied species (Lawton (1989); see also arguments in Brown & Maurer (1987)). This suggestion hinges on two lines of reasoning.

First, species that are rare locally (i.e. those species whose populations lie towards the lower bound of figure 1) tend also to have restricted geographical distributions (Brown 1984; Gaston & Lawton 1988*a*, *b*). In other words, they have small total population sizes. Second, although the data are much poorer and more difficult to interpret than is generally realized (McArdle *et al.* 1990) populations of small-bodied species may fluctuate more on average than populations of large-bodied species (scaled for generation time) (reviews in Gaston & Lawton (1988*a*, *b*); Hanski, this symposium). If they are real, these differences probably reflect the greater vulnerability of smaller organisms to density independent disturbances. Since fluctuations can drive rare populations extinct, minimum viable total population sizes should therefore be larger for small-bodied species, and large-bodied species should be able to maintain themselves in communities at lower overall densities.

A counter argument runs as follows. Although populations of rare, small-bodied species are more vulnerable to environmentally driven stochastic extinctions than similar sized populations of large-bodied species, small-bodied species have an advantage in higher intrinsic rates of increase (Gaston & Lawton 1988*a*, *b*; Pimm *et al.* 1988; Lawton 1989). Higher intrinsic rates of increase carry rare populations out of the 'danger zone' faster, and may favour the persistence of small-bodied species at lower average total population sizes and densities than are sustainable by large-bodied species.

Evidence can be adduced for both arguments (see Lawton (1989) and Pimm *et al.* (1988) for detailed discussions). For example, below a total population size of about seven pairs, large-bodied species of resident birds on British islands are less prone to extinction than are small-bodied species. Above seven pairs, these differences disappear and may possibly be reversed (Pimm *et al.* 1988). The implication is that both intrinsic rates of increase and vulnerability to environmental stochasticity play a part in determining minimum viable population sizes and densities. Additional complexities centre on the fact that not all low-density populations have restricted distributions (Rabinowitz *et al.* 1986; Gaston & Lawton 1990*a*, *b*), severing the theoretical link between determinants of minimum viable population sizes and minimum viable densities.

In brief, we have only the haziest notion how the lower bound in figure 1 is determined. My suspicion is that usually the bound slopes up as species get smaller, but I have no great confidence in the prediction. It already appears to be refuted by birds sampled throughout North America (Brown & Maurer 1987), albeit these are not data from a local community.

4. THE DENSITY OF POINTS IN BODY SIZE: ABUNDANCE SPACE

Total species richness in a community is a product of the bounds on figure 1, and the density of points within these bounds. A great deal of experimental and theoretical work in population dynamics and community ecology has been concerned with discovering how the density of points is determined, and whether there are, or are not, hard limits to the number of species that can be packed into a particular space.

(*a*) *The role of regional processes*

Work on the richness of local assemblages has focused on species interactions. Much less attention has been paid to the influence of regional (biogeographic) processes. Yet the latter are central to an understanding of local community diversity. The theory of island biogeography (MacArthur & Wilson 1967) tells us that at equilibrium, small or isolated patches of habitat hold fewer species from a regional pool than large, near patches. Individual species may be missing entirely from suitable, but isolated habitats (Harrison *et al.* 1988; Lawton & Woodroffe 1990). It is unclear whether species occupying different regions in body size: abundance space are differentially influenced by isolation. For instance, species that are common in the regional pool may generate more colonizing propagules. Body size ought also to influence dispersal abilities (Gaston & Lawton 1988*a*, *b*), although effects may be complex, not least because very small body sizes may facilitate passive dispersal and large body sizes may aid directed movements.

Familiar as these ideas are, ecologists have been slow to work out the contribution of regional processes to local diversity. Yet patterns of invasion, speciation and extinction over evolutionary time make an enormous difference to the size of the regional pool of species from which local communities are assembled (for a particularly illuminating example, see Pearson & Ghorpade (1989); for a general discussion of the problem, see Ricklefs (1987). A much neglected question is the relation between the number of species in the regional pool of potential colonists, and the richness of local assemblages (Lawton 1982; Cornell 1985*a*, *b*). There are three possibilities (figure 3*a*): all (model 1), or a constant proportion (model 2), of the species in the regional pool occur locally; or (model 3), there are constraints on the number of species able to coexist locally.

For insect herbivores on oak trees (Cornell 1985*a*, *b*) and on bracken fern (Lawton 1982, 1990; Compton *et al.* 1989; see figure 3*b*) the number of species living together locally is a roughly constant proportion of the number of species in the regional pool from which these communities are assembled (figure 3*a*, model 2). There is no evidence that the number of species is constrained in the manner envisaged by model 3. Vacant niches (unused resources that are utilized in other assemblages) are easy to identify, and there are no signs of density compensation in sparse species assemblages (see, for example, Lawton 1982). It is unclear what proportion of ecological assemblages conform to model 2; for those that do, understanding local (within community) constraints on the density of points in body size: abundance space is not very interesting. The answers lie elsewhere, in history and regional biogeography.

Model 3 communities present the other extreme, and certainly exist (Terborgh & Faaborg 1980), although how common they are is again uncertain. In model 3 communities, numbers of coexisting species are constrained; a veritable library of ecological literature has been devoted to discovering how.

(*b*) *Constraints on numbers of coexisting species: horizontal partitioning rules*

The number of coexisting species may be constrained by interspecific competition for resources such as food or space (which has been very well studied) or by 'apparent competition' for 'enemy-free space' (far less well studied) (Holt 1977, 1987; Jeffries & Lawton 1984; Lawton 1986). These constraints may have little or nothing to do with body size (Chesson, this symposium; Shorrocks & Rosewell 1986), or they may be strongly influenced by it. Here I want to focus

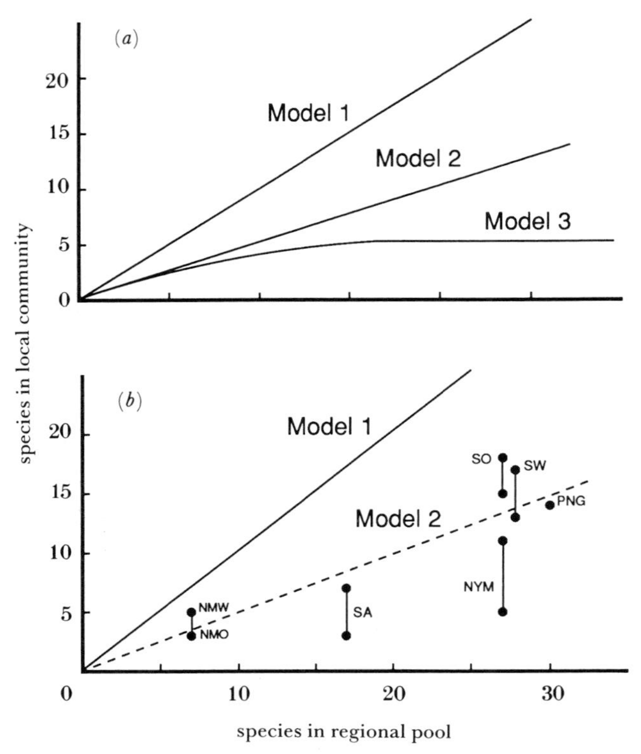

Figure 3. (*a*) Theoretical relations between regional and local species richness in ecological systems (Cornell 1985*a*, *b*). Model 1, all species in the regional pool are found in every community; Model 2, proportional sampling; Model 3, local communities saturate with species. (*b*) Relation between local and regional species richness of herbivorous arthropods (insects and mites) feeding on bracken in different parts of the world (Lawton 1982; Lawton & Gaston 1989; Compton *et al.* (1989) and references therein). Maximum and minimum values for local communities are shown for each study area. The dotted line for model 2 is for proportional sampling in which half the species in the regional pool are found locally (it is not a fitted line). (NMO, NMW, New Mexico open and woodland areas; SA, South Africa; NYM, North York Moors; SO, SW, Skipwith (York) open and woodland areas; PNG, Papua New Guinea.)

briefly on constraints on the number of coexisting species that are influenced by body size; that is, I want to examine the horizontal partitioning rules in figure 1.

The notion that there may be limits to similarity in the body sizes of competing species is usually attributed to Hutchinson (1959), although he acknowledges Julian Huxley as the originator of the idea (see Carothers 1986). Thirty years later, progress has been frustratingly slow. Schoener (1986, 1989) provides two excellent overviews. Theory based on various kinds of abstract models is equivocal on whether there are, or are not, hard limits to niche overlap (Abrams 1983). Hutchinson's much cited (albeit originally highly tentative) observation that body-length ratios of the larger to the smaller species are typically of order 1.3 may be a consequence of 'sampling' from a log-normal distribution of body sizes with small variance (Eadie *et al.* 1987). And there are still rather few convincing studies linking the observation that species are less similar in body size than expected by chance (see, for example, Schoener (1984); Diamond (1986)) with detailed studies on the mechanisms of the competitive process (Juliano & Lawton 1990*b*); Darwin's finches continue to inspire and lead our understanding (Grant 1986).

In an attempt to clarify relations between body size, limiting similarity and competitive mechanisms, Juliano & Lawton (1990*a*, *b*) examined guilds of hydroporine and laccophiline water beetles (family Dytiscidae) in habitats near York in northern England. The adults are small (largest species *ca.* 5.5 mm long), long lived, and predatory; assemblages typically contain several species that differ in body size and shape. Body size and preferred prey size are highly correlated. We compared the body size and shape of species in real assemblages with random draws from the pool of species present in the area, and with assemblages of 'pseudospecies' drawn from random points in the full range of morphological space; analyses were performed by using all the species recorded in each assemblage, or excluding rare species (defined in a standard way).

Two natural assemblages from large, well-buffered water bodies containing predatory fish were more regularly and widely spread out in morphological space than expected by chance, and these patterns were stronger when comparing just abundant species with randomly assembled 'pseudospecies'. There are theoretical reasons to expect that non-random distributions of species in body size: abundance space will be easier to detect under both these circumstances (see Juliano & Lawton (1990*a*) for further discussion). In contrast, sets of species from seven small, acid pools without fish were randomly distributed in morphological space. Armed with these results, we predicted that species in the non-random assemblages would compete for food in the manner envisaged by Hutchinson (1959), and that the random assemblages would not. Unexpectedly (Juliano & Lawton 1990*b*), we could find no evidence that adults in the two non-random assemblages competed for food, or that intensity of competition was related to differences in adult body size. In this respect, they appeared identical to the assemblages in which adults were randomly assembled in morphological space.

These results epitomize the problems that have confronted attempts to discover simple, general principles governing the horizontal partitioning rules in body size: abundance space. Pessimists would argue that ecologists have made essentially no progress with the problem in over thirty years. My view is that we have made reasonable progress with a hard problem. We now know that the question of competition and hard limits to species overlap are irrelevant for some systems. For other systems, competition is clearly important, and a series of empirical observations, experiments and models suggest that sometimes competition translates into limits on similarity in body size, and sometimes it does not. On other occasions, as in the water beetles, non-random patterns in morphology do not seem to be a result of competition for food, in the manner envisaged by Hutchinson; what the mechanism is in this particular case remains obscure. It may involve size-selective predation by fish, but we have no firm evidence on this point.

As is so often the case in ecology, theory and explanations are contingent, not absolute (Schoener 1986). The challenge now is to develop a better understanding of the contingent, horizontal partitioning rules, both in theory and in practice, building upon the rich body of data and models that already exist.

(*c*) *The vertical partitioning rules in theory*

It is clearly artificial to treat the horizontal and vertical partitioning rules as distinct, or to separate either from competitive processes that set limits to the number of coexisting species independently of body size. Interspecific competition depresses population densities on figure 1 and competitive exclusion happens when densities cross the lower bound *C–D*. Nevertheless, it is conceptually useful, for present purposes, to discuss the two sets of rules separately.

To ask about the relative abundance of species is, by definition, to ask about the vertical partitioning rules in figure 1. I am not aware that anybody has worked out in detail for more than two or three species within one trophic level why some species in a community are common and others are rare. The common species, close to the upper bound, are presumably resource limited, competitive dominants. There are several reasons why other species in the assemblage may be rarer than this, including: specializing on rare resources (small fundamental niches); competitive inferiority (realized niches much smaller than fundamental niches); the impact of natural enemies or diseases (q, in the sense of Beddington *et al.* (1978), $\ll 1$); and frequent, density-independent disturbances. It would be interesting to map out the reasons for commonness and rarity in particular species assemblages.

Although detailed information is lacking on why some species are common and others rare, even in well studied communities, we do know that despite population fluctuations around long-term averages, in assemblages where the problem has been looked at, the common species generally stay common and the rare species stay rare; that is rank abundances are statistically significantly correlated over time (Lawton & Gaston (1989) and references therein; Owen & Gilbert 1989). An alternative model of community dynamics, the 'core-satellite hypothesis' (Hanski 1982), predicts that species within assemblages switch haphazardly from common to rare or absent and back again. I know of no animal communities where this model has been verified in detail (Gaston & Lawton 1989). However, there is growing theoretical and empirical evidence that certain assemblages can exist in alternative stable states in identical physical environments, particularly where habitat fragmentation isolates the alternative states (Robinson & Dickerson 1987; Barkai & McQuaid 1988; Moss 1989; Polis *et al.* 1989; Sinclair 1989). Alternative stable states imply that species occupy alternative equilibrium positions measured on the y-axis of figure 1.

Given the rich variety of possible dynamics that exist at any points in body size: abundance space, it seems unlikely that a single model could describe the vertical partitioning rules, for species lying within a particular range of body sizes. Yet a model based on 'sequential niche breakage' (Sugihara 1980) comes intriguingly close. It may work because it stands back from the details. As Sugihara points out (p. 773): 'it is plausible to consider apportionment in a heterogeneous resource pool, involving the subdivision of several different sets of niche axes. This allows the apportionment analogy to be extended to large species ensembles which do not possess a uniform set of governing factors'.

Sugihara envisages communal niche space as a 'unit mass', sequentially split up by the component species, so that each fragment denotes relative species abundance. The initial unit mass is broken to produce two fragments; one of these is then chosen randomly and broken to yield a third, and so on. The rules for determining apportionment of niche space (the size of the fragments) between two species at any particular break are discussed by Sugihara (1980). Applying these rules, and extending the breakage sequence to a large number of species results in a distribution of fragment-lengths (species abundances) that is canonical log-normal. In other words, the theoretical species–frequency distribution generated by this minimal model is exactly the distribution of species' abundances observed in many communities, although by no means all (exceptions are reviewed by Preston (1980); Ugland & Gray (1982); Harmsen (1983); Wright (1988)).

Not unexpectedly, Sugihara's model has attracted a good deal of attention (see, for example, Ugland & Gray (1982); Harvey & Godfray (1987); Pagel *et al.* (1991) and has spawned a number of related studies (Kolasa & Strayer 1988; Kolasa 1989; Glasser 1989) exploring alternative models of niche breakage and resource partitioning between species, with no sign, as yet, of an emerging consensus. Three things, do, however, seem obvious. First, given that not all communities have the same underlying species-frequency distribution (log-normal, log-series or something else), it is very unlikely that one model will be adequate to explain species' relative abundances in all systems; once again, we need contingent theory. Secondly, we need to think very hard why minimal, phenomenological models work at all; in particular, it would seem sensible to try to derive the general partitioning rules from detailed models that explore the full range of dynamic processes known to be operating at different points in body size: abundance space (see above). Hughes (1986) has made an important start in this direction.

Thirdly, most existing models ignore body size (Harvey & Godfray (1987); Pagel *et al.* (1990) provide important exceptions). Species–frequency distributions are usually fitted to groups of similar species, moths at a light trap, for example. Although these will usually cover a limited range of body sizes, there may be sufficient differences between the largest and smallest species in the assemblage to make detection of the underlying patterns and mechanisms difficult. The way in which log-normal distributions of species' abundances are built up by combining data from species of different body sizes, incorporating figure 1, is explained in Morse *et al.* (1988).

It is unclear what range of body sizes can and should be grouped together in studies of species–frequency distributions, but suppose that figure 1 can be divided into n vertical strips of appropriate width, in which the largest species in each strip is l times the mass of the smallest. Suppose also, for the sake of argument, that the real species-frequency distribution within each strip is a log-series. If we now view the system through a cruder lens, say by combining triplets of strips (such that the largest species have l^3 the mass of the smallest) the underlying species frequency distribution in the wider strip may appear to be roughly log-normal, rather than a log-series, because combining log-series distributions is one way to generate a log-normal curve. In other words, a good deal more thought needs to be given to the way in which species of different body sizes are grouped together in empirical studies of resource partitioning and species frequency.

(*d*) *The vertical partitioning rules for North American birds*

The most important practical contribution in this genre is an analysis of biomass and energy use during the breeding season for 380 species of terrestrial birds across the whole of North America north of Mexico (Maurer & Brown 1988). Although Maurer & Brown seek continental-wide patterns, their findings may reflect what happens within smaller areas; and even if such patterns are different, the study points the way to the kinds of analyses that need doing within local assemblages. Their study combines species from different trophic levels; ideally, future work should distinguish between trophic groups.

Species were assigned to arbitrarily defined logarithmic mass classes ($n = 9$, $l = ca.\ 2$). Knowing average breeding densities for each species, Maurer & Brown (1988) could then work out patterns of energy use by each species within these nine size bands. As species within a band are approximately the same size, relative energy use is approximately the same as relative abundance within each size category (notice this is not the case when species of very different body sizes are being compared, because population energy use scales as $(\text{body mass})^{0.75}$).

In the present context, Maurer & Brown's three most important results are: (i) within each size class, population energy usages (and hence species' relative abundances) fit a log-series; (ii) the most abundant species in each size band uses the same fraction (*ca.* 10%) of the total energy used by all species in that size band; (iii) the density of species is higher in the small mass-classes; that is, resources are more finely divided, and species are more tightly packed in body size: abundance space toward the left-hand end of the ordinate in figure 1.

This last point is important, because it implies that parameters defining the exact ways in which species divide up resources are not identical for all body sizes (although the qualitative rules appear to be the same). The second point is important if we are to understand how productivity influences the upper bound in figure 1, and is dealt with briefly in the next section. The first result apparently conflicts with Sugihara's (1980) attempt to model the vertical partitioning rules as a log-normal distribution. The mismatch may, however be more apparent than real, partly because a relatively small sample of species from the right-hand tail of a log-normal distribution may look like a log-series, and partly because, as we have seen, pooling log-series may generate a log-normal distribution!

It is probably not helpful to pursue these more detailed points, mainly because the scale of Maurer & Brown's work (the avifauna of a whole continent) may, in the end, be ill-suited to studies of local communities. I have no doubt, however, that their work shows the types of data we need to gather for local species assemblages, and together with the general models of Sugihara and others, points the way towards a clearer understanding the vertical partitioning rules, and how communities are assembled in body-size: abundance space.

5. PRODUCTIVITY AND SPECIES RICHNESS

I want, now, to return briefly to the problem of the upper bound, and the relation between it, and overall species richness.

As MacArthur (1972) anticipated and Wright (1983) formalized, a number of studies have now shown geographical patterns of increasing species richness with increasing energy availability or habitat productivity (Brown 1975; Brown & Davidson 1977; Currie & Paquin 1987; Turner *et al.* 1987; Turner *et al.* 1988; Adams & Woodward 1989; Currie 1990). These results can be interpreted in terms of figure 1, via links between primary productivity and total consumer biomass.

The biomass of primary consumers increases roughly as the 1.5 power of net primary production in terrestrial ecosystems (McNaughton *et al.* 1989). It is not surprising that more energy entering the system results in a larger lump of animal biomass to be partitioned among species; it is less obvious why this increase in biomass is a power function of net primary production. It is this total biomass of animals that Sugihara's and related models seek to partition among species.

As has already been observed, Maurer & Brown (1989) find that the most abundant species in each body-size category use about the same fraction of the total energy used by all species. If these arguments extend to systems of different overall productivities (and their data for different regions of North America suggest that they do) it implies that the upper bound in figure 1 will rise as overall productivity rises, with the commonest species taking an approximately constant percentage of total resources available for species in that body-size range. It is then more or less inevitable that more productive systems will contain more species if partitioning of remaining resources stops when the rarest species in the resulting assemblage reaches the critical lower bound, C–D, in figure 1. Indeed, because the total lump of herbivore biomass to be divided up rises as the 1.5 power of net primary production, the scope for adding rare species increases quite dramatically as productivity rises, particularly if the partitioning rules are a log-series, or something similar. Interestingly, the essence of this idea is explicit in Hutchinson (1959). He wrote (p. 150): 'If the fundamental productivity of an area is limited... to such a degree that the total biomass is less than under more favourable conditions, then the rarer species in a community may be so rare that they do not exist'. This seems to me to be at least as interesting an idea as the much cited rule for limiting similarity between species and yet has received only a fraction of the attention.

6. CONCLUDING REMARKS

(*a*) *Variations in size and density*

It is clearly a gross oversimplification to treat the points representing individual species in figure 1 as average values. Real species fluctuate in density and many (e.g. fish) change markedly in size during independent life in the wild. It is unclear whether an emerging synthesis of the way in which species are assembled into communities will have to deal with these inevitable complications, or whether average values will do. The answer may depend on the taxa and upon the magnitude of population fluctuations. Changes in body size can probably be ignored in assemblages of adult birds or water beetles; they almost certainly cannot be ignored for fish, which change positions in interesting ways in body size: abundance space as they mature (Gaston & Lawton 1990*b*). Likewise, if populations display chaotic or cyclic dynamics in species-rich assemblages (Godfray & Blythe, this symposium; Sugihara & Grenfell, this symposium), it may be impossible to ignore the fluctuations as we try to understand community assembly 'from the bottom up'. My own view is that we need to keep these complexities firmly in view in the middle distance, and concentrate our immediate attention on average densities and body sizes. The problems posed by this limited perspective are already quite complicated enough.

(*b*) *Synthesis*

The full range of population dynamic processes concatenated into figure 1, and which are therefore involved in determining the relative and absolute species richness of animal assemblages, make it impossible to build a comprehensive, realistic and reasonably detailed model of local community structure. That is not the point of summarizing the processes involved. Rather, the aim of the present paper has been to draw attention to neglected details that form part of the whole picture, and that can be resolved by a combination of observations, models and experiments; examples include the 'lower bound' problem, links between local and regional species richness, and the question of how best to construct detail-independent models of the vertical partitioning rules. I view figure 1 as a framework for organizing and cataloguing our thoughts; without it, or something similar, community ecologists are in danger of failing to see the wood for the trees, and hence failing to see how microscopic theories and processes contribute to macroscopic patterns that appear, in the end, to be both simple, and general.

I am grateful to Jim Brown, Val Brown, Kevin Gaston, Paul Harvey, Mike Hassell, Mike Hochberg and Bob May for their helpful comments on the manuscript. The study forms part of the programme of work supported by the NERC Interdisciplinary Research Centre for Population Biology at Imperial College.

REFERENCES

Abrams, P. 1983 The theory of limiting similarity. *A. Rev. Ecol. Syst.* **14**, 359–376.

Adams, J. M. & Woodward, F. I. 1989 Patterns in tree species richness as a test of the glacial extinction hypothesis. *Nature, Lond.* **339**, 699–701.

Barkai, A. & McQuaid, C. 1988 Predator-prey role reversal in a marine benthic ecosystem. *Science, Wash.* **242**, 62–64.

Beddington, J. R., Free, C. A. & Lawton, J. H. 1978 Characteristics of successful natural enemies in models of biological control of insect pests. *Nature, Lond.* **273**, 513–519.

Blackburn, T. M., Harvey, P. H. & Pagel, M. D. 1990 Species number, population density and body size relationships in natural communities. *J. Anim. Ecol.* **59**, 335–345.

Brown, J. H. 1975 Geographical ecology of desert rodents. In *Ecology and evolution of communities* (ed. M. L. Cody & J. M. Diamond), pp. 315–341. Cambridge, Massachussetts: Belknap Press.

Brown, J. H. 1984 On the relationship between abundance and distribution of species. *Am. Nat.* **124**, 255–279.

Brown, J. H. & Davidson, D. W. 1977 Competition between seed-eating rodents and ants in desert ecosystems. *Science, Wash.* **196**, 880–882.

Brown, J. H. & Maurer, B. A. 1986 Body size, ecological dominance and Cope's rule. *Nature, Lond.* **324**, 248–250.

Brown, J. H. & Maurer, B. A. 1987 Evolution of species assemblages: effects of energetic constraints and species dynamics on the diversification of the North American avifauna. *Am. Nat.* **130**, 1–17.

Brown, J. H. & Maurer, B. A. 1989 Macroecology: the division of food and space among species on continents. *Science, Wash.* **243**, 1145–1150.

Clutton-Brock, T. H. & Harvey, P. H. 1977 Species differences in feeding and ranging behaviour in primates. In *Primate ecology* (ed. T. H. Clutton-Brock), pp. 557–584. London: Academic Press.

Compton, S. G., Lawton, J. H. & Rashbrook, V. K. 1989 Regional diversity, local community structure and vacant niches: the herbivorous arthropods of bracken in South Africa. *Ecol. Ent.* **14**, 365–373.

Cornell, H. V. 1985*a* Species assemblages of cynipid gall wasps are not saturated. *Am. Nat.* **126**, 565–569.

Cornell, H. V. 1985*b* Local and regional richness of cynipine gall wasps on California oaks. *Ecology* **66**, 1247–1260.

Corothers, J. H. 1986 Homage to Huxley: on the conceptual origin of minimum size ratios among competing species. *Am. Nat.* **128**, 440–442.

Currie, D. J. 1990 Energy and large-scale patterns of animal and plant species richness. *Am. Nat.* (In the press.)

Currie, D. J. & Paquin, V. 1987 Large-scale biogeographical patterns of species-richness of tree. *Nature, Lond.* **329**, 326–327.

Damuth, J. 1987 Interspecific allometry of population density in mammals and other animals: the independence of body mass and population energy use. *Biol. J. Linn. Soc.* **31**, 193–246.

Diamond, J. 1986 Evolution of ecological segregation in the New Guinea montane avifauna. In *Community ecology* (ed. J. Diamond & T. J. Case), pp. 98–125. New York: Harper & Row.

Duarte, C. M., Agusti, S. & Peters, H. 1987 An upper limit to the abundance of aquatic organisms. *Oecologia* **74**, 272–277.

Eadie, J. McA., Broekhoven, L. & Colgan, P. 1987 Size ratios and artifacts: Hutchinson's rule revisited. *Am. Nat.* **129**, 1–17.

Gaston, K. J. & Lawton, J. H. 1988*a* Patterns in the distribution and abundance of insect populations. *Nature, Lond.* **331**, 709–712.

Gaston, K. J. & Lawton, J. H. 1988*b* Patterns in body size, population dynamics, and regional distribution of bracken herbivores. *Am. Nat.* **132**, 662–680.

Gaston, K. J. & Lawton, J. H. 1989 Insect herbivores of bracken do not support the core-satellite hypothesis. *Am. Nat.* **134**, 761–777.

Gaston, K. J. & Lawton, J. H. 1990*a* Effects of scale and habitat on the relationship between regional distribution and local abundance. *Oikos*. (In the press.)

Gaston, K. J. & Lawton, J. H. 1990*b* The population ecology of rate species. In *The biology and conservation of rare fish* (ed. A. Wheeler & D. W. Sutcliffe) (*J. Fish. Biol. Suppl.* (in the press.))

Glasser, J. W. 1989 Temporal patterns in species' abundances that imply a balance between competition and predation. *Am. Nat.* **134**, 120–127.

Grant, P. R. 1986 *Darwin's finches*. Princeton, New Jersey: Princeton University Press.

Hanski, I. 1982 Dynamics of regional distribution: the core and satellite species hypothesis. *Oikos* **38**, 210–221.

Harmsen, R. 1983 Abundance distribution and evolution of community structure. *Evol. Theory* **6**, 283–292.

Harrison, S., Murphy, D. D. & Ehrlich, P. R. 1988 Distribution of the bay checkerspot, *Euphydryas editha bayensis*: evidence for a metapopulation model. *Am. Nat.* **132**, 360–382.

Harvey, P. H. & Godfray, H. C. T. 1987 How species divide resources. *Am. Nat.* **129**, 318–320.

Holt, R. D. 1977 Predation, apparent competition, and the structure of prey communities. *Theor. Popul. Biol.* **12**, 197–229.

Holt, R. D. 1987 Prey communities in patchy environments. *Oikos* **50**, 276–290.

Hughes, R. G. 1986 Theories and models of species abundances. *Am. Nat.* **128**, 879–899.

Hutchinson, G. E. 1959 Homage to Santa Rosalia or why are there so many kinds of animals? *Am. Nat.* **93**, 145–159.

Jeffries, M. J. & Lawton, J. H. 1984 Enemy free space and the structure of ecological communities. *Biol. J. Linn. Soc.* **23**, 269–286.

Juliano, S. A. & Lawton, J. H. 1990*a* The relationship between competition and morphology. I. Morphological patterns among co-occurring dytiscid beetles. *J. Anim. Ecol.* **59**, 403–419.

Juliano, S. A. & Lawton, J. H. 1990*b* The relationship between competition and morphology. II. Experiments on co-occurring dytiscid beetles. *J. Anim. Ecol.* **59**. (In the press.)

Kolasa, J. 1989 Ecological systems in hierarchical perspective: breaks in community structure and other consequences. *Ecology* **70**, 36–47.

Kolasa, J. & Strayer, D. 1988 Patterns of the abundance of species: a comparison of two hierarchical models. *Oikos* **53**, 235–241.

Lawton, J. H. 1982 Vacant niches and unsaturated communities: a comparison of bracken herbivores at sites on two continents. *J. Anim. Ecol.* **51**, 573–595.

Lawton, J. H. 1986 The effect of parasitoids on phytophagous insect communities. In *Insect parasitoids* (ed. J. Waage & D. Greathead), pp. 265–286. London: Academic Press.

Lawton, J. H. 1989 What is the relationship between population density and body size in animals? *Oikos* **55**, 429–434.

Lawton, J. H. 1990 Local and regional species-richness of bracken-feeding insects. In *Bracken biology and management* (ed. J. A. Thomson & R. T. Smith), pp. 197–202. Sydney: Australian Institute of Agricultural Science.

Lawton, J. H. & Gaston, K. J. 1989 Temporal patterns in the herbivorous insects of bracken: a test of community predictability. *J. Anim. Ecol.* **58**, 1021–1034.

Lawton, J. H. & Woodroffe, G. L. 1990 Habitat and the distribution of water voles: why are there gaps in a species' range? *J. Anim. Ecol.* **59**. (In the press.)

MacArthur, R. H. 1972 *Geographical ecology. Patterns in distribution of species*. New York: Harper and Row.

MacArthur, R. H. & Wilson, E. O. 1967 *The theory of island biogeography*. Princeton, New Jersey: Princeton University Press.

Maurer, B. A. & Brown, J. H. 1988 Distribution of energy use and biomass among species of North American terrestrial birds. *Ecology* **69**, 1923–1932.

McArdle, B. H., Gaston, K. J. & Lawton, J. H. 1990 Variation in the size of animal populations: patterns, problems and artefacts. *J. Anim. Ecol.* **59**, 439–454.

McNaughton, S. J., Oesterheld, M., Frank, D. A. & Williams, K. J. 1989 Ecosystem-level patterns of primary productivity and herbivory in terrestrial habitats. *Nature, Lond.* **341**, 142–144.

Morse, D. R., Stork, N. E. & Lawton, J. H. 1988 Species number, species abundance and body length relationships of arboreal beetles in Bornean lowland rain forest trees. *Ecol. Ent.* **13**, 25–37.

Moss, B. 1989 Water pollution and the management of ecosystems: a case study of science and scientist. In *Toward a more exact ecology* (ed. P. J. Grubb & J. B. Whittaker), pp. 401–422. Oxford: Blackwell Scientific.

Owen, J. & Gilbert, F. S. 1989 On the abundance of hoverflies (Syrphidae). *Oikos* **55**, 183–193.

Pagel, M. D., Harvey, P. H. & Godfray, H. C. J. 1991 Species abundance, biomass, and resource-use distributions. *Am. Nat.* (In the press.)

Pearson, D. L. & Ghorpade, K. 1989 Geographical distribution and ecological history of tiger beetles (Coleoptera: Cicindelidae) of the Indian subcontinent. *J. Biogeog.* **16**, 333–344.

Peters, R. H. 1983 *The ecological implications of body size*. Cambridge University Press.

Peters, R. H. & Raelson, J. V. 1984 Relations between individual size and mammalian population density. *Am. Nat.* **124**, 498–517.

Peters, R. H. & Wassenberg, K. 1983 The effect of body size on animal abundance. *Oecologia* **60**, 89–96.

Pimm, S. L., Jones, H. L. & Diamond, J. 1988 On the risk of extinction. *Am. Nat.* **132**, 757–787.

Polis, G. A., Myers, C. A. & Holt, R. D. 1989 The ecology and evolution of intraguild predation: potential competitors that eat each other. *Ann. Rev. Ecol. Syst.* **20**, 297–330.

Preston, F. W. 1980 Non canonical distributions of commonness and rarity. *Ecology* **61**, 88–97.

Rabinowitz, D., Cairns, S. & Dillon, T. 1986 Seven forms of rarity and their frequency in the flora of the British Isles. In *Conservation biology: the science of scarcity and diversity* (ed. M. E. Soelé), pp. 182–205. Sunderland, Massachussetts: Sinauer.

Ricklefs, R. E. 1987 Community diversity: relative roles of local and regional processes. *Science, Wash.* **235**, 167–171.

Robinson, J. V. & Dickerson, J. E. Jr 1987 Does invasion sequence affect community structure? *Ecology* **68**, 587–595.

Schoener, T. W. 1968 Sizes of feeding territories among birds. *Ecology* **49**, 123–141.

Schoener, T. W. 1984 Size differences among sympatric, bird-eating hawks: a worldwide survey. In *Ecological communities. Conceptual issues and the evidence* (ed. D. R. Strong Jr., D. Simberloff, L. G. Abele & A. B. Thistle), pp. 254–281. Princeton, New Jersey: Princeton University Press.

Schoener, T. W. 1986 Resource partitioning. In *Community ecology. Pattern and process* (ed. J. Kikkawa & D. J. Anderson), pp. 91–126. Oxford: Blackwell Scientific.

Schoener, T. W. 1989 The ecological niche. In *Ecological Concepts. The contribution of ecology to an understanding of the natural world* (ed. J. M. Cherrett), pp. 79–113. Oxford: Blackwell Scientific.

Shorrocks, B. & Rosewell, J. 1986 Guild size in drosophilids: a simulation model. *J. Anim. Ecol.* **55**, 527–541.

Sinclair, A. R. E. 1989 Population regulation in animals. In *Ecological Concepts. The contribution of ecology to an understanding of the natural world* (ed. J. M. Cherrett), pp. 197–241. Oxford: Blackwell Scientific.

Sugihara, G. 1980 Minimal community structure: an explanation of species abundance patterns. *Am. Nat.* **116**, 770–787.

Terborgh, J. W. & Faaborgh, J. 1980 Saturation of bird communities in the West Indies. *Am. Nat.* **116**, 178–195.

Turner, J. R. G., Gatehouse, C. M. & Corey, C. A. 1987 Does solar energy control organic diversity? Butterflies, moths and the British climate. *Oikos* **48**, 195–205.

Turner, J. R. G., Lennon, J. J. & Lawrenson, J. A. 1988 British bird species distributions and the energy theory. *Nature, Lond.* **335**, 539–541.

Ugland, K. I. & Gray, J. S. 1982 Lognormal distributions and the concept of community equilibrium. *Oikos* **39**, 171–178.

Wright, D. H. 1983 Species-energy theory: an extension of species-area theory. *Oikos* **41**, 496–506.

Wright, S. J. 1988 Patterns of abundance and the form of the species-area relation. *Am. Nat.* **131**, 401–411.

How many species?

ROBERT M. MAY

Department of Zoology, University of Oxford, South Parks Road, Oxford OX1 3PS, U.K., Imperial College, London, U.K.

SUMMARY

This paper begins with a survey of the patterns in discovering and recording species of animals and plants, from Linnaeus' time to the present. It then outlines various approaches to estimating what the total number of species on Earth might be: these approaches include extrapolation of past trends; direct assessments based on the overall fraction previously recorded among newly studied groups of tropical insects; indirect assessment derived from recent studies of arthropods in the canopies of tropical trees (giving special attention to the question of what fraction of the species found on a given host-tree are likely to be 'effectively specialized' on it); and estimates inferred from theoretical and empirical patterns in species–size relations or in food web structure. I conclude with some remarks on the broader implications of our ignorance about how many species there are.

1. INTRODUCTION

Earlier papers in this volume have surveyed our growing understanding of the factors that govern the population densities of individual species. Later papers have gone on to consider the dynamics of single populations in relation to the systems of interacting species in which they are embedded (Godfray & Blythe; Sugihara *et al.*), to their genetic structure (Travis), and to past (Valentine) and present (Hubbell; Lawton) patterns of relative abundance of species within communities. Ideally, this final paper should build on this foundation to arrive at an analytic understanding, as distinct from a mere compendium, of how many species of plants and animals we may expect to find in a given community or in a particular region. Ultimately, such analyses could aim to pyramid toward a fundamental understanding of why the total number of species on our planet is what it is, and not grossly more or less.

Needless to say, we are a long way from this goal. This paper, therefore, focuses on the plainly factual question of how many species of plants and animals there are on earth, a number currently uncertain to within a factor 10 or more, and does not ask about the underlying reasons. The paper thus differs from an earlier review with a similar title (May 1988), in which I dealt mainly with basic factors that may influence species diversity, both locally and globally.

My paper begins with an account of those species that have been named and recorded. Amazingly, this current total is not known, because there is no central catalogue or list of named species; a good estimate may be around 1.8 million (Stork 1988). I then discuss a variety of ways in which the total number of plant and animal species on earth today may be estimated: by extrapolating past trends of discovery within particular groups; by inferences from detailed studies of the arthropod canopy fauna in tropical trees; by direct estimates of the proportion of new species in studies of groups of tropical insects (often different from extrapolating past trends, which come predominately from temperate regions); from species–size relations; and so on. Different approaches yield global totals as small as 3 million or as large as 30 million or more. A penultimate section raises some broader questions about species diversity that are prompted by such estimates of local or global totals. I conclude with subjective thoughts on why it is important to know how many species there are.

2. HOW MANY RECORDED SPECIES?

It is noteworthy that Linnaeus' pioneering codification of biotic diversity came a century after Newton, in the mid-1700s. Whatever the reason for this lag between fundamental studies in physics and in biology, the legacy lingers. Today's catalogues of stars and galaxies are effectively more complete (by any reasonable measure), and vastly better funded, than catalogues of Earth's biota. And this certainly is not because stellar catalogues offer more opportunity for commercial application than do species catalogues.

The 1758 edition of Linnaeus' work records some 9000 species of plants and animals. Table 1 summarizes estimates of the numbers recorded since then up to 1970, for different groups. Table 1 also gives the time it took to record the second half of the total number of species (up to 1970) in each group, and the era when new species within each group were being discovered at the fastest rate (from Simon 1983). The table gives a sense of the differences in the attention paid to different groups, with half of all known bird species already recorded in the century after Linnaeus, while half the arachnid and crustacean species known in 1970 were recorded in the preceding 10 years; contrast figure 1 and figure 2.

The furries and featheries are, of course, very well

Table 1. *Taxonomic activity, from 1758 to 1970, for different animal groups, as revealed in patterns of recording new species (after Simon 1983)*

animal group	estimated number of species recorded up to 1970	length of time (years), prior to 1970, to record the second half of the total in the previous column	period of maximum rate of discovery of new species
Protozoa	32000	21	1897–1911
'Vermes'	41000	28	1859–1929
Arthropoda (excluding insects)	96000	10	1956–1970
Arthropoda (insects only)	790000	55	1859–1929
Coelenterata	9600	58	1899–1928
Mollusca	45000	71	1887–1899
Echinodermata	6000	63	1859–1911
Tunicata	1600	68	1900–1911
Chordata			
Pisces	21000	62	1887–1929
Amphibia	2500	60	1930–1970
Reptilia	6300	79	1859–1929
Aves	8600	125	1859–1882
Mammalia	4500	118	1859–1898

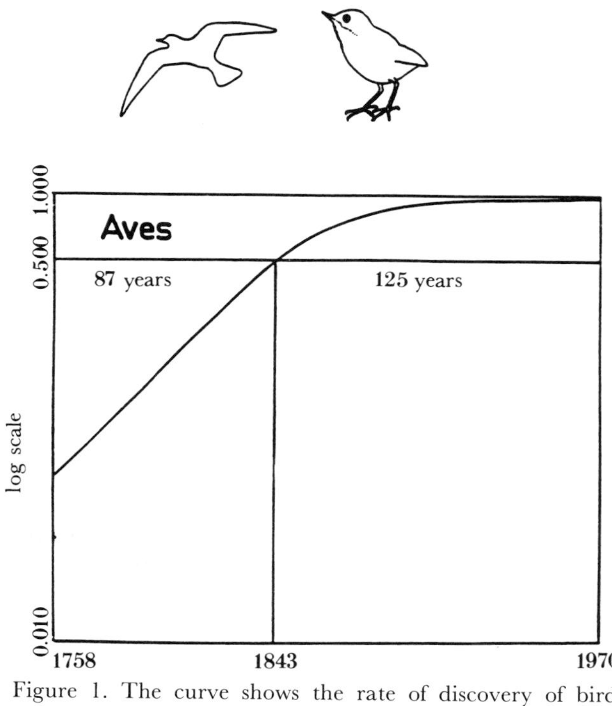

Figure 1. The curve shows the rate of discovery of bird species, from the time of Linnaeus (1758) up to 1970. Numbers of known species (expressed as a fraction of those known in 1970, on a logarithmic scale) are plotted against time. The vertical and horizontal lines show the point (1843) when half the 1970 total of species had been discovered. After Simon (1983).

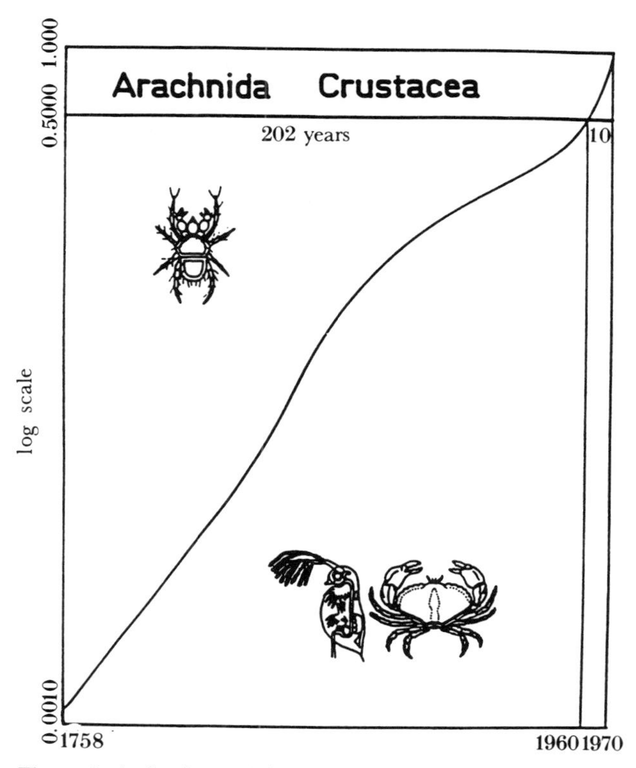

Figure 2. As for figure 1, but now showing recorded numbers of Arachnid and Crustacean species (essentially, arthropods excluding the Insecta) over time. The pattern of discovery is very different from that for bird species, with half of the total up to 1970 being recorded since 1960. After Simon (1983).

known by now. As reviewed in detail by Diamond (1985), only 134 bird species have been added to the total of just over 9000 since 1934, representing a rate of discovery of around three species per year since 1940 (most of them small, brown tropical birds). Rates of discovery are somewhat higher for mammals, with 134 of the current total of 1050 genera added since 1900, at a rate of about 1 genus per year since 1940 (most of them tropical bats, rodents or shrews, or small marsupials). As we shall see below, however, the story is altogether different for insects and other smaller creatures.

Publication rates provide another measure of the differential attention paid to different groups (see table 3, May 1988). Of papers listed in the *Zoological Record* over the past few years, mammals and birds average about 1 paper per species per year, reptiles, fish and amphibians about 0.5 papers per species per year,

Table 2. *Estimated numbers of canopy beetle species that are specific to the host tree species* Luehea seemanii, *classified into trophic groups (after Erwin 1983)*

trophic group	number of species	estimated fraction host-specific (%)	estimated number of host-specific species
herbivores	682	20	136
predators	296	5	15
fungivores	69	10	7
scavengers	96	5	5
Total	1100+	—	160

whereas insects and other groups average from 0.1 to 0.01 or fewer papers per species per year.

Even within a given class, or order, different families can show different patterns in the study they have received. Thus, looking at different families of insects, Strong *et al.* (1984) show that recorded species of whiteflies (Aleyrodidae: Hemiptera) and phytophagous thrips (Thripidae: Thysanoptera) increased dramatically this century with a peak around 1920–40; weevils (Curculionidae: Coleoptera) show a gradual rise since Linnaean times, again peaking around 1920–40; whereas the papilionoid and danaid butterflies (Papilionodea and Danainae: Lepidoptera) are more like birds, with broad peaks in recording rates in the second half of the 19th century.

Such patterns in rates of recording new species can be projected forward, group by group, by standard statistical techniques. In this way, we can obtain an estimate of the eventual global total of species of plants and animals. As mentioned earlier, however, the records of named species are scattered, so that estimates of the total number actually named by a given date are themselves insecure (three frequently cited estimates are 1.5 million (Grant 1973), 1.4 million (Southwood 1978), and the more current 1.8 million (Stork 1988)). Projected totals are correspondingly subject to wide fluctuations, depending on the views taken of trends in particular groups and the statistical procedures used. Simon (1983, table 31) surveys several estimates arrived at by these methods, which range from concluding that essentially all species have already been named(!) to Simon's own estimates of around 6–7 million animal species in total. Grant's (1973) projection of 4–5 million is often referred to, but it is essentially based on the crude procedure of multiplying the recorded total number of insect species (which he put at 0.75 million in 1973) by 5. It must be emphasized that all the estimates based on projecting rates of discovery pre-date the dramatic studies of tropical insects discussed in §3 and §4.

A cruder, but intuitively satisfying, estimate that the global total lies in the range of 3–5 million species has been provided by Raven (1985) and others. This estimate rests on two observations. First, among well-studied groups such as birds and mammals there are roughly two tropical species for each temperate or boreal species. Second, the majority of species are insects, for which temperate and boreal faunas are much better-known than tropical ones; overall, approximately two thirds of all named species are found outside the tropics. Thus if the ratio of numbers of tropical to temperate and boreal species is the same for insects as for mammals and birds, we may expect there to be something like two yet-unnamed species of tropical insects for every one named temperate or boreal species. This carries us from the recorded total of 1.4–1.8 million species to the crude estimate that the grand total may be around 3–5 million.

3. QUESTIONS RAISED BY STUDIES OF TROPICAL CANOPY FAUNAS

Recent studies of the arthropod faunas of tropical trees raise serious doubts about the validity of the above estimates, based as they are on projecting forward from past trends of discovery and recording. The essential point is that all the past trends are dominated by temperate-zone invertebrate groups, especially insects, whereas the recent work (surveyed in this section and the next two) suggests that tropical insect faunas may have very different patterns of diversity.

One frequently quoted upward revision to a global total of 30 million species or more comes from Erwin's (1982, 1983; Erwin & Scott 1980) studies of the insect fauna in the canopy of tropical trees. By using an insecticidal fog to 'knock down' the canopy arthropods, Erwin concluded that most tropical arthropod species appear to live in the treetops. This is not so surprising, because the canopy is where there is most sunshine as well as most green leaves, fruits and flowers.

Specifically, Erwin's original studies were on canopy-dwelling beetles (including weevils) collected from *Luehea seemannii* trees in Panama, over three seasons. As summarized in table 2, he found more than 1100 species of such beetles, which he partitioned among the categories of herbivore, predator, fungivore and scavenger.

To use the information in table 2 as a basis for estimating the total number of insect species in the tropics, one first needs to know what fraction of the fauna is effectively specialized to the particular tree species or genus under study. Unfortunately, there are essentially no systematic studies of this question even for temperate-zone trees, much less tropical ones. Erwin guessed 20% of the herbivorous beetles to be specific to *Luehea* (in the sense that they must use this tree species in some way for successful reproduction). As shown in table 2, his overall answer is more sensitive to this estimate than to the corresponding figures of 5%, 10%, and 5% for predator, fungivore, and scavenger beetles, respectively. In this way, Erwin arrived at an estimate of around 160 species of canopy beetles effectively specialized to a typical species of tropical tree.

Several other assumptions and guesses are needed to go from this estimate of 160 host-specific species of canopy beetles per tree to 30 million species in total. Slightly simplified, the argument runs as follows. First, Erwin noted that beetles represent 40% of all known arthropod species, leading to an estimate of around 400 canopy arthropod species per tree species. Next, Erwin suggested that for every two insect species in the canopy there may be one species elsewhere on the tree or the immediately neighbouring forest-floor, increasing the estimate to around 600 arthropod species effectively specialized to each species or genus-group of tropical tree. Finally, by using the estimate of 50000 species of tropical trees, Erwin arrived at the possibility that there are 30 million tropical arthropods in total.

As Erwin emphasizes, this chain of argument does not give an answer, but rather an agenda for research. Setting aside the question of whether *Leuhea* (along with its ensemble of associated lichens, vines, bromeliads, etc.) is a typical tropical tree, there are at least four areas of uncertainty: (i) what fraction of the beetle (or other) fauna on a given tree species are effectively specialized to it; (ii) do beetle species constitute the same fraction of insect faunas in the tropics as they do in better-studied temperate regions; (iii) for each insect species in the canopy, how many other species are found elsewhere or in or around a tree, and (iv) how do we scale up from the number of insect species associated with a given tree species in a given place to a global total? These four links in Erwin's chain of argument serve to organize the remainder of this section.

(a) *What fraction of the fauna is 'effectively specialized' to a given tree species?*

Suppose a given region contains 100 different tree species, and a total of 1000 different species of canopy beetles. At one extreme, all the beetles may be complete generalists, so that (setting aside sampling problems) all 1000 species could be found on any one tree species. At the opposite extreme, it could be that all beetles were complete specialists,with the total of 1000 made up by each of the 100 tree species contributing its particular 10 species of beetles. Now turn this around, and suppose that in a region where there are 100 tree species we find 100 species of canopy beetles in exhaustive sampling from just one tree species. How many beetle species are there likely to be in total? The above argument suggests that, if the studied tree is indeed representative, the total number of canopy beetle species lies somewhere between 100 (all generalists) and 10000 (all highly specialized).

Suppose we are investigating a region containing a total of M different tree species. The above issues can now be made precise by formally defining $p_k(i)$ to be the fraction of canopy beetles (or other insects) found on the tree species labelled k ($k = 1, 2, \ldots, M$), which utilize a total of i different tree species (including the one labelled k). Further, let N_k denote the total number of canopy beetles (or other insects) found on tree species k. Then, of the beetles found on tree species k, $N_k p_k(1)$ are found only there, $N_k p_k(2)$ are found on k and on only one other tree species, and so on. The total number of distinct canopy-beetle species, N, in the region is then given by the sum:

$$N = \sum_{k=1}^{M} \sum_{i=1}^{M} (1/i)\, N_k\, p_k(i). \tag{1}$$

That is, we sum the numbers of beetle species over all tree species, discounting each tree by the factor $1/i$ for those of its beetle species found on i tree species.

The 'effective average' number of beetle species per tree species, or overall average number of beetles 'effectively specialized' to each tree species, is then N/M, with N given by equation (1). For the tree species labelled k, the proportion of its beetle species that are 'effectively specialized' to it, f_k, is given by

$$f_k = \sum_{i=1}^{M} (1/i)\, p_k(i). \tag{2}$$

The corresponding overall proportion of beetle species 'effectively specialized', averaged over all tree species, is denoted by f:

$$f = N/\sum_k N_k. \tag{3}$$

Here N is given by equation (1). Equation (3) says that the overall average fraction of beetle species 'effectively specialized' to each tree species, f, is given by the total number of different species divided by the sum of the totals from each tree species, without discounting for overlaps.

Unfortunately, there is to my knowledge not a single systematic study of the distribution functions $p_k(i)$ for the apportionment of any faunal group among any group of plant or tree species. At best there are some studies, essentially all for temperate communities, which distinguish among monophagous, oligophagous, and polyphagous insects (roughly, those feeding on a single species or genus, on a family, or more generally, respectively). The assumptions codified in table 2 correspond to assuming that f_k is 0.20, 0.05, 0.10, 0.05 for herbivore, predator, fungivore, scavenger beetles in the canopy of *Leucea*, respectively; but these are guesses.

Although quantitative studies are lacking, the formal definitions embodied in equations (1–3) can at least alert us to some of the pitfalls in intuitive estimates of the proportion of beetle species that are 'effectively specialized'. Suppose, for example, that for a given group of insects in a particular tree species (labelled 1) the distribution function, $p_1(i)$, has the geometric-series form $p_1(i) = c\alpha^i$ (with $\alpha < 1$). This says that the fraction of species utilizing i tree species decreases geometrically as i increases. The normalization constant, c, ensures that $\Sigma\, p_1(i) = 1$ (if the total number of tree species is large, $M \to \infty$, then $c \to (1-\alpha)/\alpha$). In this event, the fraction of the insect species on tree species 1 that are effectively specialized to it, f_1, can be seen to be given

Table 3. *A very preliminary study of the distribution of beetle species among four species of British oaks, to show ideas about the distribution function, $p_k(i)$, and definitions of 'effective specialization' (from unpublished data provided by T. R. E. Southwood)*

oak species	number of beetle species found on a given oak species, N_k	number of beetle species found on a given oak species, that are found on a total of i of the 4 oak species (proportion found on i species, $p_k(i)$) $i = 1$	2	3	4	proportion of beetle species effectively specialized on this oak species, f_k, from eqn (2)
cerris	28	5 (0.18)	8 (0.29)	7 (9.25)	8 (0.29)	0.48
ilex	47	20 (0.43)	10 (0.21)	9 (0.19)	8 (0.17)	0.64
petraea	48	14 (0.29)	15 (0.31)	11 (0.23)	8 (0.17)	0.57
robur	50	15 (0.30)	15 (0.30)	12 (0.24)	8 (0.16)	0.57
totals	173	54	24	13	8	average, $f = 0.57$

(in the limit of very large M) by $f_1 \simeq [(1-\alpha)/\alpha] \ln[1/(1-\alpha)]$. But an estimate of f_1 based on biological intuition might simply ask about the 'monophagous' fraction, that are found only on that tree species or genus-group; this would lead to the estimate $f_1 \simeq p_1(1) = 1-\alpha$. Suppose that in fact $p_1(1)$, the fraction monophagous within this distribution, is 20% (that is, $\alpha = 0.8$). Then in this illustrative example the fraction 'effectively specialized' is $f_1 = 0.40$, which is twice as large as might be guessed on intuitive grounds.

What are needed, of course, are not speculative abstractions, but analytic studies of the distribution functions $p_k(i)$ for particular floral and faunal communities. By way of illustration, table 3 shows a very preliminary analysis of the distribution of beetle species among four species of oak trees, *Quercus cerris*, *Q. ilex*, *Q. petraea* and *Q. robur* in Wytham Wood, Oxford (data provided by T. R. E. Southwood). The data in table 3 represent the species 'knocked down' by insecticidal fog (specifically, from 6 foggings per tree species in the months of May and June). On this basis, the four above-mentioned oak species show totals of 28, 47, 48 and 50 beetle species, respectively, for a gross total of 173 species. But in fact only 99 distinct beetle species are present, with the pattern of overlap as shown in table 3. The table also shows that, for this example, the fraction of the beetle species that are 'effectively specialized' to a given oak species (as defined by the formulae above) is roughly around 60%. It must be emphasized that this example is presented for illustrative purposes, to make the above ideas more concrete; it is a very crude and preliminary analysis of a fragment of a much larger data set.

The definition of the distribution function $p_k(i)$ was made on biological grounds, based on the number of tree species whose resources are utilized in some way by a particular insect species. The data in table 3 are, however, surely dominated by sampling considerations. A few of the beetle species in table 3 are represented by only one individual, and many of the species that utilize these oaks are absent from the samples altogether. Moreover, the table does not distinguish between species that are known to utilize a given tree, and those that accidentally happen to be on it when the sample is collected by fogging (Moran & Southwood's (1982) 'tourist' species). In this respect the data presented in table 3 are similar to those likely to be collected in most studies of canopy and other faunas: the patterns of distribution of insect species among tree species will be a mixture of the underlying biological patterns of resource use (and of specialization and generalization), clouded or often dominated by sampling effects. I believe there is a need here both for theoretical studies (clarifying the interplay between sampling effects and the underlying distributions of resource utilization, among other things) and for empirical investigations. Such work will ultimately illuminate fundamental aspects of the structure of ecological communities, and will contribute to more reliable estimates of the global total of species. My aim in this sub-section, which obviously provides questions not answers, is to stimulate such further work.

The coarse and conventional division of phytophagous insects into monophages, oligophages and polyphages may be sharpened somewhat for the phytophagous beetles on oak trees in Southwood's study. By using standard handbooks (Fowler 1913; Joy 1931; Portevin 1935), it is possible to determine: (i) what fraction of the approximately 100 species are restricted to oaks (that is, to the genus *Quercus*); (ii) what fraction to the family Fagiacea, or (iii) the order fagaces, as opposed to (iv) those found generally on angiosperms, or (v) even more broadly. A tentative analysis of Southwood's data suggests the proportions of phytophagous beetle species in the five categories thus defined are 0.10, 0.03, 0.06, 0.77 and 0.04, respectively.

If these exceedingly rough figures were taken as representative, they would suggest that Erwin's estimate that 20% of herbivorous canopy beetles are host-specific be replaced by a figure of 10% or less (remember, the 10% above is for the oak genus, not for a particular species). This would halve the estimated global total, to 15 million species or fewer. And if tropical beetles are typically less specialized than temperate ones, as they may need to be in response to the much patchier distribution of a vastly greater diversity of tree species, then the percentage that are host-specific (and thence the estimated global species total) could be smaller yet. On essentially these grounds, Stork (1988) had suggested that the effective fraction of host-specific beetle species in tropical canopies may be closer to 5% or less than to 20% (with a corresponding global total of 7 million or fewer species); see also Beaver (1979). There is clear need both for more theory (dealing with f_k and the way sampling and other effects influence it), and for more facts.

(*b*) *In particular settings, what proportion of insect species are beetles?*

Although roughly 40 % of all recorded insect species are beetles, the existing handful of detailed studies suggest that beetle species constitute a smaller fraction of insect faunas collected by fogging tropical or temperate trees.

Southwood *et al.* (1982) sampled the invertebrate fauna (essentially all of which were insects) of six tree species in both Britain and South Africa, by using pyrethrum knockdown. In Britain, the total number of invertebrate species on the different tree species varied from 176 to 465, and the beetle species constituted, on average, 7 % of the total (ranging from a low of 5 % on *Salix* to a high of 11 % on *Quercus*; see table 1 in Southwood *et al.* (1982)). The corresponding species totals on South African trees ran from 105 to 300, of which an average of 19 % were beetle species (ranging from a low of 15 % on *Salix* to a high of 23 % on *Erythrina* and *Quercus*). Of the roughly 2800 species represented in the 24000 arthropod individuals collected in fogging samples from 10 trees in Borneo by Stork (1988, figure 3), approximately 30 % were beetles.

These numbers have led Stork to speculate that beetles may more typically constitute around 20 % of the insect species found in tree canopies, rather than the 40 % that is characteristic of beetles more generally. All other things being equal, such a revision to this link in Erwin's chain of argument would double the global total number of species. My view is that these different results from different studies speak to the complexity and contingent nature of the underlying questions. Until such time as we have a better understanding of why beetles (or any other group) make up such different fractions of the species totals in different trees at different places, I would be inclined to stay with the overall average figure of 40 % beetle species when making global estimates by 'scaling up' local studies.

(*c*) *What is the ratio of canopy insects to all insects on a tree?*

There are no studies of all the arthropod species found on the various parts of a tropical tree, but there are some studies of how the numbers of individuals are distributed.

Stork (1988) used a range of techniques to sample the fauna from each of five parts of trees in the lowland rainforest of Seram in Indonesia: canopy; tree trunk; 'ground vegetation' (from 0 to 2 m above the forest floor); leaf litter; and soil (the first few cm). In this site, Stork estimates there were approximately 4200 individuals per square metre, distributed among the five categories listed above as 1200, 50, 10, 600 and 2400. Roughly 70 % of all individuals are found in the soil and leaf litter, mostly springtails (Collembola) and mites (Acarina). In the canopy, which contains about 14 % of all individuals, ants are the most abundant arthropods (constituting roughly 43 % of the individuals). If numbers of species bore a direct relation to numbers of individuals, we can see that Erwin's estimate of one insect species elsewhere on the tree for every two species in the canopy would be out by almost a factor 10. But although ants tend to be more abundant than beetles in tropical canopies, they contribute significantly fewer species. The same seems likely to be true for the springtails in the soil and leaf litter. I have serious doubts, however, whether it is also true for the mites (moreover, the number of individual mites in the soil and leaf litter is typically an order-of-magnitude larger than the number of beetle individuals in the canopy).

Adis and colleagues (Adis & Schubart 1984; Adis & Albuqueque 1989) have made similar studies of the arthropod faunas of seven different types of Amazonian forests and plantations, using soil cores (0–7 cm), ground and arboreal traps of various kinds, and canopy fogging. Like Stork, they found the majority of individuals were Collembola and Acarina in the soil and leaf litter (with two thirds or more of these in the top 3.5 cm of the soil). The main difference between these Amazonian studies and the Indonesian ones is in the total numbers of individuals, which ranged as high as 30000 or more per square metre at some Amazonian sites. But the divisions of numbers of individuals among canopy, trunk, ground vegetation, leaf litter and soil are broadly similar in all the studies, as are the broad apportionments of individuals among taxonomic groups.

Understandably but unfortunately, all of these studies are of numbers of individuals in the different parts of the tree, not of the numbers of species. Until more is known, a consensus guess may be that for every insect species in the canopy there is one elsewhere in the typical tropical tree (rather than Erwin's 0.5). But this could lead to a serious underestimate of the global total, if the mites turn out to be as surprisingly diverse as the tropical canopy beetles.

(*d*) *How do we scale up from insect species per tree to a global total?*

Even if we knew how many insect species were effectively specialized to a truly representative species of tropical tree at a particular place, there remain serious difficulties in scaling this up to the roughly 50 000 species of tropical trees. The reason that we cannot simply multiply the number of species per tree by the number of tree species is twofold: on the one hand, the same tree species may play host to different insects in other parts of its range (so that simple multiplication underestimates the global total); on the other hand, the same insects may be effectively specialized on other tree species in other regions (so that simple multiplication could lead to overestimation).

Ward's (1977) study of Juniper in Britain is an example of the first of these two complications. Looking at each one degree band of latitude in the range of Juniper, from 50° N to 59° N, she found the recorded numbers of species of arthropods varied from 5 to 20. The total number of species associated with Juniper in Britain in this study is 27, which is larger by a factor 1.4–5.0 than the total that might be inferred from any one regional study.

The converse phenomenon, the same species specialized to different plants in different places, is shown by Thomas's (1990) analysis of the *Passiflora–Heliconius* system in South America. For this unusually well-studied association, Thomas averages over 12 sites to find a mean of 7.2 Passifloraceae species and 9.7 Heliconiinae species per site (he uses families or subfamilies, because a few Heliconiinae are not in the genus *Heliconius*, and likewise a few Passifloraceae are not in the genus *Passiflora*). The total number of recorded Passifloraceae species in the neotropics is 360. A naive scaling up may therefore suggest 360 times $(9.7/7.2) \simeq 485$ Heliconiinae species. But this total is in fact only 66 species. The complicating factor is, of course, that these relatively specialized butterflies use taxonomically different (although ecologically similar) species of Passifloraceae in different parts of the neotropics, with the result that simple multiplicative estimates are an order-of-magnitude too high. There are further complications and caveats here. For one thing, it could be that, to some extent, systematists have tended to split the Passifloraceae and lump the Heliconiinae (Gilbert & Smiley (1978) refer to one downward revision from 70 to 7 species of *Heliconius*). For another thing, only some 100–150 species of Passifloraceae are found below 1500 m, which represents an upper limit to the range of Heliconiinae species (Gilbert & Smiley 1978), so that simple multiplication might more properly suggest 135–200 Heliconiinae species. This is still significantly larger than the 66 that exist.

The essential point is that the simple procedure of multiplying the average number of species per tree by the number of tree species can be misleading for those insect species with ranges significantly larger than that of the host-tree species in question, and conversely for tree species whose ranges exceed that of the insects in a given region. We cannot begin to resolve these problems until we have a better understanding of the patterns in the distributions of geographical ranges for plant and animal species.

Despite some pioneering efforts (Rappoport 1982; Hanski 1982; Brown 1984; Root 1988), little is yet known about range-size patterns for individual species, much less for joint distributions of, for instance, herbivorous insects and their host plants. Figure 3 summarizes one recent analysis of the distributions of range sizes for all North American mammals, showing that most species have relatively restricted ranges (the median range size is about 1% of the area of North America), but that there is considerable variability. On the whole, it seems likely that for tree species with below-average ranges, we may need to acknowledge that many of the insects that are effectively specialized to them range more widely, using other trees in other places (so that Erwin's simple multiplicative procedure may overestimate totals). Conversely, trees with above-average ranges may embrace different faunas in different places, resulting in underestimated totals if one location is treated as definitive. Ultimately, such questions of range-size and geographical distribution shade into larger questions of how we define commonness and rarity, and how patchiness and regional

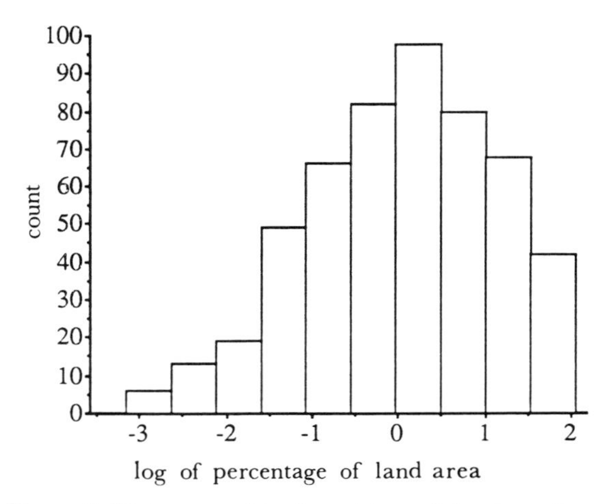

Figure 3. Histogram of the logarithms of the sizes of species's geographical ranges for 523 species of North American non-aquatic mammals. The range size is expressed as a percentage of the total land area of North America. The distribution is roughly lognormal in shape, with the proportion of species in the different (logarithmic) range-size categories falling away roughly symmetrically on either side of the median value of around 1%. After Pagel *et al.* (1990).

differentiation affect extinction probabilities (Rabinowitz *et al.* 1986; Pagel *et al.* 1990; Hubbell, this symposium).

I end this section by circling back to Erwin's method of estimating the total number of species on Earth. Suppose we accept a total of 1200 beetle species in the canopy of a tropical tree species as typical. Further assume there are 2.5 other canopy insect species for each beetle species (40% of insects are beetles), and one species elsewhere on the tree for each one in the canopy. Suppose a fraction, f, of this fauna is effectively specialized to the tree species in question. Finally, assume we can simply multiply the effective number of insect species per tree species by the number of tropical tree species (50000) to arrive at the grand total. This total is then $(300f)$ million. If, overall, 10% of the insects are effectively specialized to a given tree, we recover Erwin's 30 million. To reduce an estimate along these lines down to the previously-conventional figure of around 3 million, we need to assume that only 1% of the insect species are effectively specialized, which seems a bit low. I return to this below.

4. DIRECT ESTIMATES OF GLOBAL TOTALS

Hodkinson & Casson (1990) have presented a very direct way of estimating species totals, based essentially on determining what fraction of a thoroughly sampled group, in a particular region, have previously been described.

Specifically, they use several methods to sample the bug (Hemiptera) fauna of a moderately large and topographically diverse region of tropical rainforest in Sulawesi, in Indonesia, over a one-year period. Hodkinson & Casson (1990) estimate that their samples represent the bug fauna on the roughly 500 tree species in their study sites in Dumoga Bone

National Park in Sulawesi. They found a total of 1690 species of terrestrial bugs, of which 63% were previously unrecorded (this proportion of undescribed species is weighted according to the number of species per family within the group).

If the bugs in Sulawesi are representative of tropical insects more generally, then we may say that the total number of recorded insect species (approximately 1 million) represents 37% of the real total, leading to an estimated 2.7 million insect species in total. Hodkinson & Casson give a somewhat more detailed argument, with several intermediate steps, to reach an estimated total of 1.8 to 2.6 million, but I believe the more direct approach outlined above contains the essentials.

Hodkinson & Casson also use their data to give a different derivation of the global total number of insects. First, if they find roughly 1700 bug species on 500 tree species, we might expect 170000 bug species on the tropical total of 50000 tree species. Second, bugs comprise approximately 7.5% of all described insects (Southwood 1978), so we can estimate a grand total of 2.3 million insects (again, Hodkinson & Casson's somewhat more detailed estimate gives a total of 1.9–2.5 million). This second estimate is surprisingly close to the first, and much more direct, estimate of 2.7 million. This rough coincidence between the two estimates is not trivial or tautological; it can be seen to arise essentially because the factor of 100 in the ratio between tree species globally and in Dumoga Bone Park is roughly cancelled by the factor of 109 in the ratio between all known bug species and the number of known species in their study (or 114 if aquatic bugs are included in the global total).

This is a brief summary of an important study. There is obviously an enormous amount of work in completely sampling and classifying any group of tropical insects, and one must always worry whether the group or site is representative of more general patterns. But I see this as the simplest and most direct route to estimates of global totals. When other taxa are added, Hodkinson & Casson's estimate of insect species totals carries us back to the previously conventional global total of 3–5 million species. On the other hand, to reconcile Hodkinson & Casson's estimate with Erwin's data, we need to assume that over 400 of his roughly 1200 beetle species in the canopy of *Leuhea* have been described and recorded, which seems unlikely to me.

5. SPECIES TOTALS FROM SPECIES–SIZE RELATIONS

An altogether different approach to estimating how many species there are derives from examining patterns in the numbers of terrestrial animal species in different body-size categories (May 1978, 1988). Very roughly, as one goes from animals whose characteristic linear dimension is a few metres down to those of around 1 cm (a range spanning many orders-of-magnitude in body weight), there is an approximate empirical rule which says that for each tenfold reduction in length (1000-fold reduction in body weight) there are 100 times the number of species.

This empirical relation begins to break down at body sizes below 1 cm in characteristic length. As the relation itself is not understood, this break-down may mean nothing. But the break-down may plausibly be ascribed to our incomplete record of smaller terrestrial animals, most of which may be unrecorded tropical insects. If the observed pattern is arbitrarily extrapolated down to animals of characteristic length around 0.2 mm, we arrive at an estimated global total of around 10 million species of terrestrial animals (May 1988).

This frankly phenomenological estimate would be more interesting if we had a better understanding of the physiological, ecological or evolutionary factors generating species-size distributions (Lawton, this symposium).

6. SOME MORE GENERAL CONSIDERATIONS

This section touches briefly on some more general questions that are prompted by the above discussion.

(*a*) *Different patterns for different groups?*

We have seen that Hodkinson & Casson's direct estimates of global species totals, based on the fraction of bug species in their Sulawesi samples that were previously known, agree with estimates of around 3 million species of insects that earlier were obtained by projecting past trends of discovery. But conservative reappraisals of Erwin's estimate, based on beetle species in tropical canopies, suggest totals of at least 7 million and possibly more. To produce a total of as little as 3 million insects from Erwin's or other similar data requires that we assume only 1% of the beetle fauna are effectively specialized to a given tree species, or that more than 400 of the beetle species found in the canopy of *Leuhea* have been previously recorded; these assumptions seem a bit extreme.

One way of squaring this circle is to recognize that very different patterns may pertain to different taxonomic groups, or in different settings. We have already seen hints of this in the canopy fauna. Tobin's preliminary analysis of data from tree canopies in Manu National Park in Amazonian Peru suggests ants constitute 70% of the individuals and beetles less than 10%, but that there are many more beetle species than ant species (May 1989). Similarly, Stork's fogging samples from 10 Bornean trees were dominated by ants (4489 individuals), but they only contributed 99 of the total of 2800 species; one family of beetles contributed 1455 individuals but 739 species. In tropical canopies generally, ants contribute many individuals, typically half or more, but relatively few species, while beetles contribute relatively few individuals but many species. It seems likely to me that patterns of 'effective specialization' will also vary greatly from group to group, so that we need a much better understanding of the ecology of specific communities of organisms, before we can draw general conclusions about species diversity from limited studies of particular groups in particular places.

(*b*) *Food web structure and species totals*

Studies of the structure of food webs are beset with many difficulties of biases and inconsistencies in the way data are assembled. Provocative generalizations are nevertheless emerging (Cohen *et al.* 1990: Yodzis 1990; Nee 1990; Lawton 1989). One of these is the suggestion by Lawton and co-workers (for example, Strong *et al.* (1984)) that there may typically be something like 10 species of phytophagous insects for each plant species. Applied globally, and given that there are at least 300000 recorded species of higher plants, this would put a lower limit of around 3 million on the number of insect species. However, Lawton's ratios come mainly from studies of temperate-zone plants, and corresponding ratios may be higher for tropical plants (which would tend to support higher estimates of the global total).

Similar arguments about food web structure suggest there could be, on average, as many as five species of parasitoids (parasitic wasps and flies that lay their eggs on or in larval or pupal stages of other insects) for each phytophagous insect species, in both tropical and temperate regions (Hawkins 1990). Applied generally, this estimate could clearly escalate the global total to very high levels (Hochberg & Lawton 1990). I think such estimates are probably too extravagant, although they do suggest that closer studies of tropical hymenopterans may produce surprises to match Erwin's work on beetles.

Such estimates of global species totals, based on the number of plant species combined with the structure of food webs, are akin to the estimates in § 5 based on species–size relations. They provide independent lines of attack upon the problem, related only obliquely to direct estimates based on counting species.

(*c*) *Microbial diversity*

Throughout this chapter, I have followed the usual practice of assuming species diversity to mean the numbers of animal species, or sometimes plant and animal species. More precisely, five kingdoms are usually recognized, distinguished by different levels of cellular organization and modes of nutrition. Two of these kingdoms, the prokaryotic monerans and the eukaryotic protists, comprise microscopic unicellular organisms, and together they account for something like 5% of recorded living species. The fungal and plant kingdoms represent roughly another 22% of species. The animal kingdom thus does comprise the majority (more than 70%) of all recorded living species. The extent to which these recorded numbers of species in the different kingdoms accurately reflect their relative diversities is, however, open to question. As one moves down the size-spectrum of organisms, from the romantic large mammals and birds, through nondescript small anthropods, on down to protozoan, bacterial and viral species, not only does concern for diversity and conservation fall away, but it even changes sign. In the Smithsonian Institution in Washington, a touching label attached to Martha, the last passenger pigeon, laments her death in 1914, but no-one mourned the passing of the last smallpox virus.

Regardless of the amount of study they have received, the microorganisms that act as decomposers in the soil and leaf litter are crucial to the functioning of ecosystems. Recent work has, moreover, revealed that the diversity in natural populations of microbial organisms is far greater than that found in conventional studies of laboratory cultures (Olsen 1990). Ward *et al.* (1990) examined the ribosomal RNA sequences from a well-studied photosynthetic microbial mat from a hot spring in Yellowstone National Park, and found eight distinct sequence types, none of which were the same as any of the 12 laboratory-cultured prokaryotes believed to be characteristic of this mat. More surprisingly, only one of the eight sequences bears any close resemblance to a recognized bacterial 'phylum'. Broadly similar results were obtained by Giovannoni *et al.* (1990) from studies of ribosomal RNA gene sequences of microorganisms from samples of ocean water. These studies of natural populations of microorganisms, as distinct from laboratory cultures, are, in their own way, even more astonishing than Erwin's and others' revelations about tropical canopy faunas. They mark the advent of a new stage in our understanding of microbial ecology and diversity. We have not yet begun to address the questions that such studies of naturally occurring microbial diversity raise, for example, for the release of genetically engineered organisms.

(*d*) *What is a species?*

Up to this point, I have deliberately avoided any definition of what is meant by a species, nor will I pursue it in detail here. But a few remarks should be made.

First, some researchers recognize more species than others, even within well-studied groups. This is particularly noticeable for asexually reproducing organisms. For examples, some taxonomists see around 200 species of the parthenogenetic British blackberry, others see only around 20 (and a 'lumping' invertebrate taxonomist may concede only two or three). Some strongly inbreeding populations are almost as bad, with 'splitters' seeing an order-of-magnitude more species than do 'lumpers'; *Erophila* and *Arabidopsis*, British plants in the mustard family, are notable examples.

Second, and more fundamentally, studies of nucleotide sequences show homologies of less than 50% (as revealed by DNA hybridization) for different strains of what is currently classified as a single bacterial species, *Legionella pneumophilia* (Selander 1985). This is as large as the characteristic genetic distance between mammals and fishes. Relatively easy exchange of genetic material among different 'species' of microorganisms could mean that basic notions about what constitutes a species are significantly different for vertebrates than for bacteria.

Third, the increasing armamentarium of techniques for exploring the genetic structure of populations is uncovering further complexities in what we mean by a species. For example, *Neomachilellus scandens*, a member of the Meinertellidae, is found in inundation forests in the Rio Negro valley (which are flooded for 5–6 months each year), and in primary and secondary dryland forests near Manaus. Ecological studies show the species to be univoltine in inundation forests, with a defined reproductive period, but polyvoltine in dryland forests. These two ecologically distinct forms of *N. scandens* are effectively indistinguishable on morphological grounds, and traditional taxonomy would at most recognize them as 'biotype-specific races'. But protein analyses by Adis (1990), using electrophoresis to test 15 enzymes, shows that there is no gene flow between the univoltine and polyvoltine populations, even when the typical spatial separation is less than 50 m. Adis (1990) proposes that on ecological and genetic grounds we should recognize two species here, and that this situation may be much commoner in the tropics than is currently recognized. If so, we could be dealing with even more species than suggested by Erwin.

(*e*) *Are some species more equal than others?*

Human activities are destroying natural habitats, and the associated biota, at rates that are probably without precedent in the history of life on Earth. In particular, the clearance and burning of an area of tropical forest roughly the size of Britain each year is surely contributing to accelerating losses of tropical arthropods and other animals. Against this background, we need to go beyond knowing how many species there are, and to use this knowledge to optimize conservation efforts (either in planning National Parks and other protected regions, or in devising strategies that reconcile sustainable exploitation of resources with preservation of an appreciable fraction of the original fauna).

Such conservation efforts will pose increasingly difficult choices. Should all species be treated as equal? Or should we take the view that, for conservation purposes, a species not closely related to other living species is more important than one with many widely distributed congeners? And if the answer to this latter question is yes, how do we quantify the relative importance of different species?

Vane-Wright *et al.* (1990) have made a beginning of this task, showing how taxonomy and systematics can build from species lists toward assessment of the relative distinctness of different species and, ultimately, communities. At one extreme, we could of course regard all species as equally important. At the opposite extreme, we could take a phylogenetic or cladistic tree, which represents the hierarchical relations among the constituent species, and measure taxonomic distinctness by weighting each group equally with respect to the summed weights of their terminal taxa. This scheme has the merit of recognizing taxonomic distinctness, but it has the fault that taxonomic rank overwhelms species numbers: on this basis, the two Tuatara species would be seen as equal to all 6000 other reptile species taken together. Vane-Wright *et al.* propose an intermediate scheme, which quantifies the amount of information contained in a given hierarchical classification. Their method gives answers that depend on the topology of the hierarchy (even for a fixed number of terminal taxa), and that recognize the importance of, for example, Tuataras without amplifying it out of all proportion (see May 1990).

My purpose here is not to dwell on the details of these pioneering efforts, but rather to emphasize that a more complete understanding of how many species there are is only a first step. More and more, conservation efforts will be faced with difficult choices. It will be helpful to be guided by quantitative measures of distinctness that are based on systematic understanding of the phylogenetic relations among species.

7. CONCLUDING REMARKS

Why should we care how many species there are? One line of argument is narrowly utilitarian. Thus, for instance, essentially all modern medicines and other pharmaceutical products have been developed from natural products, and so it would seem sensible to be looking at the other shelves in the larder rather than destroying them. The triumphs of intensive agriculture have been accompanied by progressive narrowing of the genetic diversity of the plants we exploit. The likelihood of global changes in climate gives fresh emphasis to the desirability of conserving existing gene pools and exploring the possibility of utilizing new plants.

Important though such considerations are, I think a more pressing and more basic utilitarian reason for studying and cataloguing diversity is because it is a prerequisite to understanding how biological systems work. The scale and scope of human activity are now so large that they rival the natural processes that created and maintained the biosphere as a place where life can flourish. Current rates of input of carbon dioxide, chlorofluorocarbons and other gases are beginning to disturb the balance of the biosphere. Most of our uncertainties about the long-term consequences stem from uncertainties about how physical and biological processes are coupled. We need to understand the structure and functioning of ecosystems, particularly tropical ecosystems, much better than we do. And we cannot hope to do this if we do not even know what is there, and why tropical diversity is what it is.

Beyond these practical motivations, I believe we need to understand the diversity of living things for the same reasons that compel us to reach out toward understanding the origins and eventual fate of the universe, or the structure of the elementary particles that it is built from, or the sequence of molecules within the human genome that code for our own self-assembly. Unlike these other quests, understanding and conserving biological diversity is a task with a time limit. The clock ticks faster and faster as human numbers continue to grow, and each year 1–2 % of the tropical forests are destroyed. Future generations will, I believe, find it incomprehensible that Linnaeus still lags so far behind Newton, and that we continue to devote so little money and effort to understanding and conserving the other forms of life with which we share this planet.

I have been helped by Paul Harvey, Michael Hassell, Ian Hodkinson, John Lawton, Nigel Stork, Chris Thomas and many others. I am particularly grateful to Sir Richard Southwood for letting me use his unpublished data. This work was supported, in part, by the Royal Society and by NERC (through its Interdisciplinary Research Centre at Silwood Park).

REFERENCES

Adis, J. 1990 Thirty million arthropod species – too many or too few? *J. Tropical Ecol.*, **6**, 115–118.

Adis, J. & Albuqueque, M. O. 1990 Impact of deforestation on soil invertebrates from Central Amazonian inundation forests and their survival strategies to long-term flooding. *Wat. Quality Bull.* **14**, 88–98.

Adis, J. & Schubert, H. O. R. 1984 Ecological research on arthropods in Central Amazonian forest ecosystems with recommendations for study procedures. In *Trends in ecological research in the 1980s* (ed. J. H. Cooley & F. B. Golley), pp. 111–144. New York: Plenum Press.

Beaver, R. A. 1979 Host specificity of temperate and tropical animals. *Nature, Lond.* **271**, 139–141.

Brown, J. H. 1984 On the relationship between abundance and distribution of species. *Am. Nat.* **124**, 255–279.

Cohen, J. E., Briand, F. & Newman, C. M. 1990 *Community food webs: data and theory.* New York: Springer–Verlag. (*Biomathematics* **20**)

Diamond, J. M. 1985 How many unknown species are yet to be discovered? *Nature, Lond.* **315**, 538–539.

Erwin, T. L. 1982 Tropical forests: their richness in Coleoptera and other arthropod species. *Coleopt. Bull.* **36**, 74–82.

Erwin, T. L. 1983 Beetles and other insects of tropical forest canopies at Manaus, Brazil, sampled by insecticidal fogging. In *Tropical rain forest: ecology and management* (ed. S. L. Sutton, T. C. Whitmore & A. C. Chadwick), pp. 59–75. Oxford: Blackwell.

Erwin, T. L. & Scott, J. C. 1980 Seasonal and size patterns, trophic structure and richness of Coleoptera in the tropical arboreal ecosystem: the fauna of the tree *Luehea seemannii* in the Canal Zone of Panama. *Coleopt. Bull.* **34**, 305–335.

Fowler, W. W. 1913 *British Coleoptera* (6 volumes). London: Lovell, Reeve and Co.

Gilbert, L. E. & Smiley, J. T. 1978 Determinants of local diversity in phytophagous insects: host specialists in tropical environments. In *Diversity of insect faunas* (ed. L. A. Mound & N. Waloff), pp. 89–104. Oxford: Blackwell Scientific.

Giovannoni, S. J., Britschgi, T. B., Moyer, C. L. & Field, K. G. 1990 Genetic diversity in Sargasso Sea bacterioplankton. *Nature, Lond.* **345**, 60–63.

Grant, V. 1973 *The origin of adaptations.* New York: Columbia University Press.

Hanski, I. 1982 Dynamics of regional distribution: the core and satellite species hypothesis. *Oikos* **38**, 210–221.

Hawkins, B. A. 1990 Global patterns of parasitoid assemblage size. *J. Anim. Ecol.* **59**, 57–72.

Hochberg, M. E. & Lawton, J. H. 1990 Competition between kingdoms. *Trends Ecol. Evol.* (In the press.)

Hodkinson, I. D. and Casson, D. 1990 A lesser predilection for bugs: Hemiptera (Insecta) diversity in tropical rain forests. *Biol. J. Linn. Soc.* (In the press.)

Joy, N. A. 1931 *A practical handbook of British beetles.* (2 volumes.) London: Witherby.

Lawton, J. H. 1989 Food webs. In *Ecological concepts* (ed. J. M. Cherret), pp. 43–78. Oxford: Blackwell Scientific.

May, R. M. 1978 The dynamics and diversity of insect faunas. In *Diversity of insect faunas* (ed. L. A. Mound & N. Waloff), pp. 188–204. Oxford: Blackwell.

May, R. M. 1988 How many species are there on earth? *Science, Wash.* **241**, 1441–1449.

May, R. M. 1989 An inordinate fondness for ants. *Nature, Lond.* **341**, 386–387.

May, R. M. 1990 Taxonomy as destiny. *Nature, Lond.* **347**, 129–130.

Moran, V. C. & Southwood, T. R. E. 1982 The guild composition of arthropod communities in trees. *J. Anim. Ecol.* **51**, 289–306.

Nee, S. 1990 Community construction. *TREE* **5**, 337–340.

Olsen, G. J. 1990 Microbial ecology: variation among the masses. *Nature, Lond.* **345**, 20.

Pagel, M. D., May, R. M. & Collie, A. 1990 Ecological aspects of the geographical distribution and diversity of mammal species. *Am. Nat.* (In the press.)

Portevin, G. 1935 *Histoire Naturelle des Coléoptère de France* (4 volumes.) Paris: Lechavier.

Rabinowitz, D. Cairns, S. & Dillon, T. 1986 Seven forms of rarity. In *Conservation biology* (ed. M. E. Soule), pp. 182–204. Sunderland, Massachusetts: Sinauer.

Rapoport, E. H. 1982 *Areogeography: geographical strategies of species.* Oxford: Pergamon.

Raven, P. H. 1985 Disappearing species: a global tragedy. *The Futurist* **19**, 8–14.

Root, T. 1988 *Atlas of wintering North American birds.* Chicago: Chicago University Press.

Selander, R. K. 1985 Protein polymorphism and the genetic structure of natural populations of bacteria. In *Population genetics and molecular evolution* (ed. T. Ohta & K. Aoki), pp. 85–106. Berlin: Springer–Verlag.

Simon, H. R. 1983 Research and publication trends in systematic zoology. Ph.D. thesis London: The City University.

Southwood, T. R. E. 1978 The components of diversity. In *Diversity of insect faunas* (ed. L. A. Mound & N. Waloff), pp. 19–40. Oxford: Blackwell.

Southwood, T. R. E., Moran, V. C. & Kennedy, C. E. J. 1982 The richness, abundance and biomass of the arthropod communities on trees. *J. Anim. Ecol.* **51**, 635–649.

Stork, N. E. 1988 Insect diversity: facts, fiction and speculation. *Biol. J. Linn. Soc.* **35**, 321–337.

Strong, D. R., Lawton, J. H. & Southwood, T. R. E. 1984 *Insects on plants: community patterns and mechanisms.* Oxford: Blackwell Scientific.

Thomas, C. D. 1990 Estimating the number of tropical arthropod species. *Nature, Lond.* **347**, 237.

Vane-Wright, R. I., Humphries, C. J. & Williams, P. H. 1990 What to save? Systematics, the biodiversity crisis, and the agony of choice. *Biol. Conserv.* (In the press.)

Ward, D. M., Weller, R. & Bateson, M. M. 1990 16S rRNA sequences reveal numerous uncultured microorganisms in a natural community. *Nature, Lond.* **345**, 63–65.

Ward, L. K. 1977 The conservation of juniper: the associated fauna with special reference to Southern England. *J. appl. Ecol.* **14**, 81–120.

Yodzis, P. 1990 *Introduction to theoretical ecology.* New York: Harper and Row.

Discussion

R. J. H. Beverton (*Montana, Old Roman Road, Gwent, U.K.*). The 'target' of the total number of species is not, of course, static because species are continually being lost by natural and man-made processes and added by speciation. Could Professor May say something about these rates of change, and how they compare with the overall picture of species diversity?

R. M. May. It is true that throughout most of the recorded history of life on Earth, species have been going extinct and new species have been appearing. The next half-century or so is, however, likely to be singular in two respects. First, there is the scope and timescale of impending extinctions in relation to speciation. For most of the time, extinction rates and speciation rates have been roughly in balance, and both have been on geological timescales: something like half the species currently extant appeared over the past 50–100 million years or so. Over half are likely to become extinct over the next 50–100 years. Thus rates of disappearance or extinction and rates of appearance or speciation are likely to be out of balance by a factor of a million! This clearly is not what has been going on over most of geological time. Secondly, most past 'extinctions' see the basic lineage continued (at least since the Cambrian); likely current rates of extinction in the tropics will see lineages ended, not transformed.